Molecular Mechanisms of Bioactive Nutrients Promoting Health through Gut Microbiota

Molecular Mechanisms of Bioactive Nutrients Promoting Health through Gut Microbiota

Editor

Baojun Xu

MDPI

Basel • Beijing • Wuhan • Barcelona • Belgrade • Novi Sad • Cluj • Manchester

Editor
Baojun Xu
Life Sciences
BNU-HKBU United
International College
Zhuhai
China

Editorial Office
MDPI
St. Alban-Anlage 66
4052 Basel, Switzerland

This is a reprint of articles from the Special Issue published online in the open access journal *International Journal of Molecular Sciences* (ISSN 1422-0067) (available at: www.mdpi.com/journal/ijms/special_issues/Nutrients_Gut_Microbiota).

For citation purposes, cite each article independently as indicated on the article page online and as indicated below:

Lastname, A.A.; Lastname, B.B. Article Title. *Journal Name* **Year**, *Volume Number*, Page Range.

ISBN 978-3-0365-8963-3 (Hbk)
ISBN 978-3-0365-8962-6 (PDF)
doi.org/10.3390/books978-3-0365-8962-6

Contents

About the Editor

Baojun Xu

Dr. Xu is a Chair Professor at Beijing Normal University-Hong Kong Baptist University United International College (UIC, a full English teaching college in China), Fellow of the Royal Society of Chemistry, Zhuhai Scholar Distinguished Professor, Head of Department of Life Sciences, Program Director of Food Science and Technology Program, and author of over 300 peer-reviewed papers. Dr. Xu received his Ph.D. in Food Science from Chungnam National University, South Korea. He conducted postdoctoral research work in North Dakota State University (NDSU), Purdue University, and Gerald P. Murphy Cancer Foundation in the USA during 2005-2009. He conducted short-term visiting research in NDSU in 2012, and the University of Georgia in 2014, followed by visiting research during his sabbatical leave (7 months) in Pennsylvania State University in the USA in 2016. Dr. Xu is serving as the Associate Editor-in-Chief of *Food Science and Human Wellness*, Associate Editor of *Food Research International*, Associate Editor of *Food Frontiers*, and the editorial board member of around 10 international journals. He received the inaugural President's Award for Outstanding Research of UIC in 2016, and the President's Award for Outstanding Service of UIC in 2020. Dr. Xu has been listed in the world's top 2% of scientists by Stanford University in 2020, 2021, 2022, and was listed as the Best Scientist in the World in the field of Biology and Biochemistry at Research.com in 2023.

International Journal of *Molecular Sciences*

Editorial

Molecular Mechanisms of Bioactive Nutrients Promoting Health through Gut Microbiota

Baojun Xu

Food Science and Technology Program, Department of Life Sciences, BNU-HKBU United International College, Zhuhai 519087, China; baojunxu@uic.edu.cn

Many food components (such as phytochemicals, complex carbohydrates, proteins, fats, vitamins, minerals, etc.) have been found to have various biological activities. Based on dietary intake and the availability of nutrients in the intestine, human gut microbiota can produce harmful metabolites that cause human diseases or beneficial compounds that prevent host diseases [1]. Abnormal gut microbiota can produce endotoxins, exacerbating chronic inflammation and metabolic disorders [2]. Moreover, gut microbiota are crucial for maintaining metabolism and health, and dysbiosis plays a crucial role in the occurrence and development of various diseases [3].

Therefore, a promising strategy to help manage colon and host health is to regulate the composition of the gut microbiota by eating biologically active food ingredients. Bioactive ingredients obtained from dietary sources can be designed and characterized to meet human nutritional and immune needs and balance gut microbiota. To maximize knowledge on the health effect of gut microbiota on improving human health, a Special Issue titled "Molecular Mechanisms of Bioactive Nutrients Promoting Health through Gut Microbiota" was published in the *International Journal of Molecular Science*, including nine papers: six research articles and three reviews. Among these six research articles, four are animal studies [4–7], one is in vitro gut microbiota culture study [8], and one is randomized clinical study [9].

Vernocchi et al. [1] summarized the metabolism of gut microbiota and their interactions with food components, including the dietary impact on gut microbiota and metabolic composition, health effects mediated by food–microbiota metabolomes, and microbiome-based therapeutics. Finally, the author also pointed out that future research requires extensive experiments, and the role of specific nutrients needs to be evaluated in clinical trials.

The review by Mercader-Barceló et al. [10] recorded that regulating dietary factors can affect the development of idiopathic pulmonary fibrosis (IPF) through the intestinal–pulmonary axis. This review summarized evidence about the relationship between diet and IPF in human and pulmonary fibrosis animal models. The authors also discussed the biological activities of specific dietary food ingredients, including fatty acids, peptides, amino acids, carbohydrates, vitamins, minerals, and phytochemicals. Moreover, future research should aim to identify novel diet-related biomarkers of IPF, including metabolites derived from microbiota, and conduct a more in-depth analysis of lung and intestinal microbiota in IPF.

Ganesan et al. [11] summarized the treatment of colorectal cancer using diet-derived phytochemicals through colon cancer stem cells and microbiota. It also reviewed the regulation of different phytochemicals on gut microbiota through different molecular mechanisms and proposed a relationship between phytochemicals, gut microbiome, and colon cancer stem cells. The authors suggested that bioactive nutrients will improve the gut microbiota and combat colorectal cancer. In the end, the authors suggested that dietary phytochemicals-induced intestinal microbiota are still a potential research field because they have an apparent anti-tumor effect and are a new mechanism for future treatment.

Citation: Xu, B. Molecular Mechanisms of Bioactive Nutrients Promoting Health through Gut Microbiota. *Int. J. Mol. Sci.* **2023**, *24*, 13731. https://doi.org/10.3390/ijms241813731

Received: 17 August 2023
Accepted: 30 August 2023
Published: 6 September 2023

An animal study [6] has shown that pistachios can reduce inflammation and improve gut microbiota composition in mice consuming a high-fat diet. Specifically, pistachios can significantly reduce the protein levels of TNF-α and IL-1β in sera and the mRNA expression levels of IL-1β, TNF-α, F4-80, and CCL-2 in subcutaneous and visceral adipose tissues. A treatment with pistachios can reduce the ratio of *Firmicutes* to *Bacteroidetes*. Furthermore, the pistachio diet can significantly increase the abundance of healthy bacteria genera, such as *Parabacteroides, Dorea, Allobaculum, Turicibacter, Lactobacillus,* and *Anaeroplasma*. It considerably reduced the amount of inflammation-related bacteria, such as *Oscillospira, Desulfovibrio, Coprobacillus,* and *Bilophila*. Terzo et al. [6] proposed that the consumption of pistachios can alleviate inflammation caused by obesity, and this effect may be related to the regulation of microbial community composition.

Complex carbohydrates have been widely used in the food industry, but their positive effect ton gut microbiota and the related mechanisms are still unclear. Wang et al.'s [5] animal experimental studies have shown that chitosan can improve the colitis of dextran sulfate sodium-induced ulcerative disease by enhancing the internal barrier function and improving microflora. Chitosan can also improve the intestinal mucosal barrier function and affect gut microbiota. It can also better regulate the expressions of TNF-α and TJ proteins, such as claudin-1, occludin, and ZO-1. Moreover, Liang et al.'s [4] study revealed that d-Tagatose can improve constipation by modulating the composition of gut microbiota. More specifically, taking d-Tagatose can restore changes in gut microbiota caused by constipation, significantly increase level of acetylcholine (Ach) in sera, and also reduce the level of nitric oxide (NO) in sera. Nogacka et al. [8] conducted fecal culture tests in vitro on the microbiota of healthy adults with a normal weight and morbidly obese adults, with the addition of different inulintype fructans (1-kestose, actilight, P95, synergy1, and inulin) and a galactooligosaccharide to the cultures. The regulation of prebiotics leads to a significant increase in the numbers of *Bacteroides, Bifidobacterium,* and *Faecalibacterium*. The author suggests that there are differences in the tested prebiotics among different populations, indicating that research and development need to be tailored to specific populations when developing related products.

Diseases are often closely related to changes in gut microbiota composition, and unraveling the relationship may identify new molecular mechanisms for the subsequent treatment of diseases. Nagpal et al. [7] explored the relationship between gut microbiota and obesity through three commonly used mouse models of obesity or type 2 diabetes. The results indicate that obesity caused by a high-fat diet and gene mutations exhibit different gut microbiome compositions, indicating that the microbiome is sensitive to the hosts' diet, genetic background, and physiology.

When studying the relationship between diseases and gut microbiota, in addition to using animal models, it is also necessary to study human gut microbiota. Tomova et al. [9] explored the impact of food intake specificity on gut microbiota among children with autism. The dietary intake of children with autism provides the same energy as the normal, but their food choices may lead to a lack of micronutrients, such as vitamin K, B6, C, iron, copper, docosahexaenoic acid, and docosapentaenoic acid. Food selectivity and the intake of fermented milk products, total fat, omega-3, and animal/plant protein lead to similar changes in the intestinal microbiota of children with and without autism. The authors pointed out that although food intervention is difficult for children with autism, such changes may help alter gut microbiota, thereby improving their gastrointestinal and immune states.

As researchers continue to conduct studies on the role of gut microbiota and their metabolites in diseases, research has gradually shifted from non-targeted therapy to targeted microbiota therapy [2] using new technical strategies like fecal microbiota transmission (FMT), thereby treating diseases more efficiently [11]. Moreover, the latest research has been published. Wang et al. [12] utilized an FMT experiment to target the potentially critical microbiota of hyperlipidemia. The causal relationship between changes in gut microbiota and lipid metabolism has been determined, and the mechanisms of Eucommia bark extract and Eucommia leaves extract in combating hyperlipidemia have been elucidated, and feasible treatment methods for hyperlipidemia have been provided. Yang et al. [13] explored the alleviating effect of Fu brick tea thearownin on ulcerative colitis and its potential mechanisms through fecal 16S rRNA genes, metabolomics, and FMT. Chen et al. [14] used FMT, 16S rRNA sequencing, miRNA sequencing, and RNA sequencing to elucidate the role of gut microbiota/butyric acid/miR-204/ACSS2 in regulating chicken fat production and deposition.

This Special Issue has not published any papers on transformation or targeted microbiota work, but it includes one clinical randomized trial. Subsequent research on modifying gut microbiota through bioactive nutrients to achieve health promotion needs to be carried out in more research on targeted microbiota therapies and clinical randomized trials. The relevant results of these targeted microbiota therapies and clinical randomized trials can be used to develop healthy products or drugs in the food or pharmaceutical industry.

Conflicts of Interest: The author declares no conflict of interest.

References

1. Vernocchi, P.; Del Chierico, F.; Putignani, L. Gut microbiota metabolism and interaction with food components. *Int. J. Mol. Sci.* **2020**, *21*, 3688. [CrossRef] [PubMed]
2. Flaig, B.; Garza, R.; Singh, B.; Hamamah, S.; Covasa, M. Treatment of dyslipidemia through targeted therapy of gut microbiota. *Nutrients* **2023**, *15*, 228. [CrossRef] [PubMed]
3. Zhu, L.R.; Li, S.S.; Zheng, W.Q.; Ni, W.J.; Cai, M.; Liu, H.P. Targeted modulation of gut microbiota by traditional Chinese medicine and natural products for liver disease therapy. *Front. Immunol.* **2023**, *14*, 1086078. [CrossRef] [PubMed]
4. Liang, Y.-X.; Wen, P.; Wang, Y.; OuYang, D.-M.; Wang, D.; Chen, Y.-Z.; Song, Y.; Deng, J.; Sun, Y.-M.; Wang, H. The constipation-relieving property of D-tagatose by modulating the composition of gut microbiota. *Int. J. Mol. Sci.* **2019**, *20*, 5721. [CrossRef] [PubMed]
5. Wang, J.; Zhang, C.; Guo, C.; Li, X. Chitosan ameliorates DSS-induced ulcerative colitis mice by enhancing intestinal barrier function and improving microflora. *Int. J. Mol. Sci.* **2019**, *20*, 5751. [CrossRef] [PubMed]
6. Terzo, S.; Mulè, F.; Caldara, G.F.; Baldassano, S.; Puleio, R.; Vitale, M.; Cassata, G.; Ferrantelli, V.; Amato, A. Pistachio consumption alleviates inflammation and improves gut microbiota composition in mice fed a high-fat diet. *Int. J. Mol. Sci.* **2020**, *21*, 365. [CrossRef] [PubMed]
7. Nagpal, R.; Mishra, S.P.; Yadav, H. Unique gut microbiome signatures depict diet-versus genetically induced obesity in mice. *Int. J. Mol. Sci.* **2020**, *21*, 3434. [CrossRef] [PubMed]
8. Nogacka, A.M.; Salazar, N.; Arboleya, S.; Ruas-Madiedo, P.; Mancabelli, L.; Suarez, A.; Martinez-Faedo, C.; Ventura, M.; Tochio, T.; Hirano, K. In vitro evaluation of different prebiotics on the modulation of gut microbiota composition and function in morbid obese and normal-weight subjects. *Int. J. Mol. Sci.* **2020**, *21*, 906. [CrossRef] [PubMed]
9. Tomova, A.; Soltys, K.; Kemenyova, P.; Karhanek, M.; Babinska, K. The influence of food intake specificity in children with autism on gut microbiota. *Int. J. Mol. Sci.* **2020**, *21*, 2797. [CrossRef] [PubMed]
10. Mercader-Barceló, J.; Truyols-Vives, J.; Río, C.; López-Safont, N.; Sala-Llinàs, E.; Chaplin, A. Insights into the role of bioactive food ingredients and the microbiome in idiopathic pulmonary fibrosis. *Int. J. Mol. Sci.* **2020**, *21*, 6051. [CrossRef] [PubMed]
11. Ganesan, K.; Jayachandran, M.; Xu, B. Diet-derived phytochemicals targeting colon cancer stem cells and microbiota in colorectal cancer. *Int. J. Mol. Sci.* **2020**, *21*, 3976. [CrossRef] [PubMed]
12. Wang, Z.; Sun, Y.; Han, Y.; Chen, X.; Gong, P.; Zhai, P.; Yao, W.; Ba, Q.; Wang, H. Eucommia bark/leaf extract improves HFD-induced lipid metabolism disorders via targeting gut microbiota to activate the Fiaf-LPL gut-liver axis and SCFAs-GPR43 gut-fat axis. *Phytomedicine* **2023**, *110*, 154652. [CrossRef] [PubMed]
13. Yang, W.; Ren, D.; Shao, H.; Zhang, X.; Li, T.; Zhang, L.; Liu, L.; Zhao, Y.; Niu, P.; Yang, X. Theabrownin from Fu Brick tea improves ulcerative colitis by shaping the gut microbiota and modulating the tryptophan metabolism. *J. Agric. Food Chem.* **2023**, *71*, 2898–2913. [CrossRef] [PubMed]
14. Chen, C.; Chen, W.; Ding, H.; Wu, P.; Zhang, G.; Xie, K.; Zhang, T. High-fat diet-induced gut microbiota alteration promotes lipogenesis by butyric acid/miR-204/ACSS2 axis in chickens. *Poultry Sci.* **2023**, *102*, 102856. [CrossRef] [PubMed]

International Journal of
Molecular Sciences

MDPI

Review

Insights into the Role of Bioactive Food Ingredients and the Microbiome in Idiopathic Pulmonary Fibrosis

Josep Mercader-Barceló [1,2,3,*], Joan Truyols-Vives [2], Carlos Río [1], Nora López-Safont [4], Ernest Sala-Llinàs [1,5,6] and Alice Chaplin [7,*]

1 Inflammation, Repair and Cancer in Respiratory Diseases Research Group, Balearic Islands Health Research Institute (IdISBa), Ctra. Valldemossa 79, 07120 Palma, Balearic Islands, Spain; carlos.rio@ssib.es (C.R.); ernest.sala@ssib.es (E.S.-L.)
2 Department of Fundamental Biology and Health Sciences, University of the Balearic Islands (UIB), 07122 Palma, Balearic Islands, Spain; joantruyolsvives@gmail.com
3 Foners Medicina Veterinària i Innovació SLP, Foners 27, 07006 Palma, Balearic Islands, Spain
4 ADEMA School University, University of the Balearic Islands (UIB), 07009 Palma, Balearic Islands, Spain; nlopez@ademaescuelauniversitaria.com
5 Pneumology Service, Hospital Universitari Son Espases, 07010 Palma de Mallorca, Spain
6 Centro de Investigación en Red de Enfermedades Respiratorias (CIBERES), 28029 Madrid, Spain
7 Lipids in Human Pathology Research Group, Balearic Islands Health Research Institute (IdISBa), Ctra. Valldemossa 79, 07120 Palma, Balearic Islands, Spain
* Correspondence: josep.mercader@uib.es (J.M.-B.); alicemarylillian.chaplin@ssib.es (A.C.)

Received: 24 July 2020; Accepted: 19 August 2020; Published: 22 August 2020

check for updates

Abstract: Idiopathic pulmonary fibrosis (IPF) is a chronic disease mainly associated with aging and, to date, its causes are still largely unknown. It has been shown that dietary habits can accelerate or delay the occurrence of aging-related diseases; however, their potential role in IPF development has been underestimated so far. The present review summarizes the evidence regarding the relationship between diet and IPF in humans, and in animal models of pulmonary fibrosis, in which we discuss the bioactivity of specific dietary food ingredients, including fatty acids, peptides, amino acids, carbohydrates, vitamins, minerals and phytochemicals. Interestingly, many animal studies reveal preventive and therapeutic effects of particular compounds. Furthermore, it has been recently suggested that the lung and gut microbiota could be involved in IPF, a relationship which may be linked to changes in immunological and inflammatory factors. Thus, all the evidence so far puts forward the idea that the gut-lung axis could be modulated by dietary factors, which in turn have an influence on IPF development. Overall, the data reviewed here support the notion of identifying food ingredients with potential benefits in IPF, with the ultimate aim of designing nutritional approaches as an adjuvant therapeutic strategy.

Keywords: idiopathic pulmonary fibrosis; diet; microbiota; gut-lung axis

1. Introduction

Idiopathic pulmonary fibrosis (IPF) is a chronic, progressive, irreversible and heterogeneous disease characterised by the excessive deposition of extracellular matrix associated with progressive decline in lung function and respiratory failure [1]. IPF occurs primarily in older adults with a median survival of 2–3 years from diagnosis and, due to an increase in its incidence, it is becoming an economic burden on global health care [2]. The histopathological hallmark used for diagnosis is the occurrence of a usual interstitial pneumonia (UIP) pattern [2]. The current treatment options that show a modest benefit are the anti-fibrotic drugs nintedanib and pirfenidone; however, there is still no cure for IPF [1,2]. Current paradigms suggest that IPF is the consequence of an aberrant repair process, in response to subclinical epithelial injury superimposed on accelerated epithelial aging [2]. Host factors associated to the development of this condition include chronological age, the occurrence of the hallmarks of aging [3–6], such as telomere attrition, senescence, stem cell exhaustion and mitochondrial dysfunction, and genetic predisposition [7]. Furthermore, environmental factors, including exposure to wood and metal dust, pollution, tobacco smoking, gastric aspiration and viral infection, are thought to play a causative role in the initiation and progression of the disease [2]. Even though it is well known that diet has an influence on the risk of age-related diseases, the potential role of dietary factors has not been considered in this complex scenario. However, there is an increasing number of evidences in animal models suggesting a role for specific bioactive food ingredients in the promotion and mitigation of pulmonary fibrosis. Moreover, alterations in glucose and lipid metabolism are present in pulmonary fibrosis [8,9], and metabolic reprogramming in IPF patients has been described [10]. On the other hand, it has been recently suggested that the microbiome could actually influence the risk of initiation and/or progression of IPF [11–13], where gut and airway microbiota are considered relevant players. Here, we will review the studies carried out to date which investigate the role of bioactive food ingredients and microbiota in IPF, and discuss how their potential interaction could have an effect on the development and mitigation of this disease.

2. Pathogenesis of IPF

The UIP pattern characteristic of IPF includes temporal and spatially heterogeneous fibrosis, advanced scarring and microscopic honeycombing, clusters of fibroblasts and myofibroblasts (fibroblastic foci), the accumulation of hyperplastic type II alveolar epithelial cells, the reduction of type I alveolar epithelial cells and exaggerated deposition of disorganised collagen and extracellular matrix (ECM), resulting in the distortion of normal lung architecture [2]. In this context, IPF is thought to be the consequence of an aberrant repair process, in response to complex interactions between host and environmental factors. Host factors include genetic and epigenetic features that contribute to the development of an inherently susceptible and dysfunctional epithelium, to recurrent micro-injuries from cigarette smoke, wood and metal dust, gastro-oesophageal reflux, and viral infection [1]. The genetic predisposition to develop IPF has been shown in carriers of variants of genes that affect lung epithelial cells [2]. Certain epigenetic changes have also been identified in IPF patients [14], but the triggering environmental agents and the mechanisms involved in the epigenetic alterations remain elusive. The genetically and/or epigenetically predisposed epithelium displays a limited regenerative ability following recurrent injury, which is crucial for the propagation of IPF.

As a response to sustained injury, epithelial cells acquire features of mesenchymal cells, a process termed epithelial-mesenchymal transition (EMT), which is initiated by the transforming growth factor β1 (TGF-β1) [7,15]. These reprogrammed cells express α smooth muscle actin (αSMA), type I collagen and N-cadherin, and these changes trigger the loss of their polarity and tight junctions, and become more mobile [15]. Furthermore, the alveolar epithelial basement membrane is disrupted as a consequence to the damaged and dysfunctional epithelium, which favours the leakage of fibrin and fibronectin into the interstitial and alveolar spaces, as well as the intrusion of inflammatory cells and the migration of mesenchymal cells [7].

Activated epithelial and endothelial cells promote an aberrant epithelial-mesenchymal crosstalk [7,15]. Circulating fibrocytes are recruited by the alveolar epithelium, differentiate into fibroblasts or myofibroblasts, and produce collagen and cytokines that induce collagen deposition [7]. Epithelial cells release TGF-β1, which leads to the differentiation of resident pulmonary fibroblasts into myofibroblasts that secrete excessive amounts of matrix, platelet-derived growth factor (PDGF), which increases fibroblast proliferation, and Wnt proteins, that stimulate fibroblasts to produce type I collagen [7,15]. Chronically activated fibroblasts acquire an invasive phenotype, and are resistant to apoptosis, which leads to their retention in IPF lungs, allowing the fibroblasts to persistently deposit collagen [7,15]. In addition to abnormal quantities of matrix proteins, ECM remodelling is characterised by its stiffness and altered composition, which seem to stimulate the production of more collagen. ECM changes include an increase in the levels of matrix metalloproteases (MMP), such as MMP-1 and MMP-7, which are involved in ECM degradation; lysyl oxidase-like proteins, causing collagen accumulation and deposition; periostin, which promotes mesenchymal cell proliferation and fibrosis; and osteopontin, a proinflammatory cytokine involved in tissue repair [16].

Both oxidative stress and immunological factors may play a pathogenetic role in IPF [17,18]. An imbalance between oxidants and antioxidants is observed in IPF patients, who show elevated concentrations of oxidants, depleted glutathione levels, and reduced activity of the nuclear factor-erythroid-related factor 2 (Nrf-2), the antioxidant enzyme haem oxygenase 1, and superoxide dismutase (SOD) [17,18]. Furthermore, immune function is also altered in IPF, whereby changes in the innate immune response to infection and tissue injury have been observed, as well as in adaptive immunity, highlighted by the activation of CD4+ T-cells and a decrease in levels of T regulatory cells (Tregs) [16].

Contribution of Aging Hallmarks to IPF

Aging is the most evident host factor that influences IPF occurrence. The main molecular and cellular hallmarks of aging are present in IPF, which include telomere attrition, genomic instability, mitochondrial dysfunction, epigenetic alterations, cellular senescence, loss of proteostasis, dysregulated nutrient sensing, stem cell exhaustion and altered intercellular communication [19]. Furthermore, it has been shown that some of these hallmarks mediate the development of the disease [3,4,6,20]. For example, abnormal telomere shortening due to mutations in telomerase-related genes is linked to poor prognosis [5]. Aging-associated mitochondrial changes have also been identified in IPF, including an imbalance in mitochondrial dynamics, increased mitochondrial reactive oxygen species (ROS) levels and DNA mutations, leading to decreased mitochondrial respiration and ATP production. Collectively, these changes contribute to the senescent phenotype in lung fibroblasts [2,3,20]. Higher levels of senescence-associated cell cycle repressors, p21, p16 and/or p53 and beta-galactosidase activity, are found in lung cells from IPF patients, and the contributing role of senescence in IPF has been demonstrated with the administration of a senolytic drug [6]. Furthermore, bone marrow stem cells from IPF patients showed decreased stemness and capacity to prevent lung fibrosis progression [4].

However, the development of pulmonary fibrosis in individuals with no identifiable mutations, or that have not been exposed to the abovementioned environmental inputs, means that other factors must be considered. Since it is well known that diet influences the risk of age-related diseases, and that the microbiome is highly implicated in a wide range of conditions, we decided to delve into the data published so far, in order to elucidate their potential involvement in IPF.

3. The Role of Nutritional Factors in IPF

3.1. Nutrition and Aging

Nutrition is intimately linked to health and disease. Numerous epidemiological studies support the premise that unhealthy dietary habits increase the risk of age-related chronic diseases and accelerate mortality [21]. On the other hand, healthy dietary patterns could delay or prevent the occurrence of age-related diseases, thus contributing to extend a person's health and lifespan [22]. Accordingly, specific nutritional recommendations for particular subpopulations are associated with the prevention of age-related diseases; for instance, the use of calcium supplementation in post-menopausal women to reduce osteoporosis risk [23]. In addition to exerting a preventive role, the beneficial effect of particular nutrients on aging diseases is also shown in the elderly, as it happens with the intake of leucine in individuals with sarcopenia, in order to reduce muscle mass loss [24]. Furthermore, the adherence to healthy dietary patterns, such as plant-based diets, may increase overall survival in elderly individuals [25]. These data highlight the importance of precision nutrition in relation to aging, and point towards the need to design personalized evidence-based nutritional programmes to this end.

The best-known nutritional strategy to increase longevity in several animal species is the reduction of food intake without reaching malnutrition. Dietary restriction exerts a protective effect against obesity, type 2 diabetes, inflammation, hypertension and cardiovascular disease, and reduces the metabolic risk factors associated with cancer [26]. A lower intake of particular nutrients seems to be the key point for ameliorating aging, rather than calorie restriction [27]. For example, reducing protein or specific amino acids in the diet, such as methionine and tryptophan, induces protective mechanisms and extends average and maximal lifespan in experimental animals. The search of bioactive compounds able to mimic the effects of dietary restriction is of interest, given the difficulty to follow dietary restriction, particularly in the elderly. Supplementation with resveratrol increased the health and lifespan in mice, by improving mitochondrial function through the activation of sirtuin 1 (SIRT1) [28,29]. This discovery sparked an interest in molecules with an anti-aging potential, such as curcumin, epigallocatechin gallate, quercetin and resveratrol [30], which are naturally occurring, and have since demonstrated anti-aging properties in clinical trials.

However, the effects of dietary manipulations on respiratory function and age-related respiratory chronic diseases have been hardly investigated. Hegab et al. [31] recently reported that calorie restriction reversed the decrease in lung stem cells, the number and function of mitochondria, and lung inflammatory cell infiltration in old mice. Moreover, the same authors demonstrated that exposure to a high-fat diet exacerbated aging-induced lung inflammation and mitochondrial deterioration; effects that could be reversed by switching to a low-fat diet [32]. Moreover, maternal exposure to a high-fat diet negatively affected offspring lung function, and seemed to increase their susceptibility to lung diseases later in life [33]. Altogether, these results suggest that certain dietary habits may improve aging-related respiratory function.

3.2. *Bioactive Food Ingredients Influencing IPF*

Up until now, most studies on IPF have mainly focused on factors that can damage the airway. So far, the role of nutritional status in the development of IPF has been poorly evaluated in humans and, to the best of our knowledge, few epidemiological studies have analysed diet in IPF. Two case-control studies in a Japanese population found a lower fruit and higher meat consumption, together with a higher intake of saturated fat, in IPF patients [34,35]. Dietary lipid composition may contribute to the alterations observed in pulmonary surfactant phospholipids of IPF patients, whereby a lower content in polyunsaturated fatty acids at the expense of saturated fatty acids may account for their impaired biophysical surfactant function [36]. Dietary factors may influence both the risk of IPF occurrence and disease progression. On the other hand, it is also thought that IPF could affect nutritional status, since IPF patients present a high prevalence of malnutrition [37], evidencing the need for specific nutritional evaluation and counselling [38].

Interestingly, among IPF patients, cases of cardiovascular diseases, obesity, diabetes mellitus and dyslipidaemia are frequently reported [39]. Such associations suggest that dietary factors commonly involved in these metabolic diseases may also participate in the aetiology of pulmonary fibrosis, and the evidence demonstrated in experimental studies supports this rationale. In mice, neonatal overfeeding induces obesity in adulthood, and worsens airway hyperresponsiveness to methacholine, highlighted by higher amounts of collagen accumulated surrounding the bronchi and lung TGF-β1 expression [40]. Such hyperresponsiveness is also observed when adult mice are exposed to a high-fat diet [41]. The higher circulating insulin level exhibited by adult mice fed a high-fat diet seems to be the agent that enhances TGF-β1 expression in the bronchial epithelium, pointing to insulin resistance as an important player in the development of lung fibrosis. Notably, exposure to a hypercaloric diet triggers pulmonary fibrosis without the involvement of any other intentional external agent [42,43], which strengthens the hypothesis that particular bioactive food ingredients could play a causative role in the development of IPF. These observations prompted us to review the evidence regarding the participation of macronutrients, micronutrients and other bioactive food ingredients, either in the occurrence or in the attenuation of pulmonary fibrosis.

3.2.1. Macronutrients

Lipids

The exposure to diets rich in lipids triggers the occurrence of pulmonary fibrosis per se [42,43] and worsens the airway responsiveness to a challenging agent [40,41], supporting a potential role for dietary lipids as a direct causative agent of pulmonary fibrosis. The involvement of dietary lipids and lipid handling is reinforced by the fact that alterations in enzymes involved in lipid metabolism exacerbates pulmonary fibrosis [8], and that IPF patients present a decreased sphingolipid metabolism and mitochondrial β-oxidation capacity [44]. Interestingly, a recent study revealed a specific circulating lipid profile in IPF patients, and identified particular lipids that could be used as biomarkers to monitor the disease or provide prognostic information [45].

However, the effect of dietary lipids could depend on the type of fatty acid or the lipid source. Accordingly, one of the first experiments addressing these issues revealed that the intake of coconut oil or beef tallow as lipid sources triggered lower increases in lung hydroxyproline content and lipid peroxidation after the administration of bleomycin (a DNA-interactive antitumor agent commonly used to induce pulmonary fibrosis), indicating that alterations in dietary lipids can, indeed, reduce the severity of pulmonary fibrosis [46]. Here, we aim to analyse the updated evidence on the effects of different types of fatty acids on pulmonary fibrosis:

Saturated Fatty Acids (SFA)

Exposure to a high-fat diet has been shown to trigger the occurrence of incipient lung fibrosis [43], whereby the authors of this investigation noted that an increase in neutrophils in the pulmonary parenchyma may play a significant role in the development of lung fibrosis. Furthermore, in another experiment in which lung fibrosis was induced by a high-fat diet, it was observed that the levels of TGF-β1 in airway epithelial cells were increased, which was accompanied by an increased collagen deposition and expression of profibrotic factors [42], which suggest a causative role of a high intake of fatty acids in EMT. Although such a causative role could not be solely attributed to SFA in the previous experiments, Chu SG et al. [47] have recently demonstrated that the exposure to a high-fat diet rich in palmitic acid increases pulmonary fibrosis after bleomycin administration in wild-type mice. This effect was linked to the activation of the unfolded protein response and apoptosis in lung epithelial cells [47]. Moreover, these authors demonstrated the causative role of lipotoxicity on lung fibrosis by inhibiting the fatty acid transporter CD36. These results are in accordance with the higher relative contents of palmitic acid found in pulmonary surfactant phospholipids of IPF patients [36] and the positive correlation between SFA intake and IPF in a Japanese population [35], thus suggesting that a high SFA intake might increase the risk of IPF.

Polyunsaturated Fatty Acids (PUFA)

Following the observation that IPF patients show lower relative contents of oleic acid surfactant phospholipids [36], it has been hypothesized that a high PUFA intake could mitigate IPF. This question has been addressed in in vivo studies, showing that the intake of fish oil, rich in eicosapentaenoic acid (20:5, ω-3), reduces lung protein and hydroxyproline contents after a bleomycin challenge [48]. In the context of lung fibrosis, a relevant example of the importance of dietary lipids in the perinatal period for the development of diseases in adulthood comes from an experiment in which maternal diet supplementation with docosahexaenoic acid attenuated pulmonary fibrosis and improved lung function in mouse pups exposed to perinatal inflammation [49]. Furthermore, the mitigation of lung fibrosis has been demonstrated with long-chain ω-3 PUFAs (docosahexaenoic acid (22:6, ω-3)) [50], short-chain ω-3 PUFA present in flaxseed oil [51,52], and with ω-6 PUFAs, such as γ-linolenic acid (18:3, ω-6) [53]. The effects attributed to the anti-fibrotic properties of PUFA could be mediated, in part, through their conversion to resolvinD_1, which inhibits EMT in human alveolar type II cells [54], and through peroxisome proliferator-activated receptor γ (PPARγ). PPARγ agonists exhibit anti-fibrotic activity, and nitrated fatty acids, which are produced when unsaturated fatty acids react with nitric oxide (NO), are potent PPARγ agonists. Interestingly, the treatment with nitrated fatty acids reversed myofibroblast differentiation and reduced collagen deposition once lung fibrosis was established, suggesting a therapeutic role for nitrated fatty acids [55]. In all these experiments, PUFA was provided before pulmonary fibrosis was induced and, collectively, the results indicate that an adequate dietary PUFA intake might reduce the risk of lung fibrosis development.

Carbohydrates

The addition of fructose to a high-fat diet enhanced the development of pulmonary fibrosis without the participation of another intentional agent in mice [43], suggesting that high dietary contents of greatly available carbohydrates, both mono and disaccharides, might be associated with an increased risk of IPF. In fact, GLUT1-dependent glycolysis is increased in aged lungs and activates lung fibroblasts and lung fibrogenesis [9]. However, most of the studies that analyse the role of dietary carbohydrates in experimental pulmonary fibrosis attempt to mitigate fibrosis with carbohydrates, particularly specific polysaccharides, providing interesting results. For instance, barley-extracted β-glucans, which are long-chain polymers of glucose, improved hydroxyproline and oxidative stress markers, when given either in combination with N-acetylcysteine before bleomycin instillation, or alone after bleomycin administration [56]. The beneficial effects of β-glucans may rely on their ability to stimulate the immune system and antioxidant activity. Similarly, an enhanced antioxidant activity and/or immune system response also seems to explain the attenuation of pulmonary fibrosis elicited by the administration of polysaccharides from the fungi *Ophiocordyceps lanpingensis* [57] and *Ganoderma lucidum* [58]. Other polysaccharides of interest include fucoidan, a sulphated polysaccharide found in brown seaweed [59], and chitosan, a linear polysaccharide with antibacterial, antifungal, antioxidant and anti-inflammatory activities. Both these polysaccharides have shown to exert anti-fibrotic effects when administrated orally, before and after lung damage [60,61]. Anti-fibrotic effects linked to EMT inhibition were also described by the administration of plant-derived polysaccharides, including those extracted from *Dendrobium officionale* [62] or from the ancient Chinese herbal formula Yupingfeng [63].

Amino Acids, Amino Acid Derivatives and Peptides

IPF patients show an alteration in amino acid metabolism, characterised by higher amounts of proline, 4-hydroxyproline, alanine, valine, leucine, isoleucine and allysine, detected in both lung tissue and breath [64]. This has led to the question of whether the supply of functional amino acids could represent a true therapeutic strategy.

In vivo studies have addressed this issue and, in fact, oral treatment with L-arginine improved arginine metabolism in bleomycin-treated mice and reduced lung damage [65]. Similarly, in radiation-induced lung fibrosis, L-arginine reduced hydroxyproline content, procollagen I and III expression and angiotensin converting enzyme activity, which was attributed to the exogenous NO supply [66]. Likewise, pre-treatment with aerosolized arginine in experimental models of LPS-induced fibrosis reduced collagen deposition, apoptosis of alveolar cells and the expression of inflammatory cytokines and chemokines [67]. The immunomodulatory ability of certain amino acids, including L-arginine, glycine and L-norvaline, a valine isomer, seem to explain, at least in part, the protection against pulmonary fibrosis, since a reduction in the accumulation of neutrophils and macrophages and a restoration in peripheral blood Tregs were observed in animal models of lung fibrosis treated with these amino acids [65,67].

Taurine is a cysteine-derived amino acid that naturally occurs in the body. Interestingly, its anti-fibrotic potential has been documented in animal models of pulmonary fibrosis, which is explained, in part, by its antioxidant and immunomodulatory abilities [68,69]. In hamsters, diet supplementation with taurine reduced lung collagen, lipid peroxidation and SOD activity [68], and the administration of taurine both before and after bleomycin instillation reduced the amount of neutrophils in the bronchoalveolar lavage fluid (BALF) [69]. In a considerable number of experiments, the anti-fibrotic potential of taurine was studied, together with niacin, a vitamin B3 form, that when given alone, also attenuates pulmonary fibrosis [68,70], and this co-treatment exhibited a potent anti-fibrotic effect [71,72].

Recently, it was demonstrated that a small peptide isolated from the edible seaweed *Eucheuma*, named EZY-1, exhibited anti-fibrotic properties, with a superior potency and safety than pirfenidone [73], by altering the binding to PDGF receptor β and the TGF-β/Smad signal transduction. Another small peptide showing anti-fibrotic activity is the metal chelating tripeptide GHK, a naturally occurring human plasma component, the levels of which reflect regenerative capacity, and which regulates wound healing processes by inhibiting TGF-β secretion [74]. The administration of GHK inhibited collagen deposition and the inflammatory response, and suppressed EMT [75].

3.2.2. Micronutrients

Vitamins

The levels of vitamers with antioxidant activity are increased in the BALF of IPF patients, in an attempt to restore oxidative balance [76]. Such vitamers include retinol, ascorbic acid and α-tocopherol. On the contrary, these non-enzymatic antioxidants are reduced in the bleomycin model of lung fibrosis [77], highlighting the discrepancy between the bleomycin model and the human disease.

(1) Vitamin A

Retinoic acid, the bioactive metabolite of vitamin A, is an important signalling molecule during normal early lung development, and has anti-fibrotic and anti-inflammatory properties [78]. All-trans retinoic acid treatment has shown to ameliorate irradiation and bleomycin-induced lung fibrosis in mice, and its anti-fibrotic mechanisms include the decrease in TGF-β, Il-6 and Il-17A cytokine production, and a shift in the Treg/Th17 ratio [79].

(2) Vitamin B

Niacin is a vitamin B3 form involved in DNA repair, as a precursor of the coenzymes nicotinamide adenine dinucleotide and nicotinamide adenine dinucleotide phosphate. As such, its ability to enhance DNA repair caused by bleomycin has been demonstrated, being one of the first described agents able to attenuate bleomycin-induced lung fibrosis [70]. Subsequently, many studies have analysed the combination of niacin with taurine, which is expected to act through different mechanisms [68,72].

(3) Vitamin C

Vitamin C is involved in tissue repair and collagen secretion, and its deficiency leads to impaired collagen synthesis. In pulmonary fibrosis induced by chromium, paraquat or LPS, the damage was reduced by the subsequent or simultaneous administration of vitamin C [80–82]. The mechanisms of action of vitamin C, reported in the paraquat model, included reduced neutrophil recruitment and IL-17 and TGF-β secretion, and increased SOD and catalase levels [81]. Intriguingly, even though these results suggest the potential therapeutic role of vitamin C, another study reported that the depletion of ascorbate in the diet inhibits the increased neutrophilic incursion and lavage protein concentration induced by silica instillation, which may reflect the pro-oxidant function for ascorbate or the lack of ascorbate-dependent collagen hydroxylation [83].

(4) Vitamin D

Serum vitamin D has been proposed as a biomarker of IPF prognosis [84]. IPF patients show deficient circulating vitamin D levels, which correlate with disease prognosis and all-cause mortality [84,85]. Furthermore, vitamin D deficiency and bone fragility have been associated with

IPF [86]. In light of this, studies have shown a potential therapeutic role for vitamin D, such as the anti-fibrotic effect of an early vitamin D supplementation in the bleomycin model of lung fibrosis [87], and in paraquat-treated mice after a fibrotic challenge [88].

(5) Vitamin E

The role of vitamin E in the development of pulmonary fibrosis has been shown in both vitamin E deficient animals, which show a more severe interstitial pneumonitis and emphysematous lesions [89], and vitamin E treated animals, in which lung fibrosis was induced by diverse pro-fibrotic agents, including bleomycin [90], actinomycete [91], amioradone [92], radiation [93], chromium [80], and nitrogen mustard melphalan [94]. In these animal models, the mitigation of fibrosis was mainly attributed to the antioxidant activity of vitamin E. In humans, vitamin E was evaluated in combination with pentoxifylline [95], showing that chronic treatment reversed lung damage in patients with radiation-induced fibrosis, thus highlighting the anti-fibrotic potential of vitamin E.

Minerals and Salt

(1) Iron

The lungs of IPF patients show numerous aspects of dysfunctional iron metabolism, including increased iron levels, the presence of iron-laden macrophages, and iron-induced oxidative stress [96]. Fibrosis severity and pulmonary hypertension are positively associated with iron deposition in lung tissue [97] and epithelial lining fluid [98]. In BALF isolated cells, iron accumulation increases ROS generation, which is thought to play a role in macrophage activation [99]. A causative role for iron accumulation in IPF is demonstrated in experimental animals, in which fibrosis and lung function decline are prevented by the intranasal administration of an iron chelator [100]. Furthermore, animals maintained on an iron deficient diet do not accumulate lung collagen and do not increase lung lipid peroxidation after lung damage induction [83,101], which strongly supports a potential role of dietary iron in pulmonary fibrosis.

(2) Copper

Excess of copper plays an important role in the aetiology and pathogenesis of many diseases, including IPF [102], and its availability has been proposed as a distinctive factor for the development of lung fibrosis and emphysema [103]. Copper is essential for the enzymatic activity of lysyl oxidase (LOX), the cytosolic SOD isoform, and amine oxidase copper containing-3 (AOC3), proteins which are known to have an important role in IPF. LOX regulates the covalent cross-linking of ECM collagen and elastin, whereas AOC3 is a membrane-bound protein that oxidizes biogenic and dietary amines, and activates leukocyte extravasation [104]. The role of AOC3 in pulmonary fibrosis development has been shown using AOC3-deficient mice, which were protected from lung fibrosis, as well as in wild-type mice treated with an AOC3 inhibitor [105]. The reduction of copper levels achieved by a low-copper diet and the administration of a chelator agent regressed the overexpression of collagen-I, LOX, MMP2, MMP8 and TIMP1 in the lungs of fibrotic mice [106], suggesting that the management of dietary copper might influence pulmonary fibrosis development.

(3) Sodium Chloride

It has been suggested that dietary salt may influence IPF, based on the fact that its intake is associated with cardiac and renal fibrosis and, in cultured cells, promotes the trans-differentiation of monocytes into fibrocytes [107,108]. Despite the fact that a high salt diet did not exacerbate lung fibrosis after bleomycin administration in mice, a low salt diet attenuated fibrosis, which was linked to a normalization in leukocyte number and a reduction in fibrocyte number [109].

3.2.3. Phytochemicals

An increasing number of studies have analysed the anti-fibrotic effect of various phytochemicals in animal models of lung fibrosis. Quercetin, curcumin, resveratrol, epigallocatechin-3-gallate, S-allyl compounds and several lignans are the most studied phytochemicals in the context of lung fibrosis. Studies in which phytochemicals were orally administrated or supplemented in the diet are included in Table 1, together with the proposed anti-fibrotic molecular mechanisms, which involve anti-oxidant, anti-inflammatory and inhibitory effects on EMT.

Quercetin

The flavonoid quercetin is found in vegetables and fruits, particularly in onions, apples and broccoli. Quercetin presents anti-aging properties, due to its antioxidant activity and ability to induce apoptosis in senescent cells, thereby mitigating the severity of several age-related chronic diseases [110]. In experimental pulmonary fibrosis, the anti-fibrotic effect of quercetin was shown to be dependent on the activation of Nrf2-responsive genes, using Nrf2 deficient mice [111]. In addition to its antioxidant and anti-inflammatory activities, it is involved in the reversal of the resistance to apoptosis, and the reduction of the expression of senescence markers p21 and p19-ARF, and the senescence-associated secretory phenotype, which are observed in the lungs of bleomycin-treated aged mice [112]. Importantly, the removal of senescent cells with dasatinib and quercetin ameliorated pulmonary function in experimental animals [6]. This treatment was also assayed in a human clinical trial, demonstrating that IPF patients improved physical function [113], highlighting the relevant role that a natural bioactive compound such as quercetin could exert in the management of IPF.

Curcumin

The flavonoid curcumin is present in turmeric (*Curcuma longa*), and is known for its antioxidant and anti-inflammatory effects mediated by its ability to inhibit NF-kB. Curcumin arises as a potential anti-aging molecule [30] and its use against pulmonary diseases has been posed [114]. In experimental lung fibrosis, curcumin administration after lung damage induction ameliorates oxidative stress and inflammation-related parameters, changes that are accompanied by signs of reduced fibrosis [115–117]. Beneficial effects are also reported when dietary curcumin is provided before the induction of fibrosis, including the enhancement of antioxidant defenses, amelioration of fibrosis and survival improvement [118]. In addition to the enhancement of antioxidant and anti-inflammatory activities, experiments in cultured lung fibroblasts reveal curcumin's ability to decrease fibroblast proliferation and migration, increase apoptotic markers, and stimulate proteins involved in matrix degradation [119]. However, despite its potential, curcumin activity is limited by its poor bioavailability, and this feature could explain why an intraperitoneal administration exerts superior anti-fibrotic effects than oral treatment [120]. To circumvent this, novel strategies of curcumin administration are being developed. Hu Y. et al. [121] developed inhalable curcumin-loaded poly(lactic-co-glycolic) acid large porous microparticles, and showed that they entered into the lung tissue and triggered striking anti-fibrotic effects [121], serving as a potential therapy applied to IPF.

Resveratrol

Resveratrol is a stilbenoid polyphenol present in grapes, known for its antioxidant activity, its ability to enhance mitochondrial biogenesis through the activation of PPARγ coactivator-1α (PGC-1α), its ability to activate serine/threonine kinase AMP-activated protein kinase complex (AMPK) signalling, and its capacity to extend lifespan in several species, including mammals, by activating SIRT1 [28,29]. In animal models of lung fibrosis, the anti-fibrotic potential of resveratrol has been analysed, showing promising results. Resveratrol reduces collagen content when it is administered after a fibrotic challenge, a change accompanied by a reduction in markers of oxidative stress and inflammation [122,123]. Moreover, resveratrol attenuates EMT by directly acting on IPF fibroblasts, and inhibiting TGF-β-induced proliferation and differentiation into myofibroblasts [124]. SIRT1 activation and other mechanisms of action have been described for the anti-fibrotic effect of resveratrol, which are detailed in Table 1 [125–127].

Flaxseed Lignans and Schisandrin B

Plant lignans are fibre-associated polyphenolic compounds that form lignin, a constituent of the plant cell wall. They are particularly abundant in flaxseed, rye bran and oat bran, and are known for their antioxidant properties. The supplementation of the diet with flaxseeds triggers anti-fibrotic effects, both before and after lung damage [128,129], doubling the survival rate when analysed as a therapeutic agent. When flaxseeds are given as oil, this treatment also reduces lung damage, as previously mentioned, an effect that is attributed to their high short-chain ω-3 PUFA content [51,52]. However, it has been demonstrated that lignans from wholegrain flaxseed decrease inflammation, lung injury and eventual fibrosis while it improves survival after radiation exposure, evidencing that it is a bioactive component responsible for lung damage mitigation [130]. Schisandrin B, a highly abundant lignan present in *Schisandra chinensis*, known for its diverse beneficial health effects, was also shown to attenuate pulmonary fibrosis in experimental animals [131,132].

Epigallocatechin-3-Gallate

The polyphenol epigallocatechin-3-gallate (EGCG) is a major component of green tea, which aroused interest due to its anti-aging and immunomodulatory properties [133]. The administration of green tea extract reduces lung fibrosis in vivo, which is attributed to its anti-inflammatory and antioxidant activities [77,134]. Moreover, a direct inhibitory fibroblast activation is also described for EGCG [135].

S-allyl-Compounds

Garlic possesses antioxidant activity, and has been traditionally used for the treatment of ailments associated with aging. Garlic bioactive compounds include S-allyl-cysteine and S-allylmercaptocysteine, organosulphur compounds that target the Nrf2 and NF-kB pathways, which are involved in the in vivo anti-inflammatory and anti-fibrotic activities [136–140]. Notably, S-allyl compounds might not only act as therapeutic, but also as preventive, agents [138,140]. Importantly, the beneficial effect of S-allyl-compounds might be achievable from diet, since the induced damage in the lung was alleviated by the supplementation of the diet with S-allyl cysteine or the administration of a water soluble aged garlic [139]. Compared to the well-known anti-fibrotic effect of N-acetylcysteine, those elicited by S-allyl-compounds were higher, which highlight the interest of S-allyl compounds as potential dietary agents to reduce pulmonary fibrosis development.

Table 1. Effects of phytochemicals in animal models of pulmonary fibrosis.

Phytochemical Compound	Dosage	Animal Models	Main Outcomes Related to Oxidative Stress, Inflammation, EMT and Fibrosis
Quercetin [111,141–143]	10–100 mg/kg bw/day 800 mg/kg in diet	BLM and amiodarone-induced female and male mice and rats	↓ MDA levels; ↑ Nrf2, CAT and SOD levels ↓ TNFα, iNOS, IL-13/6, PDGF-β, levels; ↓ H&E staining; ↓ inflammatory cells; ↑ IFN-δ levels ↓ COL1A2, TGF-β, fibronectin 1, pERK and MMP7 levels; ↓ hydroxyproline content; ↓ Masson's trichrome staining
Curcumin [115–120,144]	74–300mg/kg bw/day 1–5% w/w in diet	Irradiation, paraquat, BLM and amiodarone-induced female and male mice and rats	↓ MPO activity; ↓ TBARS, GST and ROS levels; ↑ cathepsin K and L expression ↓ NAG, AKP and ACE levels; ↓ c-Jun expression; ↓ TNF-α, superoxide anion and NO release; ↓ mononuclear and PMN cells ↓ TGF-β1, α-SMA, hydroxyproline, type I collagen expression; ↓ Smad2-3 and ERK1/2 phosphorylation
Resveratrol [122,126,145]	50 and 100 mg/kg bw/2 days 10 mg/kg bw/day 150 mg GSE/kg bw/day	BLM, silica, and particulate matter-induced male rats and mice	↓ MDA levels; ↓ MPO activity; ↑ GSH levels ↓ IL-6/1-β, TGF-β, TNF- α, NLRP3, ASC and caspase-1 levels; ↓ neutrophils; ↓ H&E staining ↓ hydroxyproline and collagen content; ↓ Masson's trichrome staining
Schisandrin B and flaxseed lignans [51,52,128–132]	5–100 mg lignan/kg bw/day 2 mg flaxseed oil/kg bw/day 10–20% lignans, 10% flaxseed (w/w) and 15% flaxseed oil in diet	Irradiation and BLM-induced male and female mice and male rats	↓ MDA, TBARS and Nox4 levels; ↓ nitrotyrosine staining; ↑ CAT and SOD activity ↓ Alveolar PMN and macrophage influx, ↓ IL-1β/2/4/6/12/17, MIP-1α, VEGF, TNF-α and FGF levels ↓ TGF-β, MMP7, β-catenin and hydroxyproline levels; ↓ Bax, p21, Smad2 phosphorylation
SAC and SAMC [136,138–140]	25–200 mg/kg in bw/day 2 mL AGE/kg bw/2 days 0.15% in diet	TiO_2, BLM and CCl_4-induced male rats and male mice	↓ Nox4 and LPO levels; ↑ HO-1, GSH and SOD activity; ↑ Nrf2 and thiol levels ↓ TNF-α, IL-6 and iNOS levels; ↓ H&E staining; ↓ lymphocyte aggregation ↓ TGF-β1, ↓ Smad3/P-Smad3, Smad2/P-Smad2 levels; ↓ MMP-9, TIMP-1, α-SMA, fibronectin, collagen 1A1 and collagen III expression; ↓ hydroxyproline content; ↓ Azan-Mallory staining
Astaxanthin [146–148]	0.5, 1 and 2 mg/kg bw/day	BLM-induced rats	↑ SOD and CAT activity ↓ H&E staining ↓ Hydroxyproline, collagen, vimentin and α-SMA levels; ↓ Masson's trichrome staining; ↑ E-cadherin levels
Crocin [149,150]	20 and 25 mg/kg bw/day	BLM-induced male rats	↓ MDA and HO-1 levels; ↑ GSH and Nrf2 levels; ↑ GSH-px, SOD and CAT activity ↓ NO, IL-10, TLR4 and TNF-α levels; ↓ H&E staining; ↓ total inflammatory cell, lymphocyte and neutrophil ↓ Hydroxyproline content; ↓ Masson's trichrome staining
Lycopene [151]	5 mg/kg bw/day	BLM-induced male rats	↓ MDA levels; ↑ SOD activity ↓ H&E staining; NO and TNF-α levels ↓ Masson's trichrome staining

Table 1. *Cont.*

Phytochemical Compound	Dosage	Animal Models	Main Outcomes Related to Oxidative Stress, Inflammation, EMT and Fibrosis
Zingerone [152]	50 and 100 mg/kg bw/day	BLM-induced male rats	↓ MDA levels; ↑ SOD and GSH-px activity
			↓ TNFα and IL-1β levels; ↓ H&E and iNOS staining
			↓ TGF-β1 expression; ↓ hydroxyproline content
Ellagic acid [153]	15 mg/kg bw/day	BLM and cyclophosphamide-induced male rats	↓ Lipid peroxidation; ↓ protein oxidation; ↓ NADH oxidize; ↓ MPO activity; ↑ CAT, SOD and GST activity
			↓ NO production
			↓ Hydroxyproline content
Proanthocyanidin [154]	100 mg/kg bw/day	BLM-induced male rats	↓ H&E and iNOS staining; ↓ immune system cells accumulation
			↓ Hydroxyproline content

Studies in which phytochemicals were orally administrated or supplemented in the diet are included. Outcomes are categorized in anti-oxidant, anti-inflammatory and anti-EMT/fibrotic effects. ↑, increase; ↓, decrease; ACE, angiotensin converting enzyme; AGE, aged garlic extract; AKP, alkaline phosphatase; α-SMA, α smooth muscle actin; ASC, apoptosis-associated speck-like protein containing a caspase activation recruitment domain; α-SMA, α-smooth muscle actin; BLM, bleomycin; CAT, catalase; CCl_4, carbon tetrachloride; COL1A, collagenase 1A; EGCG, epigallocatechin-3-gallate; EMT, epithelial to mesenchymal transition; ERK, extracellular signal-regulated kinase; FGF, fibroblast growth factor; GSE, grape seed extract; GSH, glutathione; GSH-px, glutathione peroxidase; GST, glutathione S-transferase; H&E, hematoxylin eosin; HO-1, heme oxygenase-1; IFN-δ, interferon-δ; IL, interleukin; iNOS, inducible nitric oxide synthase; MDA, malondialdehyde; MIP-1α, macrophage inflammatory protein-1α; MMP, matrix metalloproteinase; MPO, myeloperoxidase; NAG, N-acetyl-β-D-glucosaminidase; NLRP3, nucleotide-binding domain and leucine-rich repeat protein 3; NO, nitric oxide; NOX4, NADPH oxidase 4; Nrf2, nuclear factor erythroid 2-related factor 2; PDGF-β, platelet-derived growth factor subunit B; PMN, polimorfonuclear; ROS, reactive oxygen species; SAC, S-allyl-cysteine; SAMC, S-allyl-mercaptocysteine; SOD, superoxide dismutase; TBARS, thiobarbituric acid reactive substances; TGF-β, transforming growth factor-β; TIMP1, tissue inhibitor of metalloproteinase 1; TIO_2, titanium oxide; TNFα, tumor necrosis factor α; VEGF, vascular endothelial growth factor.

4. The Role of Human Microbiota in IPF—An Emerging Therapeutic Strategy

The role of the microbiota, which refers to the microbial (bacterial) community found in the human body, in health and disease is a topic of great interest, and it is now fully established that it is crucial in human wellbeing, and how an imbalance, or dysbiosis, can lead or contribute to a wide range of diseases [155]. In this context, in the last decade, several studies have emerged, showing that lung microbiota plays an important part in airway diseases, including IPF, with a clear connection between lung dysbiosis, mortality and altered immune system pathways. As recently discussed by Dickson et al. [156], the importance of lung microbiota in IPF is evident in the ongoing, current clinical trials, in which, instead of investigating immunosuppression-associated drugs, antibiotics for microbiota manipulation are being trialed.

Historically, the lung has been considered as a sterile organ. However, with the development of non-culture dependent techniques, it has been shown that there is a diverse bacterial microenvironment [157,158]. Even though there is still evidence lacking to establish a solid causal association between altered lung microbiota and IPF progression, data so far are promising, showing that microbial composition and abundance could be predictive of disease progression and are associated to changes in alveolar cytokines and inflammatory pathways. Even though a causal relationship is still not established, it is becoming clearer that lung microbiota is essential for host homeostasis, and could be a focal point of interest in the physiopathology of IPF [159].

In line with this new area of research, it has been observed that patients presenting some pulmonary conditions, including inflammation or impaired lung function, also exhibit gastrointestinal (GIT) diseases. Furthermore, even though the GIT and the respiratory tract present different environments and functions, they have the same embryonic origin and structural similarities, suggesting a potential interaction [160]. Thus, although studies are still scarce, results published to date point towards the role of a gut-lung axis in the pathogenesis of lung diseases [161], with increasing evidence that gut microbiota may be playing a crucial part.

4.1. Bacterial Burden, Diversity and Microbial Composition in the Lung in IPF

It has been suggested that three main features regarding lung microbiota are of importance when considering IPF prognosis, pathogenesis and progression: bacterial burden, microbial diversity and overall composition [11].

On one hand, those individuals presenting decreased lung microbial diversity have a worse prognosis, and that an increased bacterial burden at diagnosis is associated with a higher progression of IPF and mortality risk [11–13]. Moreover, Molyneaux et al. [12] argue that an abundant, less diverse bacterial community in the lower airways in IPF may be the cause of repetitive alveolar injury, which is thought to be a major factor in its pathogenesis. This was further confirmed by a recent study, in which decreased lung bacterial diversity was associated with increased alveolar concentrations of proinflammatory and profibrotic cytokines and growth factors [11]. More interestingly, a very recent study showed that, even though lung bacterial burden was not associated to IPF severity, it had a strong prognostic significance, by predicting clinical outcomes, providing robust evidence that the lung microbiome could be relevant in IPF prognosis [162].

However, while these data are promising, they do not prove a causal role for lung microbiota in IPF. In order to elucidate this, a range of studies have attempted to determine bacterial composition and how its manipulation can impact IPF pathogenesis. The major bacterial phyla found in lungs is the same as in gut, thus Firmicutes and Bacteroidetes are the predominant bacteria followed by Proteobacteria and Actinobacteria [159]. It has been reported that mice with induced pulmonary fibrosis present lung dysbiosis, with an increase in the Firmicutes phylum and a decrease in Bacteroidetes, for a total of 21 days after injury. This dysbiosis was sustained through all the stages of the fibroproliferative model and chronic injury, and thus supporting the hypothesis that altered pulmonary microbiota can impact injury and repair in IPF [11]. This is in contrast with what was reported later on in a study also carried out in mice, which showed that Bacteroidetes phyla was upregulated in the lung fibrosis model. In particular, Bacteroidacea and Prevotellaceae were the most increased families within this phyla and, considering that most Bacteroidetes bacteria are gram-negative anaerobes, the authors suggest that they are responsible for the lung fibrosis pathogenesis observed in their murine model [163]. Moreover, this same study showed that there was an increase in amino acid, fatty acid and carbohydrate-associated metabolites, which serve as products for Bacteroides and Prevotella growth in the gut. They report an increase in the corresponding metabolic genes in mouse tissues, suggesting that they also promote the growth of the abovementioned phyla in the lung, which in turn promote lung-fibrotic pathogenesis [163].

At the species level, microbial communities differ significantly between the gut and lung. One of the first studies to show an association between specific bacteria or operational taxonomic units (OTUs) and disease progression in IPF reported the presence of *Streptococcus* and *Staphylococcus* in the lungs of patients with a poorer disease outcome, together with *Prevotella* spp., *Veillonella* spp. and *Escherichia* spp. [164] This was further confirmed by other studies, in which *Streptococcus*, *Haemophilus*, *Neisseria* and *Veillonella* spp. have all been observed to be more abundant in patients with IPF (vs. controls) [12,13]. A more recent study demonstrated that the honeycombed lung in IPF patients was associated with alterations in lung microbiota composition, detectable at both the species/genus and family levels of taxonomy, particularly in the *Gemella* spp., which could in turn be altering community composition and promoting injury, due to mucin overexpression and defective mucocillary clearance [165].

Manipulation of lung microbiota has been achieved in other pulmonary diseases by means of antibiotic treatment; this, however, still needs to be determined whether it occurs in IPF [156]. For this, there are ongoing trials of antibiotic treatment in IPF, which will hopefully shed some light on this matter [166,167].

4.2. *The Gut-Lung Axis: Interplay of the Gut and Lung Microbiota in IPF*

Throughout the entire life span, there is a close connection between the bacterial composition of the lung and gut microbiota, leading to the hypothesis of the existence of a host-wide network. This is evidenced by studies showing that changes in a newborn's diet alter lung microbiota composition, and that fecal transplantation in rats also modifies bacteria in lungs [168,169]. An increasing number of studies are reporting that changes in gut microbiota and its by-products could be having an effect on immune responses and inflammation linked to pulmonary diseases, referred to as the gut-lung axis. Furthermore, it has been reported that microbial by-products, such as endotoxins, metabolites, cytokines and hormones, can make their way to the lung via the bloodstream, thus suggesting a more far-reaching impact than what was previously believed. More interestingly, it seems that this axis can actually be bidirectional, and thus a lung-gut axis also takes place, whereby inflammation in the lung can lead to changes in gut microbiota [161].

In disease, an inflammatory inter-organ cross-talk seems to occur between the lung and the intestine, with a clear association between airway-related diseases and GIT conditions [159]. Even though these two organs share structural similarities and a common embryonic origin, which may account for the resemblances found in disease occurrence, it seems likely that inflammatory and immune-associated components found in these compartments are responsible for the pathological cross-talk observed [170].

To date, the microbiota present in the gut, particularly in the intestine, is the most studied by far. This is not surprising, considering that it is the largest and most diverse community of the human microbiome, with over 4×10^{13} microbial cells, which live in a mutualistic relationship with the host and carry out a wide range of functions key to maintain host metabolism and homeostasis of the immune system [171]. In comparison to gut microbiota, the number of bacteria found in other organs in the body is much lower [172].

Connecting Lung and Gut Microbiota with The Immune System and Inflammation: An Inter-organ Cross Talk

The impact of gut microbiota on the immune system has been extensively studied, demonstrating so far that bacteria and its by-products interact with the host using pro-inflammatory and regulatory signals, as well as influencing immune cell responses [159]. Currently, an area of great interest is how the gut microbiota can actually influence the immune response and inflammation in the lungs, and vice versa. Thus, interestingly, the connection between the lung microbiota and the immune system is bidirectional, by which major inflammation in the lungs can also have a big impact in lung microbiota composition [173].

The effect of gut microbiota on immunity, and its further impact on lung disease, has been researched in conditions such as COPD, asthma, cystic fibrosis and influenza virus, among others [174]. Moreover, even though data are still scarce, studies have found that the presence of certain bacterial species in the lungs correlate with immune response-associated pathways in the context of IPF, particularly Toll-like receptor signaling [174,175]. In this sense, it has been reported that *Prevotella* and *Staphylococcus* abundance in the lungs of individuals with IPF is negatively correlated with the expression of immune response genes, and that changes in the microbial community structure are associated to changes in the phenotypes of circulating leukocytes [175]. Furthermore, in the absence of lung microbiota, germ-free (GF) mice presented an increased survival rate after (lung) injury exposure [11]. This particular study presented interesting results in this sense, where GF mice were protected from mortality after antibiotic treatment, yet they still exhibited the same severity of pulmonary fibrosis as their conventional counterparts. This observation could be linked to the clinical observation that patients with IPF may not only die from progressive lung fibrosis, but due to diverse inflammatory causes and respiratory infections. Another recent study presents a relationship between specific bacteria, Bacteroides and Prevotella, and fibrotic pathogenesis through IL-17R signaling in a murine model, suggesting that the lung fibrosis occurring in IPF is promoted by specific bacteria through a profibrotic inflammatory cytokine network [163].

However, as with most data regarding host microbiota, discerning the causal relationship between pulmonary and gut microbiota and IPF progression and pathogenesis is still complicated and needs further study, in order to fully understand the mechanistic pathways. It is still unknown whether the impact observed after microbiome manipulation in IPF are due to indirect, off-target effects, mainly on gut microbiota; for this, more studies are required, that can selectively alter lung microbiota without changing gut bacterial composition [156], which is known to greatly interact with the immune system.

5. Bioactive Food Ingredients, Microbiota and IPF

The potential roles of bioactive food ingredients and microbiota on IPF may be considered as inter-dependent factors, although their purported role as independent influencing players may also be taken into account. The bioactivity of a given food ingredient cannot be solely exerted by the chemical form it is presented in food, but by a more active by-product derived from gut microbiota. Therefore, its bioactivity in host cells may be dependent on the presence or abundance of particular bacteria capable of metabolizing such compounds. The opposite relationship between food ingredients and gut microbiota could also explain the bioactivity of a compound, in which a nutrient modifies gut bacteria composition, and such changes could determine host health. In this sense, there is accumulating evidence that supports the role of interactions between bioactive food ingredients and gut microbiota, which have been primarily investigated in the context of metabolic diseases [176–178].

In the context of chronic lung diseases, the interaction between bioactive food ingredients and gut microbiota, although to a lesser extent, has also been analyzed. Thus, in individuals with cystic fibrosis, the intake of antioxidant vitamins and various flavonoids correlated with the amount of particular gut bacteria, and such variations could potentially influence immune function and inflammation, which are important in cystic fibrosis disease and co-morbidity management [179,180]. These associations pave the way to pose dietary interventions aimed at modifying gut microbiota that will eventually influence respiratory diseases. In line with this idea, a nutritional strategy was designed, whereby mice presenting a pulmonary *Pseudomonas aeruginosa* infection were fed a diet enriched in acidic oligosaccharides derived from pectin [181]. This dietary intervention stimulated the growth of intestinal bacteria species involved in immunity development, reduced the inflammatory response in the BALF, improved immune system markers, and increased pulmonary bacterial clearance, demonstrating the feasibility of such approaches in chronic lung diseases. The modulation of the immune system by diet-induced changes in gut microbiota shown in these pulmonary infected mice is also the connecting element that lead the protection against allergic airway inflammation induced by dietary fiber [182]. Thus, increases in dietary fiber induce changes in the circulating levels of short chain fatty acids produced by gut microbiota which, in turn, enhances the generation of macrophages and dendritic cells, that will eventually be responsible for the protective effect of fiber intake. In fact, nutritional-based therapies consisting of probiotic intake are proposed for their ability to enhance the pulmonary immune response [159,183]. Interestingly, the manipulation of gut microbiota in colonized mice with human selected gut bacteria affects TGF-β signaling, through the production of short chain fatty acids [184]. The modulation of TGF-β response activates Treg cells, and this could influence inflammatory conditions. Since alterations in the immune system plays an important role in IPF, it would be interesting to analyze the effects of nutritional-based strategies that improve immune function through gut microbiota for the management of this pulmonary disease.

Dietary manipulations can also impact lung microbiota composition. For example, dietary fiber is able to modify the lung microbiome [182], potentially triggered by changes in gut microbiota via the gut-lung axis, although gut microbiota independent effects elicited by fiber, or any other bioactive food ingredient, on lung microbiota composition and/or activity should not be discarded. In this regard, the potential anti-fibrotic effect of curcumin could involve the inhibition of bacterial activity with pro-fibrotic potential [185]. Likewise, the potential role of iron on IPF could be related to the iron availability to lung bacteria, which depend on the strength of the immune system, and could influence infection [186].

6. Conclusions and Future Directions

Overall, the evidence presented in this review supports a role for diet and particular bioactive food ingredients in IPF, as has been previously discussed for other chronic lung diseases [187], and suggests that nutritional approaches should be considered as potential complementary therapies. The proposed role of dietary factors and microbiota is most probably bidirectional, whereby bioactive food ingredients could be influenced by gut microbiota, and at the same time, influence both gut and lung microbiota composition. Moreover, gut and lung microbiota are interconnected through the gut-lung axis, and this crosstalk could be modulated by dietary factors. Thus, it is plausible to think that changes in this host-wide network may eventually have an impact on the development of the fibrotic lung. In addition, circulating and BALF levels of certain nutrients or their derived metabolites could be used as diagnostic and prognostic biomarkers of IPF, as has already been proposed for particular lipids and vitamin D [45,84,85]. Future studies should be aimed at identifying novel IPF biomarkers associated with the diet, which might include microbiota-derived metabolites, as well as a more in-depth profiling of the lung and gut microbiome in IPF.

Author Contributions: Conceptualization, J.M.-B.; Writing—Original Draft Preparation, J.M.-B and A.C.; Writing—Review and Editing, J.M.-B., J.T.-V., N.L.-S. and A.C.; Supervision, C.R. and E.S.-L. All authors have read and agreed to the published version of the manuscript.

Funding: Alice Chaplin is supported by a FuturMed Folium grant (Folium 17/07).

Conflicts of Interest: The authors declare no conflict of interest.

References

1. Sauleda, J.; Núñez, B.; Sala, E.; Soriano, J.B. Idiopathic Pulmonary Fibrosis: Epidemiology, Natural History, Phenotypes. *Med. Sci. (Basel)* **2018**, *6*, 110. [CrossRef] [PubMed]
2. Lederer, D.J.; Martinez, F.J. Idiopathic Pulmonary Fibrosis. *N. Engl. J. Med.* **2018**, *378*, 1811–1823. [CrossRef] [PubMed]

3. Bueno, M.; Calyeca, J.; Rojas, M.; Mora, A.L. Mitochondria dysfunction and metabolic reprogramming as drivers of idiopathic pulmonary fibrosis. *Redox Biol.* **2020**, *33*, 101509. [CrossRef] [PubMed]
4. Cárdenes, N.; Álvarez, D.; Sellarés, J.; Peng, Y.; Corey, C.; Wecht, S.; Nouraie, S.M.; Shanker, S.; Sembrat, J.; Bueno, M.; et al. Senescence of bone marrow-derived mesenchymal stem cells from patients with idiopathic pulmonary fibrosis. *Stem Cell Res. Ther.* **2018**, *9*, 257. [CrossRef] [PubMed]
5. Planas-Cerezales, L.; Arias-Salgado, E.G.; Buendia-Roldán, I.; Montes-Worboys, A.; López, C.E.; Vicens-Zygmunt, V.; Hernaiz, P.L.; Sanuy, R.L.; Leiro-Fernandez, V.; Vilarnau, E.B.; et al. Predictive factors and prognostic effect of telomere shortening in pulmonary fibrosis. *Respirology* **2019**, *24*, 146–153. [CrossRef] [PubMed]
6. Schafer, M.J.; White, T.A.; Iijima, K.; Haak, A.J.; Ligresti, G.; Atkinson, E.J.; Oberg, A.L.; Birch, J.; Salmonowicz, H.; Zhu, Y.; et al. Cellular senescence mediates fibrotic pulmonary disease. *Nat. Commun.* **2017**, *8*, 14532. [CrossRef]
7. Wolters, P.J.; Collard, H.R.; Jones, K.D. Pathogenesis of idiopathic pulmonary fibrosis. *Annu. Rev. Pathol.* **2014**, *9*, 157–179. [CrossRef]
8. Sunaga, H.; Matsui, H.; Ueno, M.; Maeno, T.; Iso, T.; Syamsunarno, M.R.; Anjo, S.; Matsuzaka, T.; Shimano, H.; Yokoyama, T.; et al. Deranged fatty acid composition causes pulmonary fibrosis in Elovl6-deficient mice. *Nat. Commun.* **2013**, *4*, 2563. [CrossRef]
9. Cho, S.J.; Moon, J.S.; Lee, C.M.; Choi, A.M.; Stout-Delgado, H.W. Glucose Transporter 1-Dependent Glycolysis Is Increased during Aging-Related Lung Fibrosis, and Phloretin Inhibits Lung Fibrosis. *Am. J. Respir. Cell Mol. Biol.* **2017**, *56*, 521–531. [CrossRef]
10. Zhao, H.; Dennery, P.A.; Yao, H. Metabolic reprogramming in the pathogenesis of chronic lung diseases, including BPD, COPD, and pulmonary fibrosis. *Am. J. Physiol. Lung Cell Mol. Physiol.* **2018**, *314*, L544–L554. [CrossRef]
11. O'Dwyer, D.N.; Ashley, S.L.; Gurczynski, S.J.; Xia, M.; Wilke, C.; Falkowski, N.R.; Norman, K.C.; Arnold, K.B.; Huffnagle, G.B.; Salisbury, M.L.; et al. Lung Microbiota Contribute to Pulmonary Inflammation and Disease Progression in Pulmonary Fibrosis. *Am. J. Respir. Crit. Care Med.* **2019**, *199*, 1127–1138. [CrossRef] [PubMed]
12. Molyneaux, P.L.; Cox, M.J.; Willis-Owen, S.A.; Mallia, P.; Russell, K.E.; Russell, A.M.; Murphy, E.; Johnston, S.L.; Schwartz, D.A.; Wells, A.U.; et al. The role of bacteria in the pathogenesis and progression of idiopathic pulmonary fibrosis. *Am. J. Respir. Crit. Care Med.* **2014**, *190*, 906–913. [CrossRef] [PubMed]
13. Molyneaux, P.L.; Willis-Owen, S.A.G.; Cox, M.J.; James, P.; Cowman, S.; Loebinger, M.; Blanchard, A.; Edwards, L.M.; Stock, C.; Daccord, C.; et al. Host-Microbial Interactions in Idiopathic Pulmonary Fibrosis. *Am. J. Respir. Crit. Care Med.* **2017**, *195*, 1640–1650. [CrossRef] [PubMed]
14. Guiot, J.; Struman, I.; Chavez, V.; Henket, M.; Herzog, M.; Scoubeau, K.; Hardat, N.; Bondue, B.; Corhay, J.L.; Moermans, C.; et al. Altered epigenetic features in circulating nucleosomes in idiopathic pulmonary fibrosis. *Clin. Epigenetics* **2017**, *9*, 84. [CrossRef] [PubMed]
15. Hill, C.; Jones, M.G.; Davies, D.E.; Wang, Y. Epithelial-mesenchymal transition contributes to pulmonary fibrosis via aberrant epithelial/fibroblastic cross-talk. *J. Lung Health Dis.* **2019**, *3*, 31–35. [CrossRef] [PubMed]
16. Drakopanagiotakis, F.; Wujak, L.; Wygrecka, M.; Markart, P. Biomarkers in idiopathic pulmonary fibrosis. *Matrix Biol.* **2018**, *68–69*, 404–421. [CrossRef]
17. Cameli, P.; Carleo, A.; Bergantini, L.; Landi, C.; Prasse, A.; Bargagli, E. Oxidant/Antioxidant Disequilibrium in Idiopathic Pulmonary Fibrosis Pathogenesis. *Inflammation* **2020**, *43*, 1–7. [CrossRef]
18. Otoupalova, E.; Smith, S.; Cheng, G.; Thannickal, V.J. Oxidative Stress in Pulmonary Fibrosis. *Compr. Physiol.* **2020**, *10*, 509–547. [CrossRef]
19. López-Otín, C.; Blasco, M.A.; Partridge, L.; Serrano, M.; Kroemer, G. The hallmarks of aging. *Cell* **2013**, *153*, 1194–1217. [CrossRef]
20. Schuliga, M.; Pechkovsky, D.V.; Read, J.; Waters, D.W.; Blokland, K.E.C.; Reid, A.T.; Hogaboam, C.M.; Khalil, N.; Burgess, J.K.; Prêle, C.M.; et al. Mitochondrial dysfunction contributes to the senescent phenotype of IPF lung fibroblasts. *J. Cell Mol. Med.* **2018**, *22*, 5847–5861. [CrossRef]
21. Jayedi, A.; Soltani, S.; Abdolshahi, A.; Shab-Bidar, S. Healthy and Unhealthy Dietary Patterns and the Risk of Chronic Disease: An Umbrella Review of Meta-Analyses of Prospective Cohort Studies. *Br. J. Nutr.* **2020**. [CrossRef] [PubMed]
22. Shlisky, J.; Bloom, D.E.; Beaudreault, A.R.; Tucker, K.L.; Keller, H.H.; Freund-Levi, Y.; Fielding, R.A.; Cheng, F.W.; Jensen, G.L.; Wu, D.; et al. Nutritional Considerations for Healthy Aging and Reduction in Age-Related Chronic Disease. *Adv. Nutr.* **2017**, *8*, 17–26. [CrossRef] [PubMed]

23. Reid, I.R.; Ames, R.W.; Evans, M.C.; Gamble, G.D.; Sharpe, S.J. Effect of calcium supplementation on bone loss in postmenopausal women. *N. Engl. J. Med.* **1993**, *328*, 460–464. [CrossRef] [PubMed]
24. Gielen, E.; Beckwée, D.; Delaere, A.; De Breucker, S.; Vandewoude, M.; Bautmans, I.; Sarcopenia Guidelines Development Group of the Belgian Society of Gerontology and Geriatrics (BSGG). Nutritional interventions to improve muscle mass, muscle strength, and physical performance in older people: An umbrella review of systematic reviews and meta-analyses. *Nutr. Rev.* **2020**. [CrossRef]
25. Bamia, C.; Trichopoulos, D.; Ferrari, P.; Overvad, K.; Bjerregaard, L.; Tjønneland, A.; Halkjaer, J.; Clavel-Chapelon, F.; Kesse, E.; Boutron-Ruault, M.C.; et al. Dietary patterns and survival of older Europeans: The EPIC-Elderly Study (European Prospective Investigation into Cancer and Nutrition). *Public Health Nutr.* **2007**, *10*, 590–598. [CrossRef]
26. Cava, E.; Fontana, L. Will calorie restriction work in humans? *Aging (Albany NY)* **2013**, *5*, 507–514. [CrossRef]
27. Fontana, L.; Partridge, L. Promoting health and longevity through diet: From model organisms to humans. *Cell* **2015**, *161*, 106–118. [CrossRef]
28. Baur, J.A.; Pearson, K.J.; Price, N.L.; Jamieson, H.A.; Lerin, C.; Kalra, A.; Prabhu, V.V.; Allard, J.S.; Lopez-Lluch, G.; Lewis, K.; et al. Resveratrol improves health and survival of mice on a high-calorie diet. *Nature* **2006**, *444*, 337–342. [CrossRef]
29. Lagouge, M.; Argmann, C.; Gerhart-Hines, Z.; Meziane, H.; Lerin, C.; Daussin, F.; Messadeq, N.; Milne, J.; Lambert, P.; Elliott, P.; et al. Resveratrol improves mitochondrial function and protects against metabolic disease by activating SIRT1 and PGC-1alpha. *Cell* **2006**, *127*, 1109–1122. [CrossRef]
30. Vaiserman, A.; Lushchak, O. Implementation of longevity-promoting supplements and medications in public health practice: Achievements, challenges and future perspectives. *J. Transl. Med.* **2017**, *15*, 160. [CrossRef]
31. Hegab, A.E.; Ozaki, M.; Meligy, F.Y.; Nishino, M.; Kagawa, S.; Ishii, M.; Betsuyaku, T. Calorie restriction enhances adult mouse lung stem cells function and reverses several ageing-induced changes. *J. Tissue Eng. Regen. Med.* **2019**, *13*, 295–308. [CrossRef] [PubMed]
32. Hegab, A.E.; Ozaki, M.; Meligy, F.Y.; Kagawa, S.; Ishii, M.; Betsuyaku, T. High fat diet activates adult mouse lung stem cells and accelerates several aging-induced effects. *Stem Cell Res.* **2018**, *33*, 25–35. [CrossRef] [PubMed]
33. Heyob, K.M.; Mieth, S.; Sugar, S.S.; Graf, A.E.; Lallier, S.W.; Britt, R.D.; Rogers, L.K. Maternal high-fat diet alters lung development and function in the offspring. *Am. J. Physiol. Lung Cell Mol. Physiol.* **2019**, *317*, L167–L174. [CrossRef] [PubMed]
34. Miyake, Y.; Sasaki, S.; Yokoyama, T.; Chida, K.; Azuma, A.; Suda, T.; Kudoh, S.; Sakamoto, N.; Okamoto, K.; Kobashi, G.; et al. Vegetable, fruit, and cereal intake and risk of idiopathic pulmonary fibrosis in Japan. *Ann. Nutr. Metab* **2004**, *48*, 390–397. [CrossRef] [PubMed]
35. Miyake, Y.; Sasaki, S.; Yokoyama, T.; Chida, K.; Azuma, A.; Suda, T.; Kudoh, S.; Sakamoto, N.; Okamoto, K.; Kobashi, G.; et al. Dietary fat and meat intake and idiopathic pulmonary fibrosis: A case-control study in Japan. *Int. J. Tuberc. Lung Dis.* **2006**, *10*, 333–339. [PubMed]
36. Schmidt, R.; Meier, U.; Markart, P.; Grimminger, F.; Velcovsky, H.G.; Morr, H.; Seeger, W.; Günther, A. Altered fatty acid composition of lung surfactant phospholipids in interstitial lung disease. *Am. J. Physiol. Lung Cell Mol. Physiol.* **2002**, *283*, L1079–L1085. [CrossRef]
37. Jouneau, S.; Lederlin, M.; Vernhet, L.; Thibault, R. Malnutrition in idiopathic pulmonary fibrosis: The great forgotten comorbidity! *Eur. Respir. J.* **2019**, *53*. [CrossRef]
38. Faverio, P.; Bocchino, M.; Caminati, A.; Fumagalli, A.; Gasbarra, M.; Iovino, P.; Petruzzi, A.; Scalfi, L.; Sebastiani, A.; Stanziola, A.A.; et al. Nutrition in Patients with Idiopathic Pulmonary Fibrosis: Critical Issues Analysis and Future Research Directions. *Nutrients* **2020**, *12*, 1131. [CrossRef]
39. Iannello, S.; Cavaleri, A.; Camuto, M.; Pisano, M.G.; Milazzo, P.; Belfiore, F. Low fasting serum triglyceride and high free fatty acid levels in pulmonary fibrosis: A previously unreported finding. *MedGenMed* **2002**, *4*, 5.
40. Ye, Z.; Huang, Y.; Liu, D.; Chen, X.; Wang, D.; Huang, D.; Zhao, L.; Xiao, X. Obesity induced by neonatal overfeeding worsens airway hyperresponsiveness and inflammation. *PLoS ONE* **2012**, *7*, e47013. [CrossRef]
41. Park, Y.H.; Oh, E.Y.; Han, H.; Yang, M.; Park, H.J.; Park, K.H.; Lee, J.H.; Park, J.W. Insulin resistance mediates high-fat diet-induced pulmonary fibrosis and airway hyperresponsiveness through the TGF-β1 pathway. *Exp. Mol. Med.* **2019**, *51*, 1–12. [CrossRef] [PubMed]

42. Ge, X.N.; Greenberg, Y.; Hosseinkhani, M.R.; Long, E.K.; Bahaie, N.S.; Rao, A.; Ha, S.G.; Rao, S.P.; Bernlohr, D.A.; Sriramarao, P. High-fat diet promotes lung fibrosis and attenuates airway eosinophilia after exposure to cockroach allergen in mice. *Exp. Lung Res.* **2013**, *39*, 365–378. [CrossRef] [PubMed]
43. Vedova, M.C.D.; Soler Garcia, F.M.; Muñoz, M.D.; Fornes, M.W.; Gomez Mejiba, S.E.; Gómez, N.N.; Ramirez, D.C. Diet-Induced Pulmonary Inflammation and Incipient Fibrosis in Mice: A Possible Role of Neutrophilic Inflammation. *Inflammation* **2019**, *42*, 1886–1900. [CrossRef] [PubMed]
44. Zhao, Y.D.; Yin, L.; Archer, S.; Lu, C.; Zhao, G.; Yao, Y.; Wu, L.; Hsin, M.; Waddell, T.K.; Keshavjee, S.; et al. Metabolic heterogeneity of idiopathic pulmonary fibrosis: A metabolomic study. *BMJ Open Respir. Res.* **2017**, *4*, e000183. [CrossRef]
45. Yan, F.; Wen, Z.; Wang, R.; Luo, W.; Du, Y.; Wang, W.; Chen, X. Identification of the lipid biomarkers from plasma in idiopathic pulmonary fibrosis by Lipidomics. *BMC Pulm. Med.* **2017**, *17*, 174. [CrossRef]
46. Kennedy, J.I.; Chandler, D.B.; Fulmer, J.D.; Wert, M.B.; Grizzle, W.E. Effects of dietary fats on bleomycin-induced pulmonary fibrosis. *Exp. Lung Res.* **1987**, *12*, 149–161. [CrossRef]
47. Chu, S.G.; Villalba, J.A.; Liang, X.; Xiong, K.; Tsoyi, K.; Ith, B.; Ayaub, E.A.; Tatituri, R.V.; Byers, D.E.; Hsu, F.F.; et al. Palmitic Acid-Rich High-Fat Diet Exacerbates Experimental Pulmonary Fibrosis by Modulating Endoplasmic Reticulum Stress. *Am. J. Respir. Cell Mol. Biol.* **2019**, *61*, 737–746. [CrossRef]
48. Kennedy, J.I.; Chandler, D.B.; Fulmer, J.D.; Wert, M.B.; Grizzle, W.E. Dietary fish oil inhibits bleomycin-induced pulmonary fibrosis in the rat. *Exp. Lung Res.* **1989**, *15*, 315–329. [CrossRef]
49. Velten, M.; Britt, R.D.; Heyob, K.M.; Tipple, T.E.; Rogers, L.K. Maternal dietary docosahexaenoic acid supplementation attenuates fetal growth restriction and enhances pulmonary function in a newborn mouse model of perinatal inflammation. *J. Nutr.* **2014**, *144*, 258–266. [CrossRef]
50. Chen, J.; Zeng, T.; Zhao, X.; Xiea, K.; Bi, Y.; Zhong, Z. Docosahexaenoic acid (DHA) ameliorates paraquat-induced pulmonary fibrosis in rats possibly through up-regulation of Smad 7 and SnoN. *Food Chem. Toxicol.* **2013**, *57*, 330–337. [CrossRef]
51. Lawrenz, J.; Herndon, B.; Kamal, A.; Mehrer, A.; Dim, D.C.; Baidoo, C.; Gasper, D.; Nitz, J.; Molteni, A.; Baybutt, R.C. Dietary Flaxseed Oil Protects against Bleomycin-Induced Pulmonary Fibrosis in Rats. *Pulm. Med.* **2012**, *2012*, 457031. [CrossRef] [PubMed]
52. Abidi, A.; Kourda, N.; Feki, M.; Ben Khamsa, S. Protective Effect of Tunisian Flaxseed Oil against Bleomycin-Induced Pulmonary Fibrosis in Rats. *Nutr. Cancer* **2020**, *72*, 226–238. [CrossRef] [PubMed]
53. Ziboh, V.A.; Yun, M.; Hyde, D.M.; Giri, S.N. gamma-Linolenic acid-containing diet attenuates bleomycin-induced lung fibrosis in hamsters. *Lipids* **1997**, *32*, 759–767. [CrossRef] [PubMed]
54. Zheng, S.; Wang, Q.; D'Souza, V.; Bartis, D.; Dancer, R.; Parekh, D.; Gao, F.; Lian, Q.; Jin, S.; Thickett, D.R. ResolvinD. *Lab. Invest.* **2018**, *98*, 130–140. [CrossRef]
55. Reddy, A.T.; Lakshmi, S.P.; Zhang, Y.; Reddy, R.C. Nitrated fatty acids reverse pulmonary fibrosis by dedifferentiating myofibroblasts and promoting collagen uptake by alveolar macrophages. *FASEB J.* **2014**, *28*, 5299–5310. [CrossRef]
56. Iraz, M.; Bilgic, S.; Samdanci, E.; Ozerol, E.; Tanbek, K. Preventive and early therapeutic effects of β-glucan on the bleomycin-induced lung fibrosis in rats. *Eur. Rev. Med. Pharmacol. Sci.* **2015**, *19*, 1505–1516.
57. Zhou, S.; Zhou, Y.; Yu, J.; Du, Y.; Tan, Y.; Ke, Y.; Wang, J.; Han, B.; Ge, F. Ophiocordyceps lanpingensis polysaccharides attenuate pulmonary fibrosis in mice. *Biomed. Pharmacother.* **2020**, *126*, 110058. [CrossRef]
58. Chen, J.; Shi, Y.; He, L.; Hao, H.; Wang, B.; Zheng, Y.; Hu, C. Protective roles of polysaccharides from Ganoderma lucidum on bleomycin-induced pulmonary fibrosis in rats. *Int. J. Biol. Macromol.* **2016**, *92*, 278–281. [CrossRef]
59. Yu, H.H.; Chengchuan Ko, E.; Chang, C.L.; Yuan, K.S.; Wu, A.T.H.; Shan, Y.S.; Wu, S.Y. Fucoidan Inhibits Radiation-Induced Pneumonitis and Lung Fibrosis by Reducing Inflammatory Cytokine Expression in Lung Tissues. *Mar. Drugs* **2018**, *16*, 392. [CrossRef]
60. Kim, Y.S.; Li, Q.; Youn, H.Y.; Kim, D.Y. Oral Administration of Chitosan Attenuates Bleomycin-induced Pulmonary Fibrosis in Rats. *In Vivo* **2019**, *33*, 1455–1461. [CrossRef]
61. Zhou, L.L.; He, X.Y.; Xu, F.Y.; Du, B.X.; Zou, Z.; Shi, X.Y. Chitosan aerosol inhalation alleviates lipopolysaccharide-induced pulmonary fibrosis in rats. *Exp. Lung Res.* **2014**, *40*, 467–473. [CrossRef] [PubMed]
62. Chen, J.; Lu, J.; Wang, B.; Zhang, X.; Huang, Q.; Yuan, J.; Hao, H.; Chen, X.; Zhi, J.; Zhao, L.; et al. Polysaccharides from Dendrobium officinale inhibit bleomycin-induced pulmonary fibrosis via the TGFβ1-Smad2/3 axis. *Int. J. Biol. Macromol.* **2018**, *118*, 2163–2175. [CrossRef] [PubMed]

63. Xu, L.; Li, L.C.; Zhao, P.; Qi, L.W.; Li, P.; Gao, J.; Fei, G.H. Total polysaccharide of Yupingfeng protects against bleomycin-induced pulmonary fibrosis via inhibiting transforming growth factor-β1-mediated type I collagen abnormal deposition in rats. *J. Pharm. Pharmacol.* **2014**, *66*, 1786–1795. [CrossRef] [PubMed]
64. Gaugg, M.T.; Engler, A.; Bregy, L.; Nussbaumer-Ochsner, Y.; Eiffert, L.; Bruderer, T.; Zenobi, R.; Sinues, P.; Kohler, M. Molecular breath analysis supports altered amino acid metabolism in idiopathic pulmonary fibrosis. *Respirology* **2019**, *24*, 437–444. [CrossRef] [PubMed]
65. Gao, L.; Zhang, J.H.; Chen, X.X.; Ren, H.L.; Feng, X.L.; Wang, J.L.; Xiao, J.H. Combination of L-Arginine and L-Norvaline protects against pulmonary fibrosis progression induced by bleomycin in mice. *Biomed. Pharmacother* **2019**, *113*, 108768. [CrossRef] [PubMed]
66. Song, L.; Wang, D.; Cui, X.; Hu, W. The protective action of taurine and L-arginine in radiation pulmonary fibrosis. *J. Environ. Pathol. Toxicol. Oncol.* **1998**, *17*, 151–157.
67. Ma, X.; Zhang, Y.; Jiang, D.; Yang, Y.; Wu, G.; Wu, Z. Protective Effects of Functional Amino Acids on Apoptosis, Inflammatory Response, and Pulmonary Fibrosis in Lipopolysaccharide-Challenged Mice. *J. Agric. Food Chem.* **2019**, *67*, 4915–4922. [CrossRef]
68. Giri, S.N.; Wang, Q. Taurine and niacin offer a novel therapeutic modality in prevention of chemically-induced pulmonary fibrosis in hamsters. *Adv. Exp. Med. Biol.* **1992**, *315*, 329–340. [CrossRef]
69. Wang, Q.J.; Giri, S.N.; Hyde, D.M.; Nakashima, J.M. Effects of taurine on bleomycin-induced lung fibrosis in hamsters. *Proc. Soc. Exp. Biol. Med.* **1989**, *190*, 330–338. [CrossRef]
70. Wang, Q.J.; Giri, S.N.; Hyde, D.M.; Nakashima, J.M.; Javadi, I. Niacin attenuates bleomycin-induced lung fibrosis in the hamster. *J. Biochem. Toxicol.* **1990**, *5*, 13–22. [CrossRef]
71. Gurujeyalakshmi, G.; Hollinger, M.A.; Giri, S.N. Regulation of transforming growth factor-beta1 mRNA expression by taurine and niacin in the bleomycin hamster model of lung fibrosis. *Am. J. Respir. Cell Mol. Biol.* **1998**, *18*, 334–342. [CrossRef] [PubMed]
72. Blaisdell, R.J.; Giri, S.N. Mechanism of antifibrotic effect of taurine and niacin in the multidose bleomycin-hamster model of lung fibrosis: Inhibition of lysyl oxidase and collagenase. *J. Biochem. Toxicol.* **1995**, *10*, 203–210. [CrossRef] [PubMed]
73. Yu, H.; Zhang, Z.; Huang, H.; Wang, Y.; Lin, B.; Wu, S.; Ma, J.; Chen, B.; He, Z.; Wu, J.; et al. Inhibition of bleomycin-induced pulmonary fibrosis in mice by the novel peptide EZY-1 purified from Eucheuma. *Food Funct.* **2019**, *10*, 3198–3208. [CrossRef] [PubMed]
74. Gruchlik, A.; Chodurek, E.; Dzierzewicz, Z. Effect of GLY-HIS-LYS and its copper complex on TGF-β secretion in normal human dermal fibroblasts. *Acta Pol. Pharm.* **2014**, *71*, 954–958.
75. Zhou, X.M.; Wang, G.L.; Wang, X.B.; Liu, L.; Zhang, Q.; Yin, Y.; Wang, Q.Y.; Kang, J.; Hou, G. GHK Peptide Inhibits Bleomycin-Induced Pulmonary Fibrosis in Mice by Suppressing TGFβ1/Smad-Mediated Epithelial-to-Mesenchymal Transition. *Front. Pharmacol.* **2017**, *8*, 904. [CrossRef]
76. Markart, P.; Luboeinski, T.; Korfei, M.; Schmidt, R.; Wygrecka, M.; Mahavadi, P.; Mayer, K.; Wilhelm, J.; Seeger, W.; Guenther, A.; et al. Alveolar oxidative stress is associated with elevated levels of nonenzymatic low-molecular-weight antioxidants in patients with different forms of chronic fibrosing interstitial lung diseases. *Antioxid. Redox Signal.* **2009**, *11*, 227–240. [CrossRef]
77. Sriram, N.; Kalayarasan, S.; Sudhandiran, G. Enhancement of antioxidant defense system by epigallocatechin-3-gallate during bleomycin induced experimental pulmonary fibrosis. *Biol. Pharm. Bull.* **2008**, *31*, 1306–1311. [CrossRef]
78. Tabata, C.; Kadokawa, Y.; Tabata, R.; Takahashi, M.; Okoshi, K.; Sakai, Y.; Mishima, M.; Kubo, H. All-trans-retinoic acid prevents radiation- or bleomycin-induced pulmonary fibrosis. *Am. J. Respir. Crit. Care Med.* **2006**, *174*, 1352–1360. [CrossRef]
79. Dong, Z.; Tai, W.; Yang, Y.; Zhang, T.; Li, Y.; Chai, Y.; Zhong, H.; Zou, H.; Wang, D. The role of all-trans retinoic acid in bleomycin-induced pulmonary fibrosis in mice. *Exp. Lung Res.* **2012**, *38*, 82–89. [CrossRef]
80. Hemmati, A.A.; Nazari, Z.; Ranjbari, N.; Torfi, A. Comparison of the preventive effect of vitamin C and E on hexavalent chromium induced pulmonary fibrosis in rat. *Inflammopharmacology* **2008**, *16*, 195–197. [CrossRef]
81. Rodrigues da Silva, M.; Schapochnik, A.; Peres Leal, M.; Esteves, J.; Bichels Hebeda, C.; Sandri, S.; Pavani, C.; Ratto Tempestini Horliana, A.C.; Farsky, S.H.P.; Lino-Dos-Santos-Franco, A. Beneficial effects of ascorbic acid to treat lung fibrosis induced by paraquat. *PLoS ONE* **2018**, *13*, e0205535. [CrossRef] [PubMed]

82. Mohamed, H.A.; Elbastawisy, Y.M.; Elsaed, W.M. Attenuation of lipopolysaccharide-induced lung inflammation by ascorbic acid in rats: Histopathological and ultrastructural study. *SAGE Open Med.* **2019**, *7*, 2050312119828260. [CrossRef] [PubMed]
83. Ghio, A.J.; Kennedy, T.P.; Crissman, K.M.; Richards, J.H.; Hatch, G.E. Depletion of iron and ascorbate in rodents diminishes lung injury after silica. *Exp. Lung Res.* **1998**, *24*, 219–232. [CrossRef] [PubMed]
84. Tzilas, V.; Bouros, E.; Barbayianni, I.; Karampitsakos, T.; Kourtidou, S.; Ntassiou, M.; Ninou, I.; Aidinis, V.; Bouros, D.; Tzouvelekis, A. Vitamin D prevents experimental lung fibrosis and predicts survival in patients with idiopathic pulmonary fibrosis. *Pulm. Pharmacol. Ther.* **2019**, *55*, 17–24. [CrossRef]
85. Gao, Y.; Zhao, Q.; Qiu, X.; Zhuang, Y.; Yu, M.; Dai, J.; Cai, H.; Yan, X. Vitamin D levels are prognostic factors for connective tissue disease associated interstitial lung disease (CTD-ILD). *Aging (Albany NY)* **2020**, *12*, 4371–4378. [CrossRef]
86. Ma, D.; Peng, L. Vitamin D and pulmonary fibrosis: A review of molecular mechanisms. *Int. J. Clin. Exp. Pathol.* **2019**, *12*, 3171–3178.
87. Zhang, Z.; Yu, X.; Fang, X.; Liang, A.; Yu, Z.; Gu, P.; Zeng, Y.; He, J.; Zhu, H.; Li, S.; et al. Preventive effects of vitamin D treatment on bleomycin-induced pulmonary fibrosis. *Sci. Rep.* **2015**, *5*, 17638. [CrossRef]
88. Schapochnik, A.; da Silva, M.R.; Leal, M.P.; Esteves, J.; Hebeda, C.B.; Sandri, S.; de Fátima Teixeira da Silva, D.; Farsky, S.H.P.; Marcos, R.L.; Lino-Dos-Santos-Franco, A. Vitamin D treatment abrogates the inflammatory response in paraquat-induced lung fibrosis. *Toxicol. Appl. Pharmacol.* **2018**, *355*, 60–67. [CrossRef]
89. Nakamura, H.; Sato, S.; Takahashi, K. Effects of vitamin E deficiency on bleomycin-induced pulmonary fibrosis in the hamster. *Lung* **1988**, *166*, 161–176. [CrossRef]
90. Kilinç, C.; Ozcan, O.; Karaöz, E.; Sunguroğlu, K.; Kutluay, T.; Karaca, L. Vitamin E reduces bleomycin-induced lung fibrosis in mice: Biochemical and morphological studies. *J. Basic Clin. Physiol. Pharmacol.* **1993**, *4*, 249–269. [CrossRef]
91. Denis, M. Antioxidant therapy partially blocks immune-induced lung fibrosis. *Inflammation* **1995**, *19*, 207–219. [CrossRef] [PubMed]
92. Card, J.W.; Leeder, R.G.; Racz, W.J.; Brien, J.F.; Bray, T.M.; Massey, T.E. Effects of dietary vitamin E supplementation on pulmonary morphology and collagen deposition in amiodarone- and vehicle-treated hamsters. *Toxicology* **1999**, *133*, 75–84. [CrossRef]
93. Bese, N.S.; Munzuroglu, F.; Uslu, B.; Arbak, S.; Yesiladali, G.; Sut, N.; Altug, T.; Ober, A. Vitamin E protects against the development of radiation-induced pulmonary fibrosis in rats. *Clin. Oncol. (R Coll Radiol)* **2007**, *19*, 260–264. [CrossRef] [PubMed]
94. Wigenstam, E.; Rocksén, D.; Ekstrand-Hammarström, B.; Bucht, A. Treatment with dexamethasone or liposome-encapsuled vitamin E provides beneficial effects after chemical-induced lung injury. *Inhal. Toxicol.* **2009**, *21*, 958–964. [CrossRef]
95. Delanian, S.; Balla-Mekias, S.; Lefaix, J.L. Striking regression of chronic radiotherapy damage in a clinical trial of combined pentoxifylline and tocopherol. *J. Clin. Oncol.* **1999**, *17*, 3283–3290. [CrossRef]
96. Ogger, P.P.; Byrne, A.J. Lung fibrosis enters the iron age. *J. Pathol.* **2020**. [CrossRef]
97. Kim, K.H.; Maldonado, F.; Ryu, J.H.; Eiken, P.W.; Hartman, T.E.; Bartholmai, B.J.; Decker, P.A.; Yi, E.S. Iron deposition and increased alveolar septal capillary density in nonfibrotic lung tissue are associated with pulmonary hypertension in idiopathic pulmonary fibrosis. *Respir. Res.* **2010**, *11*, 37. [CrossRef]
98. Puxeddu, E.; Comandini, A.; Cavalli, F.; Pezzuto, G.; D'Ambrosio, C.; Senis, L.; Paci, M.; Curradi, G.; Sergiacomi, G.L.; Saltini, C. Iron laden macrophages in idiopathic pulmonary fibrosis: The telltale of occult alveolar hemorrhage? *Pulm. Pharmacol. Ther.* **2014**, *28*, 35–40. [CrossRef]
99. Lee, J.; Arisi, I.; Puxeddu, E.; Mramba, L.K.; Amicosante, M.; Swaisgood, C.M.; Pallante, M.; Brantly, M.L.; Sköld, C.M.; Saltini, C. Bronchoalveolar lavage (BAL) cells in idiopathic pulmonary fibrosis express a complex pro-inflammatory, pro-repair, angiogenic activation pattern, likely associated with macrophage iron accumulation. *PLoS ONE* **2018**, *13*, e0194803. [CrossRef]
100. Ali, M.K.; Kim, R.Y.; Brown, A.C.; Donovan, C.; Vanka, K.S.; Mayall, J.R.; Liu, G.; Pillar, A.L.; Jones-Freeman, B.; Xenaki, D.; et al. Critical role for iron accumulation in the pathogenesis of fibrotic lung disease. *J. Pathol.* **2020**, *251*, 49–62. [CrossRef]
101. Chandler, D.B.; Barton, J.C.; Briggs, D.D.; Butler, T.W.; Kennedy, J.I.; Grizzle, W.E.; Fulmer, J.D. Effect of iron deficiency on bleomycin-induced lung fibrosis in the hamster. *Am. Rev. Respir. Dis.* **1988**, *137*, 85–89. [CrossRef] [PubMed]

102. Baldari, S.; Di Rocco, G.; Toietta, G. Current Biomedical Use of Copper Chelation Therapy. *Int. J. Mol. Sci.* **2020**, *21*, 1069. [CrossRef]
103. Janssen, R.; de Brouwer, B.; von der Thüsen, J.H.; Wouters, E.F.M. Copper as the most likely pathogenic divergence factor between lung fibrosis and emphysema. *Med. Hypotheses* **2018**, *120*, 49–54. [CrossRef] [PubMed]
104. Carpéné, C.; Boulet, N.; Chaplin, A.; Mercader, J. Past, Present and Future Anti-Obesity Effects of Flavin-Containing and/or Copper-Containing Amine Oxidase Inhibitors. *Medicines (Basel)* **2019**, *6*, 9. [CrossRef]
105. Marttila-Ichihara, F.; Elima, K.; Auvinen, K.; Veres, T.Z.; Rantakari, P.; Weston, C.; Miyasaka, M.; Adams, D.; Jalkanen, S.; Salmi, M. Amine oxidase activity regulates the development of pulmonary fibrosis. *FASEB J.* **2017**, *31*, 2477–2491. [CrossRef] [PubMed]
106. Ovet, H.; Oztay, F. The copper chelator tetrathiomolybdate regressed bleomycin-induced pulmonary fibrosis in mice, by reducing lysyl oxidase expressions. *Biol. Trace. Elem. Res.* **2014**, *162*, 189–199. [CrossRef] [PubMed]
107. Gomer, R.H. New approaches to modulating idiopathic pulmonary fibrosis. *Curr. Allergy Asthma Rep.* **2013**, *13*, 607–612. [CrossRef] [PubMed]
108. Cox, N.; Pilling, D.; Gomer, R.H. NaCl potentiates human fibrocyte differentiation. *PLoS ONE* **2012**, *7*, e45674. [CrossRef]
109. Chen, W.; Pilling, D.; Gomer, R.H. Dietary NaCl affects bleomycin-induced lung fibrosis in mice. *Exp. Lung Res.* **2017**, *43*, 395–406. [CrossRef]
110. Martel, J.; Ojcius, D.M.; Wu, C.Y.; Peng, H.H.; Voisin, L.; Perfettini, J.L.; Ko, Y.F.; Young, J.D. Emerging use of senolytics and senomorphics against aging and chronic diseases. *Med. Res. Rev.* **2020**. [CrossRef]
111. Boots, A.W.; Veith, C.; Albrecht, C.; Bartholome, R.; Drittij, M.J.; Claessen, S.M.H.; Bast, A.; Rosenbruch, M.; Jonkers, L.; van Schooten, F.J.; et al. The dietary antioxidant quercetin reduces hallmarks of bleomycin-induced lung fibrogenesis in mice. *BMC Pulm. Med.* **2020**, *20*, 112. [CrossRef] [PubMed]
112. Hohmann, M.S.; Habiel, D.M.; Coelho, A.L.; Verri, W.A.; Hogaboam, C.M. Quercetin Enhances Ligand-induced Apoptosis in Senescent Idiopathic Pulmonary Fibrosis Fibroblasts and Reduces Lung Fibrosis In Vivo. *Am. J. Respir. Cell Mol. Biol.* **2019**, *60*, 28–40. [CrossRef] [PubMed]
113. Justice, J.N.; Nambiar, A.M.; Tchkonia, T.; LeBrasseur, N.K.; Pascual, R.; Hashmi, S.K.; Prata, L.; Masternak, M.M.; Kritchevsky, S.B.; Musi, N.; et al. Senolytics in idiopathic pulmonary fibrosis: Results from a first-in-human, open-label, pilot study. *EBioMedicine* **2019**, *40*, 554–563. [CrossRef] [PubMed]
114. Lelli, D.; Sahebkar, A.; Johnston, T.P.; Pedone, C. Curcumin use in pulmonary diseases: State of the art and future perspectives. *Pharmacol. Res.* **2017**, *115*, 133–148. [CrossRef] [PubMed]
115. Thresiamma, K.C.; George, J.; Kuttan, R. Protective effect of curcumin, ellagic acid and bixin on radiation induced toxicity. *Indian J. Exp. Biol.* **1996**, *34*, 845–847.
116. Venkatesan, N. Pulmonary protective effects of curcumin against paraquat toxicity. *Life Sci.* **2000**, *66*, PL21–PL28. [CrossRef]
117. Punithavathi, D.; Venkatesan, N.; Babu, M. Protective effects of curcumin against amiodarone-induced pulmonary fibrosis in rats. *Br. J. Pharmacol.* **2003**, *139*, 1342–1350. [CrossRef]
118. Lee, J.C.; Kinniry, P.A.; Arguiri, E.; Serota, M.; Kanterakis, S.; Chatterjee, S.; Solomides, C.C.; Javvadi, P.; Koumenis, C.; Cengel, K.A.; et al. Dietary curcumin increases antioxidant defenses in lung, ameliorates radiation-induced pulmonary fibrosis, and improves survival in mice. *Radiat. Res.* **2010**, *173*, 590–601. [CrossRef]
119. Zhang, D.; Huang, C.; Yang, C.; Liu, R.J.; Wang, J.; Niu, J.; Brömme, D. Antifibrotic effects of curcumin are associated with overexpression of cathepsins K and L in bleomycin treated mice and human fibroblasts. *Respir. Res.* **2011**, *12*, 154. [CrossRef]
120. Smith, M.R.; Gangireddy, S.R.; Narala, V.R.; Hogaboam, C.M.; Standiford, T.J.; Christensen, P.J.; Kondapi, A.K.; Reddy, R.C. Curcumin inhibits fibrosis-related effects in IPF fibroblasts and in mice following bleomycin-induced lung injury. *Am. J. Physiol. Lung Cell Mol. Physiol.* **2010**, *298*, L616–L625. [CrossRef]
121. Hu, Y.; Li, M.; Zhang, M.; Jin, Y. Inhalation treatment of idiopathic pulmonary fibrosis with curcumin large porous microparticles. *Int. J. Pharm.* **2018**, *551*, 212–222. [CrossRef] [PubMed]
122. Sener, G.; Topaloğlu, N.; Sehirli, A.O.; Ercan, F.; Gedik, N. Resveratrol alleviates bleomycin-induced lung injury in rats. *Pulm. Pharmacol. Ther.* **2007**, *20*, 642–649. [CrossRef] [PubMed]

123. Zhang, Y.Q.; Liu, Y.J.; Mao, Y.F.; Dong, W.W.; Zhu, X.Y.; Jiang, L. Resveratrol ameliorates lipopolysaccharide-induced epithelial mesenchymal transition and pulmonary fibrosis through suppression of oxidative stress and transforming growth factor-β1 signaling. *Clin. Nutr.* **2015**, *34*, 752–760. [CrossRef]
124. Fagone, E.; Conte, E.; Gili, E.; Fruciano, M.; Pistorio, M.P.; Lo Furno, D.; Giuffrida, R.; Crimi, N.; Vancheri, C. Resveratrol inhibits transforming growth factor-β-induced proliferation and differentiation of ex vivo human lung fibroblasts into myofibroblasts through ERK/Akt inhibition and PTEN restoration. *Exp. Lung Res.* **2011**, *37*, 162–174. [CrossRef] [PubMed]
125. Rong, L.; Wu, J.; Wang, W.; Zhao, R.P.; Xu, X.W.; Hu, D. Sirt 1 activator attenuates the bleomycin-induced lung fibrosis in mice via inhibiting epithelial-to-mesenchymal transition (EMT). *Eur. Rev. Med. Pharmacol. Sci.* **2016**, *20*, 2144–2150.
126. Ding, S.; Wang, H.; Wang, M.; Bai, L.; Yu, P.; Wu, W. Resveratrol alleviates chronic "real-world" ambient particulate matter-induced lung inflammation and fibrosis by inhibiting NLRP3 inflammasome activation in mice. *Ecotoxicol. Environ. Saf.* **2019**, *182*, 109425. [CrossRef]
127. Wang, J.; He, F.; Chen, L.; Li, Q.; Jin, S.; Zheng, H.; Lin, J.; Zhang, H.; Ma, S.; Mei, J.; et al. Resveratrol inhibits pulmonary fibrosis by regulating miR-21 through MAPK/AP-1 pathways. *Biomed. Pharmacother.* **2018**, *105*, 37–44. [CrossRef]
128. Lee, J.C.; Krochak, R.; Blouin, A.; Kanterakis, S.; Chatterjee, S.; Arguiri, E.; Vachani, A.; Solomides, C.C.; Cengel, K.A.; Christofidou-Solomidou, M. Dietary flaxseed prevents radiation-induced oxidative lung damage, inflammation and fibrosis in a mouse model of thoracic radiation injury. *Cancer Biol. Ther.* **2009**, *8*, 47–53. [CrossRef]
129. Christofidou-Solomidou, M.; Tyagi, S.; Tan, K.S.; Hagan, S.; Pietrofesa, R.; Dukes, F.; Arguiri, E.; Heitjan, D.F.; Solomides, C.C.; Cengel, K.A. Dietary flaxseed administered post thoracic radiation treatment improves survival and mitigates radiation-induced pneumonopathy in mice. *BMC Cancer* **2011**, *11*, 269. [CrossRef]
130. Pietrofesa, R.; Turowski, J.; Tyagi, S.; Dukes, F.; Arguiri, E.; Busch, T.M.; Gallagher-Colombo, S.M.; Solomides, C.C.; Cengel, K.A.; Christofidou-Solomidou, M. Radiation mitigating properties of the lignan component in flaxseed. *BMC Cancer* **2013**, *13*, 179. [CrossRef]
131. Wang, Y.; Dong, X.; Zhao, N.; Su, X.; Li, Y.; Wen, M.; Li, Z.; Wang, C.; Chen, J.; Zhuang, W. Schisandrin B attenuates bleomycin-induced pulmonary fibrosis in mice through the wingless/integrase-1 signaling pathway. *Exp. Lung Res.* **2020**, *46*, 185–194. [CrossRef] [PubMed]
132. Zhang, D.; Liu, B.; Cao, B.; Wei, F.; Yu, X.; Li, G.F.; Chen, H.; Wei, L.Q.; Wang, P.L. Synergistic protection of Schizandrin B and Glycyrrhizic acid against bleomycin-induced pulmonary fibrosis by inhibiting TGF-β1/Smad2 pathways and overexpression of NOX4. *Int. Immunopharmacol.* **2017**, *48*, 67–75. [CrossRef]
133. Pae, M.; Wu, D. Immunomodulating effects of epigallocatechin-3-gallate from green tea: Mechanisms and applications. *Food Funct.* **2013**, *4*, 1287–1303. [CrossRef] [PubMed]
134. Donà, M.; Dell'Aica, I.; Calabrese, F.; Benelli, R.; Morini, M.; Albini, A.; Garbisa, S. Neutrophil restraint by green tea: Inhibition of inflammation, associated angiogenesis, and pulmonary fibrosis. *J. Immunol.* **2003**, *170*, 4335–4341. [CrossRef] [PubMed]
135. Sriram, N.; Kalayarasan, S.; Manikandan, R.; Arumugam, M.; Sudhandiran, G. Epigallocatechin gallate attenuates fibroblast proliferation and excessive collagen production by effectively intervening TGF-β1 signalling. *Clin. Exp. Pharmacol. Physiol.* **2015**, *42*, 849–859. [CrossRef]
136. Li, C.; Sun, X.; Li, A.; Mo, M.; Zhao, Z. S-Allylmercaptocysteine attenuates Bleomycin-induced pulmonary fibrosis in mice via suppressing TGF-β1/Smad and oxidative stress pathways. *Int. Immunopharmacol.* **2020**, *79*, 106110. [CrossRef]
137. Nie, Y.; Yu, K.; Li, B.; Hu, Y.; Zhang, H.; Xin, R.; Xiong, Y.; Zhao, P.; Chai, G. S-allyl-l-cysteine attenuates bleomycin-induced pulmonary fibrosis and inflammation via AKT/NF-κB signaling pathway in mice. *J. Pharmacol. Sci.* **2019**, *139*, 377–384. [CrossRef]
138. Mizuguchi, S.; Takemura, S.; Minamiyama, Y.; Kodai, S.; Tsukioka, T.; Inoue, K.; Okada, S.; Suehiro, S. S-allyl cysteine attenuated CCl4-induced oxidative stress and pulmonary fibrosis in rats. *Biofactors* **2006**, *26*, 81–92. [CrossRef]
139. Moustafa, G.G.; Hussein, M.M.A. New insight on using aged garlic extract against toxic impacts of titanium dioxide bulk salt triggers inflammatory and fibrotic cascades in male rats. *Biomed. Pharmacother.* **2016**, *84*, 687–697. [CrossRef]

140. Tsukioka, T.; Takemura, S.; Minamiyama, Y.; Mizuguchi, S.; Toda, M.; Okada, S. Attenuation of Bleomycin-Induced Pulmonary Fibrosis in Rats with S-Allyl Cysteine. *Molecules* **2017**, *22*, 543. [CrossRef]
141. Oka, V.O.; Okon, U.E.; Osim, E.E. Pulmonary Responses Following Quercetin Administration in Rats After Intratracheal Instillation of Amiodarone. *Niger. J. Physiol. Sci.* **2019**, *34*, 63–68. [PubMed]
142. Verma, R.; Kushwah, L.; Gohel, D.; Patel, M.; Marvania, T.; Balakrishnan, S. Evaluating the Ameliorative Potential of Quercetin against the Bleomycin-Induced Pulmonary Fibrosis in Wistar Rats. *Pulm. Med.* **2013**, *2013*, 921724. [CrossRef] [PubMed]
143. Impellizzeri, D.; Talero, E.; Siracusa, R.; Alcaide, A.; Cordaro, M.; Maria Zubelia, J.; Bruschetta, G.; Crupi, R.; Esposito, E.; Cuzzocrea, S.; et al. Protective effect of polyphenols in an inflammatory process associated with experimental pulmonary fibrosis in mice. *Br. J. Nutr.* **2015**, *114*, 853–865. [CrossRef] [PubMed]
144. Punithavathi, D.; Venkatesan, N.; Babu, M. Curcumin inhibition of bleomycin-induced pulmonary fibrosis in rats. *Br. J. Pharmacol.* **2000**, *131*, 169–172. [CrossRef]
145. Hemmati, A.A.; Nazari, Z.; Samei, M. A comparative study of grape seed extract and vitamin E effects on silica-induced pulmonary fibrosis in rats. *Pulm. Pharmacol. Ther.* **2008**, *21*, 668–674. [CrossRef]
146. Song, X.; Wang, B.; Lin, S.; Jing, L.; Mao, C.; Xu, P.; Lv, C.; Liu, W.; Zuo, J. Astaxanthin inhibits apoptosis in alveolar epithelial cells type II in vivo and in vitro through the ROS-dependent mitochondrial signalling pathway. *J. Cell Mol. Med.* **2014**, *18*, 2198–2212. [CrossRef]
147. Wang, M.; Zhang, J.; Song, X.; Liu, W.; Zhang, L.; Wang, X.; Lv, C. Astaxanthin ameliorates lung fibrosis in vivo and in vitro by preventing transdifferentiation, inhibiting proliferation, and promoting apoptosis of activated cells. *Food Chem. Toxicol.* **2013**, *56*, 450–458. [CrossRef]
148. Zhang, J.; Xu, P.; Wang, Y.; Wang, M.; Li, H.; Lin, S.; Mao, C.; Wang, B.; Song, X.; Lv, C. Astaxanthin prevents pulmonary fibrosis by promoting myofibroblast apoptosis dependent on Drp1-mediated mitochondrial fission. *J. Cell Mol. Med.* **2015**, *19*, 2215–2231. [CrossRef]
149. Mehrabani, M.; Goudarzi, M.; Mehrzadi, S.; Siahpoosh, A.; Mohammadi, M.; Khalili, H.; Malayeri, A. Crocin: A protective natural antioxidant against pulmonary fibrosis induced by bleomycin. *Pharmacol. Rep.* **2020**. [CrossRef]
150. Zaghloul, M.S.; Said, E.; Suddek, G.M.; Salem, H.A. Crocin attenuates lung inflammation and pulmonary vascular dysfunction in a rat model of bleomycin-induced pulmonary fibrosis. *Life Sci.* **2019**, *235*, 116794. [CrossRef]
151. Zhou, C.; Han, W.; Zhang, P.; Cai, M.; Wei, D.; Zhang, C. Lycopene from tomatoes partially alleviates the bleomycin-induced experimental pulmonary fibrosis in rats. *Nutr. Res.* **2008**, *28*, 122–130. [CrossRef] [PubMed]
152. Gungor, H.; Ekici, M.; Onder Karayigit, M.; Turgut, N.H.; Kara, H.; Arslanbas, E. Zingerone ameliorates oxidative stress and inflammation in bleomycin-induced pulmonary fibrosis: Modulation of the expression of TGF-β1 and iNOS. *Naunyn Schmiedebergs Arch. Pharmacol.* **2020**. [CrossRef] [PubMed]
153. Saba; Khan, S.; Parvez, S.; Chaudhari, B.; Ahmad, F.; Anjum, S.; Raisuddin, S. Ellagic acid attenuates bleomycin and cyclophosphamide-induced pulmonary toxicity in Wistar rats. *Food Chem. Toxicol.* **2013**, *58*, 210–219. [CrossRef] [PubMed]
154. Agackiran, Y.; Gul, H.; Gunay, E.; Akyurek, N.; Memis, L.; Gunay, S.; Sirin, Y.S.; Ide, T. The efficiency of proanthocyanidin in an experimental pulmonary fibrosis model: Comparison with taurine. *Inflammation* **2012**, *35*, 1402–1410. [CrossRef]
155. Lynch, S.V.; Pedersen, O. The Human Intestinal Microbiome in Health and Disease. *N. Engl. J. Med.* **2016**, *375*, 2369–2379. [CrossRef]
156. Dickson, R.P.; Harari, S.; Kolb, M. Making the case for causality: What role do lung microbiota play in idiopathic pulmonary fibrosis? *Eur. Respir. J.* **2020**, *55*. [CrossRef]
157. Dickson, R.P.; Huffnagle, G.B. The Lung Microbiome: New Principles for Respiratory Bacteriology in Health and Disease. *PLoS Pathog* **2015**, *11*, e1004923. [CrossRef]
158. Dickson, R.P.; Erb-Downward, J.R.; Martinez, F.J.; Huffnagle, G.B. The Microbiome and the Respiratory Tract. *Annu. Rev. Physiol.* **2016**, *78*, 481–504. [CrossRef]
159. Enaud, R.; Prevel, R.; Ciarlo, E.; Beaufils, F.; Wieërs, G.; Guery, B.; Delhaes, L. The Gut-Lung Axis in Health and Respiratory Diseases: A Place for Inter-Organ and Inter-Kingdom Crosstalks. *Front. Cell. Infect. Microbiol.* **2020**, *10*, 9. [CrossRef]
160. Budden, K.F.; Gellatly, S.L.; Wood, D.L.; Cooper, M.A.; Morrison, M.; Hugenholtz, P.; Hansbro, P.M. Emerging pathogenic links between microbiota and the gut-lung axis. *Nat. Rev. Microbiol.* **2017**, *15*, 55–63. [CrossRef]

161. Zhang, D.; Li, S.; Wang, N.; Tan, H.Y.; Zhang, Z.; Feng, Y. The Cross-Talk Between Gut Microbiota and Lungs in Common Lung Diseases. *Front. Microbiol.* **2020**, *11*, 301. [CrossRef] [PubMed]
162. Invernizzi, R.; Barnett, J.; Rawal, B.; Nair, A.; Ghai, P.; Kingston, S.; Chua, F.; Wu, Z.; Wells, A.U.; Renzoni, E.R.; et al. Bacterial burden in the lower airways predicts disease progression in idiopathic pulmonary fibrosis and is independent of radiological disease extent. *Eur. Respir. J.* **2020**, *55*. [CrossRef] [PubMed]
163. Yang, D.; Chen, X.; Wang, J.; Lou, Q.; Lou, Y.; Li, L.; Wang, H.; Chen, J.; Wu, M.; Song, X.; et al. Dysregulated Lung Commensal Bacteria Drive Interleukin-17B Production to Promote Pulmonary Fibrosis through Their Outer Membrane Vesicles. *Immunity* **2019**, *50*, 692–706.e697. [CrossRef] [PubMed]
164. Han, M.K.; Zhou, Y.; Murray, S.; Tayob, N.; Noth, I.; Lama, V.N.; Moore, B.B.; White, E.S.; Flaherty, K.R.; Huffnagle, G.B.; et al. Lung microbiome and disease progression in idiopathic pulmonary fibrosis: An analysis of the COMET study. *Lancet Respir. Med.* **2014**, *2*, 548–556. [CrossRef]
165. Dickson, R.P.; Huffnagle, G.B.; Flaherty, K.R.; White, E.S.; Martinez, F.J.; Erb-Downward, J.R.; Moore, B.B.; O'Dwyer, D.N. Radiographic Honeycombing and Altered Lung Microbiota in Patients with Idiopathic Pulmonary Fibrosis. *Am. J. Respir. Crit. Care Med.* **2019**, *200*, 1544–1547. [CrossRef] [PubMed]
166. Anstrom, K.J.; Noth, I.; Flaherty, K.R.; Edwards, R.H.; Albright, J.; Baucom, A.; Brooks, M.; Clark, A.B.; Clausen, E.S.; Durheim, M.T.; et al. Design and rationale of a multi-center, pragmatic, open-label randomized trial of antimicrobial therapy—The study of clinical efficacy of antimicrobial therapy strategy using pragmatic design in Idiopathic Pulmonary Fibrosis (CleanUP-IPF) clinical trial. *Respir. Res.* **2020**, *21*, 68. [CrossRef]
167. Hammond, M.; Clark, A.B.; Cahn, A.P.; Chilvers, E.R.; Fraser, W.D.; Livermore, D.M.; Maher, T.M.; Parfrey, H.; Swart, A.M.; Stirling, S.; et al. The Efficacy and Mechanism Evaluation of Treating Idiopathic Pulmonary fibrosis with the Addition of Co-trimoxazole (EME-TIPAC): Study protocol for a randomised controlled trial. *Trials* **2018**, *19*, 89. [CrossRef]
168. Madan, J.C.; Koestler, D.C.; Stanton, B.A.; Davidson, L.; Moulton, L.A.; Housman, M.L.; Moore, J.H.; Guill, M.F.; Morrison, H.G.; Sogin, M.L.; et al. Serial analysis of the gut and respiratory microbiome in cystic fibrosis in infancy: Interaction between intestinal and respiratory tracts and impact of nutritional exposures. *mBio* **2012**, *3*. [CrossRef]
169. Liu, T.; Yang, Z.; Zhang, X.; Han, N.; Yuan, J.; Cheng, Y. 16S rDNA analysis of the effect of fecal microbiota transplantation on pulmonary and intestinal flora. *3 Biotech.* **2017**, *7*, 370. [CrossRef]
170. Keely, S.; Talley, N.J.; Hansbro, P.M. Pulmonary-intestinal cross-talk in mucosal inflammatory disease. *Mucosal Immunol.* **2012**, *5*, 7–18. [CrossRef]
171. Illiano, P.; Brambilla, R.; Parolini, C. The mutual interplay of gut microbiota, diet and human disease. *FEBS J.* **2020**, *287*, 833–855. [CrossRef] [PubMed]
172. Sender, R.; Fuchs, S.; Milo, R. Revised Estimates for the Number of Human and Bacteria Cells in the Body. *PLoS Biol.* **2016**, *14*, e1002533. [CrossRef] [PubMed]
173. Molyneaux, P.L.; Mallia, P.; Cox, M.J.; Footitt, J.; Willis-Owen, S.A.; Homola, D.; Trujillo-Torralbo, M.B.; Elkin, S.; Kon, O.M.; Cookson, W.O.; et al. Outgrowth of the bacterial airway microbiome after rhinovirus exacerbation of chronic obstructive pulmonary disease. *Am. J. Respir. Crit. Care Med.* **2013**, *188*, 1224–1231. [CrossRef] [PubMed]
174. McAleer, J.P.; Kolls, J.K. Contributions of the intestinal microbiome in lung immunity. *Eur. J. Immunol.* **2018**, *48*, 39–49. [CrossRef] [PubMed]
175. Huang, Y.; Ma, S.F.; Espindola, M.S.; Vij, R.; Oldham, J.M.; Huffnagle, G.B.; Erb-Downward, J.R.; Flaherty, K.R.; Moore, B.B.; White, E.S.; et al. Microbes Are Associated with Host Innate Immune Response in Idiopathic Pulmonary Fibrosis. *Am. J. Respir. Crit. Care Med.* **2017**, *196*, 208–219. [CrossRef] [PubMed]
176. Chaplin, A.; Carpéné, C.; Mercader, J. Resveratrol, Metabolic Syndrome, and Gut Microbiota. *Nutrients* **2018**, *10*, 1651. [CrossRef]
177. Ly, N.P.; Litonjua, A.; Gold, D.R.; Celedón, J.C. Gut microbiota, probiotics, and vitamin D: Interrelated exposures influencing allergy, asthma, and obesity? *J. Allergy. Clin. Immunol.* **2011**, *127*, 1087–1094. [CrossRef]
178. Delzenne, N.M.; Rodriguez, J.; Olivares, M.; Neyrinck, A.M. Microbiome response to diet: Focus on obesity and related diseases. *Rev. Endocr. Metab Dis.* **2020**. [CrossRef]
179. Li, L.; Krause, L.; Somerset, S. Associations between micronutrient intakes and gut microbiota in a group of adults with cystic fibrosis. *Clin Nutr.* **2017**, *36*, 1097–1104. [CrossRef]

180. Li, L.; Somerset, S. Associations between Flavonoid Intakes and Gut Microbiota in a Group of Adults with Cystic Fibrosis. *Nutrients* **2018**, *10*, 1264. [CrossRef]
181. Bernard, H.; Desseyn, J.L.; Bartke, N.; Kleinjans, L.; Stahl, B.; Belzer, C.; Knol, J.; Gottrand, F.; Husson, M.O. Dietary pectin-derived acidic oligosaccharides improve the pulmonary bacterial clearance of Pseudomonas aeruginosa lung infection in mice by modulating intestinal microbiota and immunity. *J. Infect. Dis.* **2015**, *211*, 156–165. [CrossRef] [PubMed]
182. Trompette, A.; Gollwitzer, E.S.; Yadava, K.; Sichelstiel, A.K.; Sprenger, N.; Ngom-Bru, C.; Blanchard, C.; Junt, T.; Nicod, L.P.; Harris, N.L.; et al. Gut microbiota metabolism of dietary fiber influences allergic airway disease and hematopoiesis. *Nat. Med.* **2014**, *20*, 159–166. [CrossRef] [PubMed]
183. Fabbrizzi, A.; Amedei, A.; Lavorini, F.; Renda, T.; Fontana, G. The lung microbiome: Clinical and therapeutic implications. *Intern. Emerg. Med.* **2019**, *14*, 1241–1250. [CrossRef] [PubMed]
184. Atarashi, K.; Tanoue, T.; Oshima, K.; Suda, W.; Nagano, Y.; Nishikawa, H.; Fukuda, S.; Saito, T.; Narushima, S.; Hase, K.; et al. Treg induction by a rationally selected mixture of Clostridia strains from the human microbiota. *Nature* **2013**, *500*, 232–236. [CrossRef]
185. Lee, H.S.; Hua, H.S.; Wang, C.H.; Yu, M.C.; Chen, B.C.; Lin, C.H. induces connective tissue growth factor expression through the TLR2-JNK-AP-1 pathway in human lung fibroblasts. *FASEB J.* **2019**, *33*, 12554–12564. [CrossRef]
186. Ali, M.K.; Kim, R.Y.; Karim, R.; Mayall, J.R.; Martin, K.L.; Shahandeh, A.; Abbasian, F.; Starkey, M.R.; Loustaud-Ratti, V.; Johnstone, D.; et al. Role of iron in the pathogenesis of respiratory disease. *Int. J. Biochem. Cell. Biol.* **2017**, *88*, 181–195. [CrossRef]
187. Scoditti, E.; Massaro, M.; Garbarino, S.; Toraldo, D.M. Role of Diet in Chronic Obstructive Pulmonary Disease Prevention and Treatment. *Nutrients* **2019**, *11*, 1357. [CrossRef]

International Journal of
Molecular Sciences

MDPI

Review

Diet-Derived Phytochemicals Targeting Colon Cancer Stem Cells and Microbiota in Colorectal Cancer

Kumar Ganesan [1,2,†], Muthukumaran Jayachandran [1,†] and Baojun Xu [1,*]

1 Food Science and Technology Programme, Beijing Normal University-Hong Kong Baptist University United International College, Zhuhai 519087, China; kumarg@hku.hk (K.G.); jmkbio1@gmail.com (M.J.)
2 Laboratory and Clinical Research Institute for Pain, Department of Anesthesiology, The University of Hong Kong, Hong Kong SAR, China
* Correspondence: baojunxu@uic.edu.hk; Tel.: +86-7563620636; Fax: +86-7563620882
† These authors contributed equally to the article as first authors.

Received: 20 April 2020; Accepted: 30 May 2020; Published: 1 June 2020

check for updates

Abstract: Colorectal cancer (CRC) is a fatal disease caused by the uncontrolled propagation and endurance of atypical colon cells. A person's lifestyle and eating pattern have significant impacts on the CRC in a positive and/or negative way. Diet-derived phytochemicals modulate the microbiome as well as targeting colon cancer stem cells (CSCs) that are found to offer significant protective effects against CRC, which were organized in an appropriate spot on the paper. All information on dietary phytochemicals, gut microbiome, CSCs, and their influence on CRC were accessed from the various databases and electronic search engines. The effectiveness of CRC can be reduced using various dietary phytochemicals or modulating microbiome that reduces or inverses the progression of a tumor as well as CSCs, which could be a promising and efficient way to reduce the burden of CRC. Phytochemicals with modulation of gut microbiome continue to be auspicious investigations in CRC through noticeable anti-tumorigenic effects and goals to CSCs, which provides new openings for cancer inhibition and treatment.

Keywords: phytochemicals; gut microbiota; colon cancer stem cells; CRC therapy

1. Introduction

Colorectal cancer (CRC) is one of the most fatal diseases and foremost causes of death globally, representing the third most common malignancy. The American Cancer Society estimated that the rough sum of CRC incidences in the United States in 2018 alone was 97,220 (colon cancer), and 43,030 (rectal cancer), which had a great influence on curative care, which exceeded $17 billion in the medical care system [1]. The CRC develops (70%) via a serious transformation of specific morphological traits, denoted as adenoma to a carcinoma sequence [2]. About 30% of CRC cases are caused due to hereditary disorder, often connected with familial adenomatous polyposis and/or hereditary non-polyposis [3]. Chronic inflammatory bowel diseases (IBD) or family history of CRC are the primary causes of CRC [4]. In economically developed countries, the mortality connected to CRC is greater than the economically developing nations, and it affects over a million people annually [5]. Several epidemiological studies have also shown different risk factors to CRC including age, family history, IBD, obesity, smoking, lack of exercise, alcohol consumption, and diet [6]. Unfortunately, the present treatments are inadequate, owing to its effective treatment, and besides, have various side effects, chemo-resistance, and recurrence of the illness.

The growing oncogenic study provides awareness about the malignancies in humans that could have a history of stem cell diseases. Rendering to the cancer stem cell (CSC) study, CRC originates from a minor portion of tumor cells in the colon that demonstrate self-renewal, pluripotency, and can recruit and sustain tumor development [7]. The cancer-developing cells or CSC were initially recognized in blood cancer, which is copious in most of the hard tumors, especially in CRC. The smaller fraction of CSC can develop the spread of tumorous tissues, in analog to target tissues that produce effective histological units, and organs. CSCs are generally tumor-initiating, self-renewal, long-lasting cells that divide asymmetrically and harvest aggressively thriving cancer progenitor cells. These cells are resistant to cytotoxic conditions, divide into the manifold, and create endless copies, characterizing clinically relevant CRC development [8]. Nowadays, the prevalence of CRC is increasing even in historically low-risk nations, including Korea, Japan, China, and Eastern Europe. A high-frequency rate of CRC has been reported in these topographical areas, which is due to the outcomes of western diets, microbiota alterations in the gut, and cancer-causing dietary components [2,9]. Being overweight and obesity are also recognized risk factors of CRC. High consumption of red meat and reduced intake of fruits and vegetables are additional key factors to the increase f the menace of CRC [10].

To alleviate the effects of CRC and understanding the colon CSCs proliferation, there is an urgent requirement to develop an innovative and safer drug for treating CRC and preventing CSCs growth. Recently, diet-derived phytochemicals or bioactive compounds have the potential to reduce the effects of CRC that upsurge many interests among researchers [7,11]. Recently, the impact of phytochemicals in decreasing the risk of CRC and the connection with CSCs are well-documented in the literature [12–14]. The actions of bioactive compounds are varied depending on distinct chemicals by targeting diverse pathways and beneficial to human health. Various preclinical investigations have been examined related to anti-cancer activities of phytocompounds in CRC models. The results suggest several novel compounds such as apigenin, betanin, α, and β-carotene, diallyl sulfide, ethyl gallate, gallic acid, resveratrol, quercetin, luteolin, silymarin [15,16]. These compounds are harmless and can be employed in synergistic treatment to decrease cancer cell growth via chemotherapeutic mediators [15].

The microbiome is comprised of the main inhabitants in the human gut, comprising of 100 trillion microbes with diverse actions that maintain the integrity of a healthy colon [17]. Undigested dietary residues in the colonic lumen are the prime energy sources for the gut microbiota, which digest those dietary residues, resulting in the formation of several active metabolites with favorable functions. Imbalance of gut microbiota or dysbiosis can lead to several pathologies, including infectious diseases, gastrointestinal cancers, inflammatory bowel disease, and even obesity and diabetes. Dysbiosis may cause chronic inflammation, recognized as one of the prime causes of CRC. Earlier, our publications have also summarized the functions of gut microbiota, particularly, short-chain fatty acid synthesis with their benefits to the hosts in regulating various diseases such as diabetes, cardiovascular diseases, and cancer [18–20]. Dietary interventions or the consumption of phytochemicals is the beneficial component, which has been proved as effective in treating CRC [21–28]. Taking this into account, we aimed to review in-depth analysis of various diet-derived phytochemicals mediating the gut microbiome and its role in CRC prevention and treatment. In addition, we intend to review the dietary phytochemical interventions targeting colon CSCs on CRC prevention.

2. Diet-Derived Phytochemicals Modulate the Gut Microbiome

Earlier studies suggested the gut microbiota (*Bacteroides fragilis*, *Escherichia coli* strain NC101, *Desulfovibrio*, *Helicobacter hepaticus*, *Clostridium ramosum*, Fusobacterium, Campylobacter, *Prevotella*, etc.) in humans play a significant role and alter the immune function through pro-carcinogenic markers resulting in the etiology of CRC [20]. Altering the immune system in the gut normally enhances tumor microhabitats, and inflammation, ensuing the CRC development [19]. In recent research has also recommended genetically reformed colon bacteria, which are beneficial and are currently employed in experimental cases that outcomes are promising [29]. Furthermore, they can be greatly beneficial to the host as probiotics that inhibit CRC through alterations of microbiota and colon environment.

The consumption of natural products produces essential bioeffects in the body through multifaceted relations with gut microbiota [30,31]. Natural phytochemicals normally have fiber-rich glycosides that exist as complex molecules with the properties of lower bioavailability and lesser solubility [32].

The nature of the phytochemicals could be altered during microbial fermentation in the colon, ensuring high quantities of various byproducts with greater pharmacological activity [33]. Numerous metabolites that derived from gut microbiota may further be subject to various enzymatic cleavage by methylation, glucuronidation, glycination, or sulfation in the hepatocytes, which are then trafficked into the tissues and finally excreted into the gut [32,34]. Gut microbiota converts glucuronides to aglycones by β-glucuronidases, which can be immediately reabsorbed in the colon. Thus, the synthesis of microbial β-glucuronidase and its enterohepatic passage have possible steps to extend the holding period of phytochemicals in the host [32,34]. Rising data suggested the dietary phytometabolites derived from gut microbiota, which are capable of enhancing the bioavailability, antioxidant properties, detoxification of xenobiotics, and prebiotics function [34,35]. Furthermore, these compounds can eliminate gut pathogenic organisms, reduce oxidative DNA damage and pro-inflammatory mediators, and thus regulate normal cell division and apoptosis [36,37]. The effects of phytochemicals on gut microbiota and their anti-inflammatory effects are presented in Table 1.

2.1. Polyphenols

Polyphenols are one of the prime classes of chemicals in plants, extensively studied for their health-promoting properties [38–40]. Human diets contain varieties of polyphenols and have significant protective activities against various cancer types. Scavenging of free radicals and reducing oxidative stress are the key mechanisms by which a polyphenol can achieve [38]. Several studies confirmed the actions of polyphenols on CRC inhibition, which often interconnected with the relationship of gut microbiota [41–43]. For instance, an animal study was conducted related to cranberry polyphenols on *Akkermansia* (mucin-degrading bacterium), which protected the host from obesity, diabetes, and gut inflammation. In this study, the mice were administered with high fat and sugar diet and cranberry extract (CE) (200 mg/kg/day) for eight weeks, and the various gut microbiota were analyzed by the methods of 16S rRNA and 454 pyrosequencing. The outcomes of the study revealed the administration of CE reduced body weight, visceral fat obesity, triglyceride accumulation, and inflammation, and elevated antioxidant properties and insulin sensitivity. Furthermore, the metagenomics study of CE treatment exhibited an increased percentage of *Akkermansia* [42].

The anti-carcinogenic properties of the gut microbiota are generally attributed based on the two properties, (a) either by improving the host's immune system or (b) by generating the metabolites, which can interfere with the pathways involving CRC formation. A study demonstrated that the presence of amines, bile acids, and high consumption of meat can reduce some bacterial growth such as *Clostridium*, which inhibits the development of CRC [43]. By using the alimentary metabolites, gut microbiota produces biologically active short-chain fatty acids. The *Rosburia faecis* and *Eubacterium rectale* group of bacteria can normally produce the butyrate, which involves reducing cell apoptosis and diversity [41]. A study showed the polyphenol metabolites modulated microbiota that directly restricted the growth/proliferation of CRC [44]. Another study has also related intestinal metabolites, quercetin, chlorogenic, and caffeic acids to interfering in cyclooxygenase-2 expression resulting in the prevention of DNA damage in the colon [45]. The polyphenols-mediated gut microbiota changes are a potential technique for inhibiting colon cancer, although insufficient trials have been piloted, in which, wine [46], blueberry [47], and cocoa [48] displayed a bifidogenic outcome.

Table 1. Effects of phytochemicals on gut microbiota and their anti-inflammatory effects.

Phytochemicals	Compounds	Model	Effect on Gut Microbiota	Anti-Inflammatory Effect	References
Anthocyanins	Anthocyanins	C57BL/6J mice	Feces of gut microbiota-deficient mice showed an increase in anthocyanins and a decrease in their phenolic acid metabolites, while a corresponding increase was observed in jejunum tissue	Decreased the inflammatory status of mice	[49]
Anthocyanins	Anthocyanins	C57BL/6J mice	Treatment modified the gut microbiota composition	Effectively reduced the expression levels of IL-6 and TNFα genes, markedly increased SOD and GPx activity	[50]
Catechins	Epigallocatechin-3-gallate	C57BL/6J mice	The Firmicutes/Bacteroidetes ratio significantly lowered in HFD + EGCG, but higher in control diet + EGCG	Potential use for prevention, or therapy, for oxidative stress-induced health risks	[51]
Catechins	Epigallocatechin-3-gallate	C57BL/6J mice	Maintained the microbial ecology balance and prevented dysbiosis	Suppressed the activation of NF-κB and decrease expression of iNOS	[52]
Catechins	Epigallocatechin-3-gallate	Wistar rats	Affected the growth of certain species of gut microbiota	Suppressed the activation of NF-κB	[53]
Catechins	Quercetin	C57BL/6J mice	Increased Firmicutes/Bacteroidetes ratio and gram-negative bacteria and increased Helicobacter. Regulated gut microbiota balance	Reverted dysbiosis-mediated TLR-4, NF-κB signaling pathway activation, and related endotoxemia, with subsequent inhibition of inflammasome response and reticulum stress pathway activation	[54]
Catechins	Quercetin	Wistar rats	Attenuated Firmicutes/Bacteroidetes ratio, inhibiting the growth of bacterial species associated with diet-induced obesity (Erysipelotrichaceae, Bacillus, *Eubacterium cylindroides*). Quercetin was effective in lessening high-fat sucrose diet-induced gut microbiota dysbiosis	Suppressed the activation of NF-κB	[55]
Catechins	Quercetin	Fischer 344 rats	Exerted prebiotic properties by decreased pH, increased butyrate production, and altered gut microbiota	Suppressed the activation of NF-κB	[56]
Catechins	Kaempferol	3 T3-L1 adipocytes	Treatment modified the gut microbiota composition	Reduced LPS pro-inflammatory action, promoted anti-inflammatory and antioxidant effects	[57]
Flavonones	Baicalein	C57BL/6J mice	Firmicutes/Bacteroidetes ratio significantly lowered and regulated dysbiosis	Suppressed the activation of NF-κB and decreased the expression of iNOS and TGF-β	[58]
Organosulfur compounds	Garlic essential oil and Diallyl disulfide	C57BL/6 mice	Treatment modified the gut microbiota composition	Significantly decreased the release of pro-inflammatory cytokines in the liver, accompanied by elevated antioxidant capacity via inhibition of cytochrome P450 2E1 expression	[59]
Phenolic acid	Curcumin	Mice	A direct effect of bioactive metabolites reaching the adipose tissue rather than from changes in gut microbiota composition	Nutritional doses of *Curcuma longa* decreased proinflammatory cytokine expression in subcutaneous adipose tissue	[60]

Table 1. *Cont.*

Phytochemicals	Compounds	Model	Effect on Gut Microbiota	Anti-Inflammatory Effect	References
Phenolic acid	Curcumin	LDLR−/− mice	Improved intestinal barrier function and prevented the development of metabolic diseases	Significantly attenuated the Western diet-induced increase in plasma LPS levels	[61]
Phenolic acid	Curcumin	Human IEC lines Caco-2 and HT-29	Modulated chronic inflammatory diseases by reducing intestinal barrier dysfunction despite poor bioavailability	Significantly attenuated LPS-induced secretion of master cytokine IL-1β from IEC and macrophages. Reduced IL-1β-induced activation of p38 MAPK in IEC and subsequent increase in the expression of myosin light-chain kinase	[62]
Polyphenols	Polyphenols	C57BL/6J ApcMin mice	Bacterial diversity was higher in the bilberry group than in the other groups	Attenuation of inflammation in cloudberry-fed mice	[63]
Stilbenes	Resveratrol	Kunming mice	HF microbiomes were different from those in CT and HF-RES mice. After treatment, Lactobacillus and Bifidobacterium were significantly increased, whereas *Enterococcus faecalis* was significantly decreased, resulting in a higher abundance of Bacteroidetes and a lower abundance of Firmicutes	Decreased the inflammatory status of mice	[64]
Stilbenes	Resveratrol	Glp1r−/− mice	Treatment modified the gut microbiota composition	Decreased the inflammatory status of mice	[65]
Stilbenes	Resveratrol	Wistar rats	Trans-resveratrol supplementation alone or in combination with quercetin scarcely modified the gut microbiota profile but acted at the intestinal level, altering mRNA expression of tight-junction proteins and inflammation-associated genes	Altered mRNA expression of tight-junction proteins and inflammation-associated genes	[55]
Stilbenes	Resveratrol	Adipocytes	Treatment modified the gut microbiota composition	Resveratrol opposed the effect induced by LPS, functioning as an ameliorating factor in disease state	[66]
Stilbenes	Resveratrol	Human	Steroid metabolism of the affected gut microbiota was studied	-	[67]
Stilbenes	Piceatannol	C57BL/6 mice	Altered the composition of the gut microbiota by increasing Firmicutes and Lactobacillus and decreasing Bacteroidetes	Decreased the inflammatory status of mice	[68]
Stilbenes	Piceatannol	Zucker obese rats	It did not modify the profusion of the most abundant phyla in gut microbiota, though slight changes were observed in the abundance of several Lactobacillus, Clostridium, and Bacteroides species belonging to Firmicutes and Bacteroidetes	Showed a tendency to reduce plasma LPS by 30%	[69]

Abbreviation: Caco-2—human epithelial colorectal adenocarcinoma cells; CT—control diet; EGCG—Epigallocatechin-3-gallate; GPx—glutathione peroxidase; HF-RES—high-fat diet supplemented with resveratrol; HFD—high-fat diet; IEC—intestinal epithelial cells; IL 6—interleukin 6; iNOS—inducible nitric oxide synthase; LPS—lipopolysaccharides; MAPK—mitogen-activated protein kinase; mRNA—messenger ribonucleic acids; NF-κB—nuclear factor kappa B; SOD—superoxide dismutase; TGF β—transforming growth factor-beta; TLR-4—toll-like receptor 4; TNFα—tumor necrosis factor-alpha; P450 2E1—cytochrome P450 2E1.

4. Effect of Diet-Derived Phytochemicals on the CSCs

Signal transduction pathways, namely, Hh, Wnt/β-catenin, and Notch, contribute to a variety of usual stem cells and provide striking strategies to CSC [101,102]. Irregular cascade signaling of Wnt/β-catenin causes the majority of malignancy in most individuals [103]. Preclinical investigations have been undertaken to find small molecules, which are capable of distracting the pathway of Wnt/β-catenin [104,105]. Based on the findings, monoclonal antibodies and siRNA are promising blockers against the Wnt1/2 pathway [104,105]. However, targeting Wnt1/2 is still a primitive stage and no beneficial mediators have yet been permitted for patient practice until today [106]. Numerous bioactive chemicals have been studied in inhibiting the above-stated pathways. For example, Corn lily-derived cyclopamine that targeted hedgehog signaling [107]. Epigallocatechin gallate (EGCG) inhibited Wnt/β-catenin signaling and was found to influence CSC self-renewal and invasive abilities [108,109]. Retinoic acid is an active molecule derived from vitamin A, can downregulate the Notch signaling, and differentiate CSCs or reduce their development [110]. The Akt/mTOR signaling pathway is one of the significant pathways intricate in the CSC. This CSC existence and invasion of the stimulation of Akt/mTOR is very decisive. Declining motility and apoptosis commencement of CSC occurs repetitively, owing to Akt deterrence [35] (Figure 1).

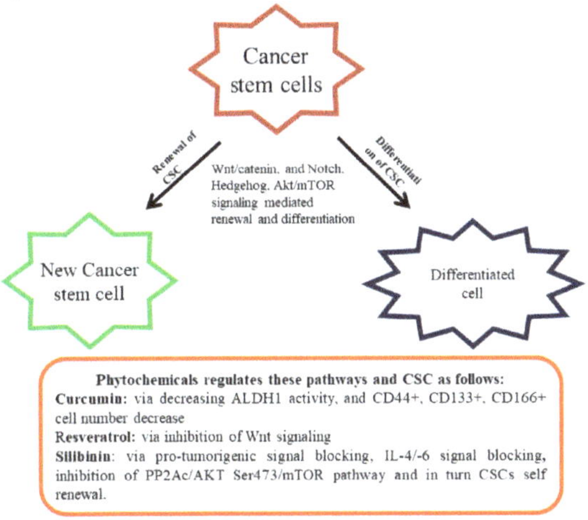

Figure 1. Renewal and differentiation of cancer stem cells (CSC). Diet-derived phytochemicals generally attenuate various signaling mediated renewal and differentiation and thereby regulate CSC proliferation.

The anti-cancer effect of polyphenols is normally achieved by the inhibition of tumor cell proliferation, and stimulation of caspase-3-dependent apoptosis via the Akt/mTOR pathway [111]. An assortment of the investigation suggested that polyphenols and flavonoids can affect various CSCs and inhibit proliferation and thus the outcomes exhibited phytochemicals are promising anti-cancer agents targeting CSCs [112]. There are several colon CSCs markers with varying functions comprising a cluster of differentiation 44 (CD44, a receptor of hyaluronic acid), CD133 (unidentified), CD166 (fixative substances), and aldehyde dehydrogenase-1 (Aldh-1 an enzyme). In tumors, CD133 is recognized as a colon cancer-originating cell. The markers of CD166 along with CD4435 or CD24/CD29 identified the populace of colorectal CSC. Aldh1 is also accepted as a novel indicator of CSCs in humans. Curcumin contributed to the control of colon CSCs and standardized the several markers of CRC stem cells. It reduced Aldh1, CD44+, CD133+, CD166+ cell numbers, and enhanced apoptosis in tumors [113]. In another study, curcumin-enhanced G2/M phase arrested and downregulated β-catenin expression [114].

An interesting study on the *Sasa quelpaertensis* extract (SQE) showed the induction of CSC variation and inhibited Wnt signaling. SQE contains high quantities of polyphenol, including *p*-coumaric acid and tricin that inhibited the renewal and differentiation of CSC [115]. In this study associated with colon, HCT116, and HT29 CSCs were labeled with respective markers (CD133+ and CD44+) and introduced into the nude mice to develop the CRC. The nude mice were administered with the SQE extract (300 mg/kg b.w) that reduced signaling of CSC marker expression and Wnt/β-catenin, as well as the hypoxia-inducible factor-1α [115]. Resveratrol, a renowned phytochemical present in several dietary sources inhibited the effect on colon CSCs through the hindrance of Wnt signaling [116]. Ellagic acid is an active principle of walnut displayed to inhibit CRC by regulating the colon CSCs [117]. Silibinin is another imperative phytochemical revealed to regulate colon CSCs via blocking of pro-tumorigenic signaling, including, IL-4/IL-6 [118]. By overwhelming the PP2Ac/AKT Ser473/ mTOR pathway, silibinin impeded colon CSCs self-renewal [92]. In an interesting study connected to the colon CSC, cinnamic acid found to reduce the CSC markers connected with HT-29 colon cancer cells [119].

5. The Anti-Tumorigenic Potential of Phytochemicals through Various Molecular Goals in Colon CSC

Globally, diet-derived phytochemicals lead to reduced CRC incidences. For incidence, Mediterranean people generally have a low prevalence of CRC, because of the high consumption of olive oil and tomato [120]. Both olive oil and tomato have phytochemical-rich dietary materials that can reduce CRC in Mediterranean individuals [120]. Various in vitro and in vivo studies showed that phytochemicals inhibit cell propagation, differentiation, angiogenesis, and anti-apoptotic activities in the colon (Figure 2 and Table 3). These diet-derived phytochemicals offered a significant success rate in numerous medical trials of CRC individuals [121–123]. A beneficial efficacy of diet-derived phytochemicals in the CRC management especially targeting of CSCs increases greater interests among researchers [124,125]. Antitumor effects of diet-derived phytochemicals are presented via four molecular targets as below.

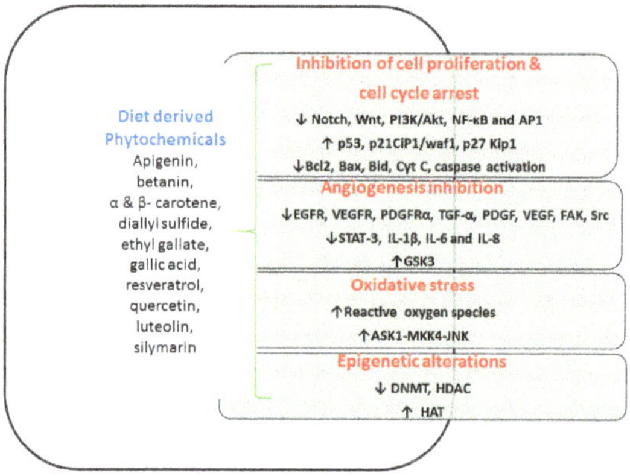

Figure 2. Various in vitro and in vivo studies showed the phytochemicals inhibit cell propagation, differentiation, angiogenesis, and anti-apoptotic activities in the colon. Abbreviation: Akt-serine/threonine-specific protein kinase; AP1—Activator protein 1; ASK1—apoptosis signal-regulating kinase 1; Bax—bcl-2-like protein; Bcl 2-B-cell lymphoma 2; Bid—BH3 Interacting Domain Death Agonist; CIP1/waf1—cyclin-dependent kinase inhibitor 1; Cyt C—cytochrome C; DNMT—DNA methyltransferase; EGFR—epidermal growth factor receptor; FAK—Focal adhesion kinase; GSK3-glycogen synthase kinase-3; HAT—histone acetyltransferases; HDAC—histone deacetylase; IL-interleukin; JNK—c-Jun N-terminal kinases; Kip1—kinesin-like protein1; MKK4—mitogen-activated protein kinase kinase 4; NF-κB—nuclear factor kappa-B; PDGF—platelet-derived growth factor; PDGFRα—platelet-derived growth factor receptor A; PI3K—Phosphoinositide 3-kinases; SrC—protooncogene c; STAT3—signal transducer and activator of transcription 3; TGFα—Transforming Growth Factor-alpha; VEGF—vascular endothelial growth factor; VEGFR—vascular endothelial growth factor receptor.

Table 3. List of phytochemicals and their anti-tumorigenic effect on colon CSC.

Dietary Phytochemical	Sources	Molecular Mechanistic Action	References
(+)-catechin, chlorogenic acid, ellagic acid, and gallic acid	Walnut phenolic extract (WPE)	WPE down-regulated the CSC markers such as CD133, DLK1, CD44, and Notch1. WPE downregulated the β-catenin/p-GSK3β signaling pathway. The CSC's self-renewing capacity was suppressed by WPE. Overall, WPE regulated the characteristics of colon CSCs.	[117]
Cinnamic acid	Fruits, vegetables, and whole grains	Cinnamic acid reduced the CSC markers associated with HT-29 colon cancer cells.	[119]
Curcumin	Turmeric	Curcumin decreased the ALDH1 activity, decreases CD44+, CD133+, CD166+ cell numbers, and induces apoptosis. Induces G2/M phase arrest, and downregulates the expression of β-catenin.	[113,114]
EGCG	Apple skin, green and black tea, onions, carob, plums, hazelnuts, and pecans.	EGCG suppressed glycoprotein; reduced the expression Wnt signaling, cell cycle, Hedgehog, Akt/mTOR, NF-κB, and VEGF pathways; Induced apoptosis.	[126]
Lycopene	Olive, tomatoes, watermelon, pink grapefruit, pink guava, papaya, seabuckthorn, wolfberry, and rosehip	Downregulated Akt/mTOR, and VEGF, Epigenetic alterations	[127]
p-Coumaric Acid and tricin	Sasa quelpaertensis extract (SQE)	Induced CSC differentiation and inhibited Wnt signaling. Suppressed the expression of CSC markers, hypoxia-inducible factor-1α (HIF-1α) signaling, and Wnt/β-catenin signaling.	[115]
Quercetin	Leafy vegetables, broccoli, red onions, peppers, apples, grapes, black and green tea, red wine	Induced apoptosis, and downregulated Wnt, Hedgehog, NF-κB, PI3K/Akt, MRP1, 4, and 5	[128]
Resveratrol	Peanuts, pistachios, grapes, wine, blueberries, cranberries, cocoa, and dark chocolate	Resveratrol acted on colon CSCs via inhibition of Wnt signaling	[116]
Silibinin	Milk thistle seeds	Silibinin acted via pro-tumorigenic signaling blocking and IL-4/-6 signal blocking; Suppressed the activation of the PP2Ac/AKT Ser473/mTOR pathway; Inhibited tumor formation rate, tumor growth, and colon CSLCs self-renewal.	[92,118]
Sulforaphane	Broccoli Sprouts, Cauliflower, Cabbage, Brussels Sprout, Bok Choy, Collards	Reduced the expression of NF-κB, Akt/mTOR, ALDH1, Wnt signaling, Induced apoptosis, downregulated epithelial-mesenchymal transition	[129]

Abbreviation: Akt—serine/threonine-specific protein kinase; Aldh1—Aldehyde Dehydrogenase 1; CD—a cluster of differentiation; CSCs—colon cancer stem cells; DLK1—Delta Like Non-Canonical Notch Ligand 1; EGCG—epigallocatechin gallate; GSK3β—glycogen synthase kinase 3 beta; HT-29—human colorectal adenocarcinoma cells; IL—interleukin; MRP- Multidrug resistance-associated protein; mTOR—mammalian target of rapamycin; NF-κB—nuclear factor kappa-B; PP2Ac- Protein phosphatase 2A homologs, catalytic domain; VEGF—vascular endothelial growth factor.

5.1. *Inhibition of Cell Multiplication and Cell Cycle Progression*

Colon CSCs have the ability of proliferation and metastatic effect with atypical maintenance of numerous signaling pathways, accountable for malignancy. Diet-derived phytocompounds that are connected to multiple signalings, such as PI3K/Akt, Hh, Wnt, and Notch could be beneficial healing approaches in managing CSCs induced malignancy. The unusual stimulus of NF-κB signaling normally accelerates malignant cell proliferation that averts apoptosis [130]. Phytocompounds contribute to the initiation of this apoptosis, prevent cell division with cell cycle growth, and hence phytocompounds are a great attractive drug candidature for tumor therapy. Various cancer models connected with phytochemicals that have established with the upregulation of proapoptotic proteins (Bax, and Cyt C), triggers caspase cascade, and cleavage of poly (ADP-Ribose) polymerase and thus regulates cancer development [105,131]. Diet-derived phytochemicals such as curcumin, EGCG, and lycopene demonstrated an ability to increase apoptosis via induction of p53-dependent Bax, upregulating p21waf1/Cip1, and p27Kip1 CDK inhibitors and thus repressed the normal cell cycle [132,133].

Likewise, isothiocyanates exhibited a reduction in the incidence of CRC through elevated apoptosis, cessation of the cell cycle, and self-renewal of CSCs [15]. Curcumin, gingerol, EGCG, and resveratrol inhibited the signaling of Notch, Wnt signaling, β-catenin/TCF transcription as well as targets to avert

CSC self-renewal [134,135]. Sulforaphane is generally acquired from broccoli, which is effective in preventing colon CSCs proliferation through modulation of multiple signaling pathways, comprising PI3K-Akt, NF-κB, Hh, Wnt/β-catenin [136,137].

5.2. Inhibition of Angiogenesis Mechanism

Angiogenesis supports CRC initiation, development, and metastasis and its suppression provides a fascinating strategy for the treatment of CRC. Diet-derived phytochemicals reduce angiogenesis through several pathways. Curcumin gingerol, and EGCG inhibited Wnt signaling with various receptors of the epidermal growth factor (EGFR), vascular endothelial growth factors (VEGFR-1, VEGFR-2, and VEGFR-3) and downregulated IL-1β, IL-6, and IL-8 and thus these compounds inhibited chemoresistance, angiogenesis, and invasion [138]. Experiments validated that dose-dependent manners of curcumin prevented interleukin from the gut and inhibited angiogenesis and CSCs stimulation [138]. EGCG impeded angiogenesis and growth of the tumors through the activation of receptors of EGFR and platelet-derived growth factor receptor-α (PDGFRα) [139]. Studies established that capsaicin inhibited CRC-provoked angiogenesis through the reduction of the STAT-3 facilitated downstream mechanism [140]. Isoflavones also suppressed Wnt signaling by augmenting glycogen synthase kinase expression, fixes with β-catenin resulting in elevated phosphorylation, and successively decreased CRC development [141].

5.3. Oxidative Stress and Anti-Tumorigenic Effect

Investigators established that CSCs in many tumor cells contain a negligible concentration of reactive oxygen species (ROS) and these quantities are dynamic for preserving normal stem cell functions [142]. These ROS conservations in normal cells as well as CSCs are greatly important. The beneficial outcome of elevated ROS eradicates CSCs, which can be one of the vital goals for CRC treatment. Hence, the increased ROS plays as "double-edged sword", which is not only an illness maker but also as a missile in tumor treatment. Curcumin has contradictory roles in hunting and creating ROS, and however, the consumption of dietary curcumin possesses a potential anticancer activity. Curcumin-induced ROS generation and their oxidative stress that largely induced cell apoptosis in HT29 cell lines through the activation of signaling cascade ASK1-MKK4-JNK [143]. Studies revealed the twin function of lycopene as a ROS scavenger and creator based on its dose-dependent manner. Ribeiro et al. [144] established oxidative stress in the HT29 cell line, resulting in functional DNA impairment, which was greatly secured by lycopene (1-3 μM); however, DNA damage is amplified while lycopene treatment in higher concentrations (4-10 μM). Capsaicin-stimulated apoptosis in human CRC cell lines, which is connected with an upsurge production of ROS and disruption of membrane potential in mitochondria [145].

5.4. Epigenetic Alterations

There are three enzymes viz., DNA methyltransferases (DNMTs), histone acetyltransferases (HATs), and histone deacetylases (HDACs) play an energetic function in chromatin organization and direction of transcription. HAT activity is connected to dynamic chromatin in transcription, while, DNMTs and HDACs induce silencing of the gene. The disparity of DNA methylation and histone acetylation/deacetylation often contributes to cancer. Multiple signaling pathways in CRC comprise Wnt/β-catenin, Hh, Notch, and TGF-β/BMP provide self-renewal, and variation in stem cells that are regularly modulated by epigenetic mechanisms. The mechanism of HDAC inhibition is an extensive platform of anti-tumorigenic effects comprising cell cycle arrest, apoptosis, and cell differentiation that have fascinated new consideration as possible anticancer candidates. Various researchers have recommended that curcumin, phenyl isothiocyanate, EGCG have anti-tumorigenic properties that are possibly mediated through an epigenetic mechanism by DNMTs and HATs inhibition [146,147].

6. Effect of the Gut Microbiome on Colon CSCs and CRC

The key factors contributing to CRC are colon CSCs and diet, which is a renowned and significant environmental factor connected to CRC. The metabolites from gut microbiota have the potential to be either tumorigenic or anti-tumorigenic agents. Intestinal microbiota produces short-chain fatty acids (SCFAs) from the dietary fibers through fermentation in the host. SCFAs are aliphatic carbon-based acids, in which the most abundant SCFAs are acetate (C2), propionate (C3), and butyrate (C4) [148]. These SCFAs are shown to exert numerous beneficial effects on the host's energy metabolism. The dietary fibers reach into the large intestine without undergoing any course of digestion; the reason is owing to the absence of dietary digestive enzymes in the upper intestinal tract. The gut microbiota present in the large intestine is accountable for the breakdown of these dietary fibers into active metabolites [149]. One noteworthy beneficial effect of SCFAs with over the host immune system is butyrate. Butyrate is a metabolic product of dietary fiber and resistant starch by the bacterial action (*Faecalibacterium prausnitzii* and *Eubacterium rectale*) in the colonic lumen. The literature claimed that the butyrate can induce G1 phase-cell cycle arrest, cell differentiation, and apoptosis in CRC [150]. As we stated earlier only a small populace of cells is accountable for the initial generation of malignancy cells and referred to as CSCs. The method of traditional malignancy therapy is proven ineffective against these CSCs. The CSC holds a specific cluster of differentiation (CD) markers on their surface, hence targeting those CD markers may be the best way to target the CSCs [151]. In a recent study on the effects of butyrate on the colon CSCs, sodium butyrate (NaB) employed in CRC stem cells (human) type HCT116. The investigation was carried out by analyzing the expression profiles of a definite marker for CRC stem cells, including CD24, CD133, and CD44. These results revealed that the SCFA-NaB had variable impacts on HCT116 stem cells (CD24, CD44, and CD133). The results still varied bestowing on the concentration of NaB and incubation time. Overall, this study offers some interesting information on NaB and whether it is possible to develop it as a novel therapeutic drug targeting cancer stem cells [152] (Figure 3).

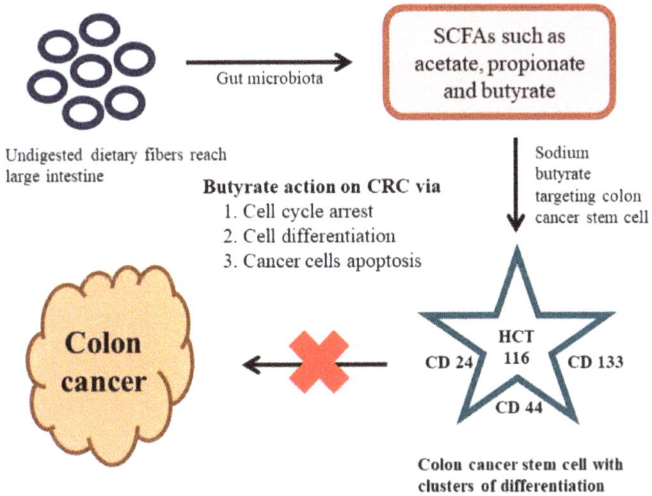

Figure 3. Effect of the gut microbiome on colon cancer stemm cells (CSCs) and colorectal cancer (CRC).

Currently, the studies on the complexity of microbiota associated with CRC revealed that microbiota unconditionally influences CRC at high risk due to the range and complexity of the gut microbiota. Despite the contemporary debate, regarding whether alterations in the microbiota give rise to colon carcinogenesis, in which, some noteworthy explanations have been made to recommend a causative function of the gut microbiota in the CRC. The rodent model investigations employed as natural, chemically-induced, or genetically predisposed CRC, which revealed the enhancing tumorigenic properties of microbiota and their effects on the development of CRC [97,100]. These tumorigenicity effects are attributed to inflammation, which plays as a cancer inducer in animals. Dysbiosis has also detected during the exposure of subsequent radiation in animals, which represented the vulnerability of the microbiota; additionally, dysbiosis may play as a pilot and facilitate the formation of CRC [153] (Figure 4). Eubiosis is referred to as the balance microbiome status maintaining healthy human body conditions [154]. Upon the impaired eubiosis, macrophages generally produce TGF-β, IL-6, and TNF; and T cells produce the pro-inflammatory Th17 cells by the differentiation of CD4 T cells; and thus, cause an adaptive immune response. The Th17 cell is abundant in the mucosal inflammation, which leads to CRC development [155]. The commensal bacteria, *Clostridia* species, can promote the overproduction of Th17 cells, leads to increased IL-17 generation in the epithelial cells. It is well established that Th17 acted as a driving force for the initiation of CRC Min-mouse models exposing the animals to enterotoxigenic *Bacteroides fragilis* [156] (Figure 4). These findings suggest that the inflammatory process plays a pivotal role among the gut microbiota and CRC. The pathogenic bacteria stimulate cancer formation through diverse mechanisms, including (a) dysbiosis and inflammation induced by a microorganism-associated molecular pattern (MAMP) triggering toll-like receptor (TLR) and additional pattern recognition receptors (PRR); (b) detrimental effects are intervened by bacterial toxins such as colibactin and CDT, and (c) acetaldehydes and nitrosamines by activating toxins through metabolic activities [157].

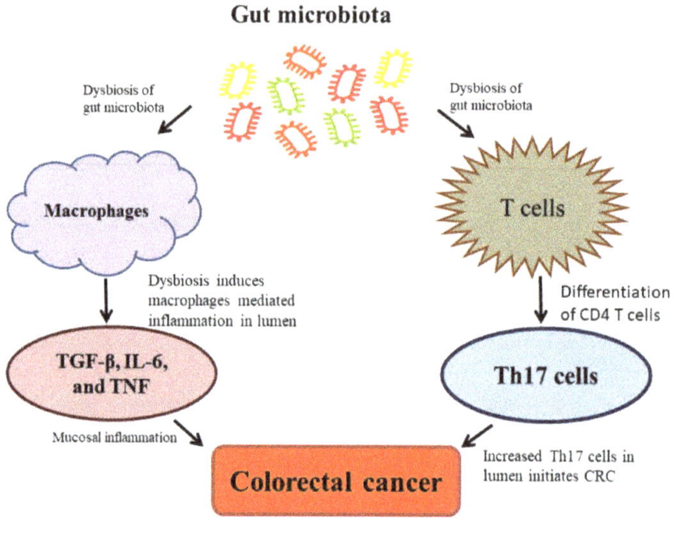

Figure 4. Dysbiosis of gut microbiota causes a high risk of colorectal cancer (CRC).

Nevertheless, the stimulus of the innate immunity achieved by toll-like receptor and agonists of the NOD-like receptor has been established as possible innate immunity and increase of anti-tumor activity [158]. Taking this into account, these explanations recommend an ambiguous role of the gut microbiota in carcinogenesis that may be reliant on the grade and mechanism. Microbial diversity is considerably poorer in tumor tissues matched with noncancerous tissues, proposing that a more appropriate microhabitat occurs in the vicinity to gut tissue. In CRC patients, the higher abundances of Erysipelotrichaceae, Prevotellaceae, Coriobacteriaceae, Lactobacillales, *Fusobacterium, Porphyromonas, Peptostreptococcus, Mogibacterium, Escherichia- Shigella, Prevotella* and lower loads of *Bifidobacterium, Faecalibacterium, Blautia, Staphylococcus*, and *Bacillus* were found [156–158]. These observations recommend certain bacteria may well compete in the converted niche and disclose novel steps in which the microbiota influences CRC development.

Gut Microbiome Regulates Wnt/β-catenin Signaling Pathways

Adult CSC has the properties of self-renewal and targeting for cancer-originating mutation [159]. Elevated mutations in colon CSC ensuing the changes in variation/plasticity and site of the stem cell/propagation are the most represented primary sign for colon tumorigenesis [160]. The initiation of colon tumorigenesis is frequently determined by mutations in the Wnt signaling. Wnt is generally a secreted signaling protein. Conversely, the loss of function of adenomatous polyposis coli or gain of function of β-catenin causes the balance of unrestricted β-catenin that provides abnormal Wnt signaling leads to tumorigenesis [161]. Notably, mutation triggering of the Wnt pathway in G-protein-coupled receptor ($Lgr5^+$) cells contributes to intestinal tumors with high competence compared to other colonic cell tumors [162]. According to the CSC hypothesis, the populace of colon cells can propagate tumor generation, measured as multipotent resulting in the cell of cancer [162,163]. Current data shows that multiple CSC hierarchies occur in the colon, facilitate cell fate in the account for various extrinsic factors including, diet, inflammation, and body anxiety [164]. Additionally, a function of diet in the maintenance of colon CSC has also described [165].

The Wnt signaling generally occurs in an upstream of the β-catenin pathway [166] (Figure 5). Briefly, Wnt ligands largely fix with the complex of the Frizzled/LRP co-receptor, which triggers the canonical pathway. Axin, a Wnt signaling inhibitor protein is employed to the cell membrane, resulting in the inactivation of the adenomatous polyposis coli complex succeeding in the equilibrium of β-catenin. When Wnt is triggered, β-catenin is instantly soothed, allowing transfer to the nucleus and fixes with T cell factor and eventually elicits the expression of target genes. Among them, Leucine-rich repeat-containing $Lgr5^+$ genes participated in stem cell proliferation [167].

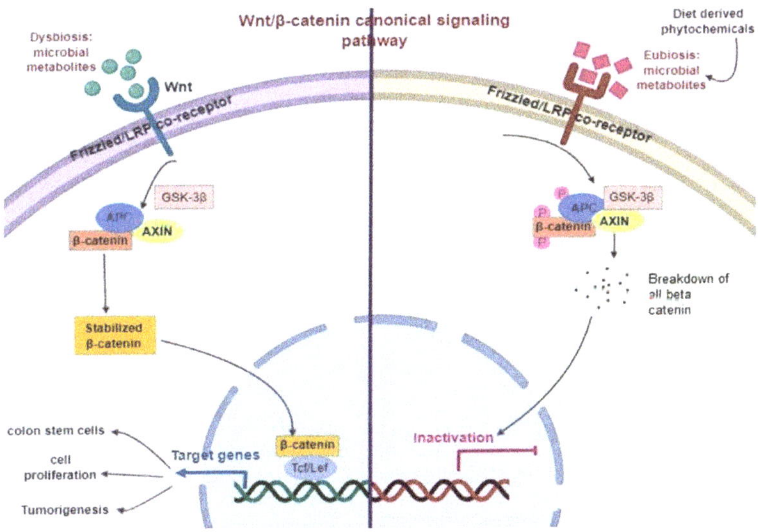

Figure 5. Diet-derived phytochemicals stabilize the microbiome status (Eubiosis) that inhibits Wnt/β-catenin signaling pathways successively prevent intestinal infection and inflammation.

The adenomatous polyposis coli is normally a tumor-suppressor protein that is mutated in almost 80% of CRC. Thus, the stimulation of Wnt/β-catenin is a primary biomarker of colitis-related CRC [168]. Diet-derived phytochemicals balance the microbiome status (Eubiosis), which inhibits Wnt/β-catenin signaling pathways and successively prevent intestinal infection and inflammation [96,169].

7. The Triangular Relationship between Phytochemicals, Gut Microbiome, and CSCs

Gut microbiota is chiefly affected by the dietary phytochemicals that can disturb its physiological relations in the host [170]. Through their alimentary canal route, phytochemicals are digested by colonic bacteria and produce several by-products [171]. These phytochemicals are rich in various active principles comprising polyphenols and flavonoids that upsurge the *Firmicutes, Bacteroidetes, Actinobacteria,* and *Proteobacteria* [30,171], which alters the pH of the colon environment and maintains the balance of the colonic microbiome [172]. Therefore, the effect of colonic bacteria on the dietary phytochemicals targeting dietary intervention which may contribute to host well-being [171]. The phytochemicals facilitate colonic bacteria, which may influence as adjuvants to treat cancer, obesity, diabetes, and chronic inflammatory diseases and prove as potentially prophylactics and candidates for the treatment of these diseases [30,171]. Furthermore, diet-derived phytochemical modulates colonic microbiota that targeting CSCs recognized as capable of decreasing the burden of CRC by triangle relationship (Figure 6).

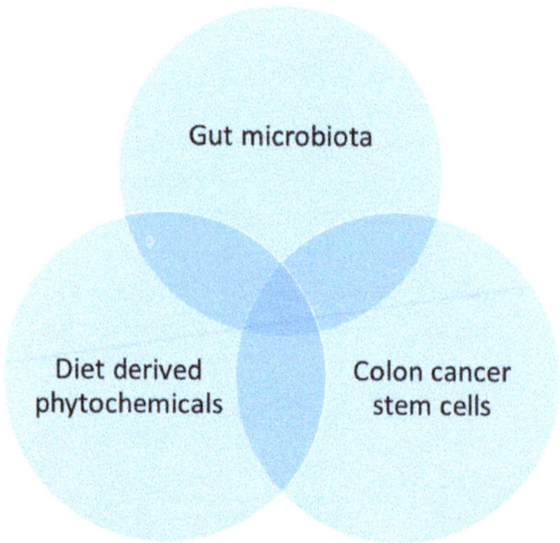

Figure 6. Triangular relationship between phytochemicals, gut microbiome, and cancer stem cells (CSCs).

The accumulating data put forward to the etiology of CRC, which is linked through the actions of colonic bacteria not only because of the pro-carcinogenic actions of particular pathogens but also other bacterial communities, especially their metabolome [173]. The multipotent colon CSCs undergo self-renewal during the asymmetric cell division and produce a populace of transit magnifying cells in CRC [174]. These cells undergo migration, proliferation, and differentiation to produce mature tumors and cancer progenitors. Uncontrolled proliferation or cell division of CSCs can repopulate [175].

Several phytochemicals including cinnamic acid, curcumin, EGCG, lycopene, quercetin, resveratrol, silibinin have been described to interfere with various regulatory pathways in the preservation of CSCs or to modify the CSC phenotype [12]. Notch, Hedgehog, and Wnt/β-catenin signaling pathways are the central signaling pathways, and they are involving in the self-renewal and differentiation of CSCs [102]. Thus, synergistic activities are anticipated when the CSC-directing phytochemicals and modulating colonic bacteria. By considering the above facts, CSCs influence a significant function in the tumor formation, targeting various signaling pathways and involves in the cancer development that may gain much interest in the field of cancer prevention via phytochemicals modulated colonic microbiota [176]. It is clearly understood that diet-derived phytochemicals undergo various alteration in the colonic bacteria and vice versa, various phytochemicals could regulate the colonic CSCs has also found to modify the gut microbiota population through triangular rapport, which may benefit to the host in combating the CRC.

8. Conclusions

CRC remains a significant threat to human society. Several investigations have elucidated the actions of several phytochemicals on the colon carcinogenesis via regulating several pathways, new insights into the relationships among the phytochemicals and colonic bacteria seem interesting and promising. Phytochemicals are a concoction of various bioactive compounds directing various cell signaling pathways that altered gut microbiota composition. This may support to destroying malignant cells with minor risks of emerging drug resistance. Dietary phytochemicals, or bioactive compounds and their analogs offer the advance of better-quality drugs that may ultimately provide the resolution to eradicate CSCs. These bioactive compounds would reinforce gut microbiota and combat against dreaded CRC. Dietary phytochemical-induced gut microbiota continues to be an encouraging and dynamic research niche in the upcoming days with evident anti-tumorigenesis effects and goals of abolishing the CSCs; propose novel opportunities for CRC prevention and treatment.

Author Contributions: All authors conceived and designed the review; K.G. and M.J. wrote the paper and B.X. critically revised and improved the manuscript. All authors have read and agreed to the published version of the manuscript.

Funding: This study is jointly supported by two research grants R201714 and R201914 from Beijing Normal University-Hong Kong Baptist University United International College, Zhuhai, China.

Acknowledgments: The authors sincerely thank the faculty members and technical staff in the Food Science and Technology Programme, Beijing Normal University-Hong Kong Baptist University United International College, Zhuhai, China who provided a great arrangement of support and cooperation in the study.

Conflicts of Interest: The authors declare no conflict of interest. The funders had no role in the design of the study; in the collection, analyses, or interpretation of data; in the writing of the manuscript; or in the decision to publish the results.

Abbreviations

Akt	serine/threonine-specific protein kinase
Aldh1	aldehyde Dehydrogenase 1
AOM	azoxymethane
AP1	activator protein 1
APC	adenomatous polyposis coli
Apc^{Min}	adenomatous polyposis coli/ multiple intestinal neoplasia
ASK1	apoptosis signal-regulating kinase 1
b.w.	body weight
Bax	bcl-2-like protein
Bcl 2	B-cell lymphoma 2
Bid	BH3 Interacting domain death agonist
Caco-2	human epithelial colorectal adenocarcinoma cells
CD	cluster of differentiation
Cdx2	human caudal type homeobox 2
CIP1/waf1	cyclin-dependent kinase inhibitor 1
c-MYC	myc protein
CRC	colorectal cancer
CSCs	colon cancer stem cells
Cyt C	cytochrome C
DLK1	delta like non-canonical notch ligand 1
DMH	1,2-Dimethylhydrazine
DNA	deoxyribonucleic acid
DNMT	DNA methyltransferase
DSS	dextran sodium sulfate

EGCG	Epigallocatechin-3-gallate
EGFR	epidermal growth factor receptor
EpCAM	epithelial cell adhesion molecule
EphB	ephrin B
ESA	epithelial surface antigen
FAK	Focal adhesion kinase
GPx	glutathione peroxidase
GSK3	glycogen synthase kinase-3
HAT	histone acetyltransferases
HDAC	histone deacetylase
HFD	high-fat diet
Hh	hedgehog
HT-29	human colorectal adenocarcinoma cells
IBD	chronic inflammatory bowel disease
IEC lines	intestinal epithelial cell lines
IL	interleukin
iNOS	inducible nitric oxide synthase
JNK	c-Jun N-terminal kinases
Kip1	kinesin-like protein1
LDLR	low-density lipoprotein receptor
LGR5	leucine-rich repeat-containing G protein-coupled receptor 5
LPS	lipopolysaccharides
MAM-GlcUA	methyl azoxy methanol-beta-D-glucosiduronic acid
MAPK	mitogen-activated protein kinase
MKK4	mitogen-activated protein kinase kinase 4
mRNA	messenger ribonucleic acid
MRP	Multidrug resistance-associated protein
mTOR	mammalian target of rapamycin
NF-κB	nuclear factor kappa-B
NO	nitric oxide
Nod	nucleotide-binding oligomerization domain-containing protein
PDGF	platelet-derived growth factor
PDGFRα	platelet-derived growth factor receptor A
PI3K	Phosphoinositide 3-kinases
PP2Ac	Protein phosphatase 2A homologs, catalytic domain
siRNA	Small interfering RNA
SOD	superoxide dismutase
SrC	protooncogene c
STAT3	signal transducer and activator of transcription 3
Tcf4	T-cell factor 4
TGFα	Transforming Growth Factor-alpha
TLR-4	Toll-like receptor 4
TNF-α	tumor necrosis factor-alpha
VEGF	vascular endothelial growth factor
VEGFR	vascular endothelial growth factor receptor

References

1. Tarver, T. Cancer Facts & Figures 2012. American Cancer Society (ACS). *J. Consum. Health. Internet.* **2012**, *16*, 366–367. [CrossRef]
2. Jemal, A.; Center, M.M.; DeSantis, C.; Ward, E.M. Global patterns of cancer incidence and mortality rates and trends. *Cancer Epidemiol. Biomarkers Preven.* **2010**, *19*, 1893–1907. [CrossRef] [PubMed]
3. Chen, E.; Xu, X.; Liu, T. Hereditary nonpolyposis colorectal cancer and cancer syndromes: Recent basic and clinical discoveries. *J. Oncol.* **2018**, *2018*, 1–11. [CrossRef] [PubMed]

4. Kim, N.H.; Jung, Y.S.; Park, J.H.; Park, D.I.; Sohn, C.I. Association between family history of colorectal cancer and the risk of metachronous colorectal neoplasia following polypectomy in patients aged <50 years. *J. Gastroenterol. Hepatol.* **2018**, *34*, 383–389. [CrossRef]
5. Sahni, M.; Goel, A.; Pande, P.; Kumar, K. Multiple gastrointestinal cancers in a single patient—A rare clinical entity. *Indian, J. Surg. Oncol.* **2018**, *9*, 633–635. [CrossRef]
6. Ganesan, K.; Sukalingam, K.; Xu, B. Impact of consumption of repeatedly heated cooking oils on the incidence of various cancers- A critical review. *Crit. Rev. Food Sci. Nutr.* **2017**, 1–18. [CrossRef]
7. Ganesan, K.; Xu, B. Molecular targets of vitexin and isovitexin in cancer therapy: A critical review. *Ann. N.Y. Acad. Sci.* **2017**, *1401*, 102–113. [CrossRef]
8. Hatano, Y.; Fukuda, S.; Hisamatsu, K.; Hirata, A.; Hara, A.; Tomita, H. Multifaceted interpretation of colon cancer stem cells. *Int. J. Mol. Sci.* **2017**, *18*, 1446. [CrossRef]
9. Whalen, K.A.; McCullough, M.; Flanders, W.D.; Hartman, T.J.; Judd, S.; Bostick, R.M. Paleolithic and Mediterranean diet pattern scores and risk of incident, sporadic colorectal adenomas. *Am. J. Epidemiol.* **2014**, *180*, 1088–1097. [CrossRef]
10. Conlon, M.; Bird, A. The impact of diet and lifestyle on gut microbiota and human health. *Nutrients* **2014**, *7*, 17–44. [CrossRef]
11. Kuppusamy, P.; Yusoff, M.M.; Maniam, G.P.; Ichwan, S.J.A.; Soundharrajan, I.; Govindan, N. Nutraceuticals as potential therapeutic agents for colon cancer: A review. *Acta Pharma. Sin. B* **2014**, *4*, 173–181. [CrossRef] [PubMed]
12. Scarpa, E.-S.; Ninfali, P. Phytochemicals as innovative therapeutic tools against cancer stem cells. *Int. J. Mol. Sci.* **2015**, *16*, 15727–15742. [CrossRef] [PubMed]
13. Yoshida, T.; Maoka, T.; Das, S.K.; Kanazawa, K.; Horinaka, M.; Wakada, M.; Satomi, Y.; Nishino, H.; Sakai, T. Halocynthiaxanthin and Peridinin sensitize colon cancer cell lines to tumor necrosis factor-related apoptosis-inducing ligand. *Mol. Cancer Res.* **2007**, *5*, 615–625. [CrossRef] [PubMed]
14. Singh, R.K.; Chang, H.-W.; Yan, D.; Lee, K.M.; Ucmak, D.; Wong, K.; Abrouk, M.; Farahnik, B.; Nakamura, M.; Zhu, T.H.; et al. Influence of diet on the gut microbiome and implications for human health. *J. Translat. Med.* **2017**, *15*, 73. [CrossRef]
15. Singh, S.V.; Singh, K. Cancer chemoprevention with dietary isothiocyanates mature for clinical translational research. *Carcinogenesis* **2012**, *33*, 1833–1842. [CrossRef]
16. Chen, C.-Y.; Kao, C.-L.; Liu, C.-M. The cancer prevention, anti-inflammatory and anti-oxidation of bioactive phytochemicals targeting the TLR4 signaling pathway. *Int. J. Mol. Sci.* **2018**, *19*, 2729. [CrossRef]
17. Czank, C.; Cassidy, A.; Zhang, Q.; Morrison, D.J.; Preston, T.; Kroon, P.A.; Botting, N.P.; Kay, C.D. Human metabolism and elimination of the anthocyanin, cyanidin-3-glucoside: A 13C-tracer study. *Am. J. Clin. Nutr.* **2013**, *97*, 995–1003. [CrossRef]
18. Jayachandran, M.; Xiao, J.; Xu, B. A Critical review on health promoting benefits of edible mushrooms through gut microbiota. *Int. J. Mol. Sci.* **2017**, *18*, 1934. [CrossRef]
19. Ganesan, K.; Chung, S.K.; Vanamala, J.; Xu, B. Causal relationship between diet-induced gut microbiota changes and diabetes: A novel strategy to transplant *Faecalibacterium prausnitzii* in preventing diabetes. *Int. J. Mol. Sci.* **2018**, *19*, 3720. [CrossRef]
20. Ganesan, K.; Guo, S.; Fayyaz, S.; Zhang, G.; Xu, B. Targeting programmed *Fusobacterium nucleatum* Fap2 for colorectal cancer therapy. *Cancers* **2019**, *11*, 1592. [CrossRef]
21. Ganesan, K.; Xu, B. Anti-diabetic effects and mechanisms of dietary polysaccharides. *Molecules* **2019**, *24*, 2556. [CrossRef] [PubMed]
22. Ganesan, K.; Xu, B. Anti-obesity effects of medicinal and edible mushrooms. *Molecules* **2018**, *23*, 2880. [CrossRef] [PubMed]
23. Ganesan, K.; Jayachandran, M.; Xu, B. A critical review on hepatoprotective effects of bioactive food components. *Crit. Rev. Food Sci. Nutrit.* **2017**, *58*, 1165–1229. [CrossRef]
24. Ganesan, K.; Xu, B. A critical review on polyphenols and health benefits of black soybeans. *Nutrients* **2017**, *9*, 455. [CrossRef]
25. Ganesan, K.; Mickymaray, S.; Alfaiz, F.A.; Thatchinamoorthi, R.; Aboody, M.S.A.; Xu, B. Immunomodulatory and antineoplastic efficacy of common spices and their connection with phenolic antioxidants. *Bioact. Compd. Health Dis.* **2020**, *3*, 15–31. [CrossRef]

26. Islam, T.; Ganesan, K.; Xu, B. New insight into mycochemical profiles and antioxidant potential of edible and medicinal mushrooms: A review. *Int. J. Med. Mushrooms* **2019**, *21*, 237–251. [CrossRef]
27. Ganesan, K.; Xu, B. Telomerase inhibitors from natural products and their anticancer potential. *Int. J. Mol. Sci.* **2017**, *19*, 13. [CrossRef]
28. Sukalingam, K.; Ganesan, K.; Xu, B. *Trianthema portulacastrum* L. (giant pigweed): Phytochemistry and pharmacological properties. *Phytochem. Rev.* **2017**, *16*, 461–478. [CrossRef]
29. Abreu, M.T.; Peek, R.M. Gastrointestinal malignancy and the microbiome. *Gastroenterol.* **2014**, *146*, 1534–1546.e1533. [CrossRef]
30. Eid, H.M.; Wright, M.L.; Anil Kumar, N.V.; Qawasmeh, A.; Hassan, S.T.S.; Mocan, A.; Nabavi, S.M.; Rastrelli, L.; Atanasov, A.G.; Haddad, P.S. Significance of microbiota in obesity and metabolic diseases and the modulatory potential by medicinal plant and food ingredients. *Front. Pharmacol.* **2017**, *8*, 387. [CrossRef]
31. Espín, J.C.; González-Sarrías, A.; Tomás-Barberán, F.A. The gut microbiota: A key factor in the therapeutic effects of (poly)phenols. *Biochem. Pharmacol.* **2017**, *139*, 82–93. [CrossRef] [PubMed]
32. Bohn, T.; McDougall, G.J.; Alegría, A.; Alminger, M.; Arrigoni, E.; Aura, A.-M.; Brito, C.; Cilla, A.; El, S.N.; Karakaya, S.; et al. Mind the gap-deficits in our knowledge of aspects impacting the bioavailability of phytochemicals and their metabolites-a position paper focusing on carotenoids and polyphenols. *Mol. Nutr. Food Res.* **2015**, *59*, 1307–1323. [CrossRef] [PubMed]
33. Marín, L.; Miguélez, E.M.; Villar, C.J.; Lombó, F. Bioavailability of dietary polyphenols and gut microbiota metabolism: Antimicrobial properties. *BioMed Res. Int.* **2015**, *2015*, 1–18. [CrossRef]
34. Louis, P.; Hold, G.L.; Flint, H.J. The gut microbiota, bacterial metabolites and colorectal cancer. *Nat. Rev. Microbiol.* **2014**, *12*, 661–672. [CrossRef] [PubMed]
35. Ozdal, T.; Sela, D.A.; Xiao, J.; Boyacioglu, D.; Chen, F.; Capanoglu, E. The reciprocal interactions between polyphenols and gut microbiota and effects on bioaccessibility. *Nutrients* **2016**, *8*, 78. [CrossRef] [PubMed]
36. Del Rio, D.; Rodriguez-Mateos, A.; Spencer, J.P.E.; Tognolini, M.; Borges, G.; Crozier, A. Dietary (poly)phenolics in human health: Structures, bioavailability, and evidence of protective effects against chronic diseases. *Antioxidants Redox Signal.* **2013**, *18*, 1818–1892. [CrossRef] [PubMed]
37. Etxeberria, U.; Fernández-Quintela, A.; Milagro, F.I.; Aguirre, L.; Martínez, J.A.; Portillo, M.P. Impact of polyphenols and polyphenol-rich dietary sources on gut microbiota composition. *J. Agric. Food Chem.* **2013**, *61*, 9517–9533. [CrossRef]
38. Ganesan, K.; Xu, B. Polyphenol-rich dry common beans (*Phaseolus vulgaris* L.) and their health benefits. *Int. J. Mol. Sci.* **2017**, *18*, 2331. [CrossRef]
39. Ganesan, K.; Xu, B. Polyphenol-rich lentils and their health promoting effects. *Int. J. Mol. Sci.* **2017**, *18*, 2390. [CrossRef]
40. Ganesan, K.; Xu, B. A critical review on phytochemical profile and health promoting effects of mung bean (*Vigna radiata*). *Food Sci. Hum. Wellness* **2018**, *7*, 11–33. [CrossRef]
41. Dueñas, M.; Cueva, C.; Muñoz-González, I.; Jiménez-Girón, A.; Sánchez-Patán, F.; Santos-Buelga, C.; Moreno-Arribas, M.; Bartolomé, B. Studies on modulation of gut microbiota by wine polyphenols: From isolated cultures to omic approaches. *Antioxidants* **2015**, *4*, 1–21. [CrossRef] [PubMed]
42. Anhê, F.F.; Roy, D.; Pilon, G.; Dudonné, S.; Matamoros, S.; Varin, T.V.; Garofalo, C.; Moine, Q.; Desjardins, Y.; Levy, E.; et al. A polyphenol-rich cranberry extract protects from diet-induced obesity, insulin resistance and intestinal inflammation in association with increased *Akkermansia spp.* population in the gut microbiota of mice. *Gut* **2014**, *64*, 872–883. [CrossRef] [PubMed]
43. Pahle, J.; Menzel, L.; Niesler, N.; Kobelt, D.; Aumann, J.; Rivera, M.; Walther, W. Rapid eradication of colon carcinoma by *Clostridium perfringens* enterotoxin suicidal gene therapy. *BMC Cancer* **2017**, *17*, 129. [CrossRef] [PubMed]
44. Duda-Chodak, A.; Tarko, T.; Satora, P.; Sroka, P. Interaction of dietary compounds, especially polyphenols, with the intestinal microbiota: A review. *Eur. J. Nutr.* **2015**, *54*, 325–341. [CrossRef]
45. Miene, C.; Weise, A.; Glei, M. Impact of polyphenol metabolites produced by colonic microbiota on expression of COX-2 and GSTT2 in human colon cells (LT97). *Nutrit. Cancer* **2011**, *63*, 653–662. [CrossRef]
46. Barron, C.C.; Moore, J.; Tsakiridis, T.; Pickering, G.; Tsiani, E. Inhibition of human lung cancer cell proliferation and survival by wine. *Cancer Cell Int.* **2014**, *14*, 6. [CrossRef]
47. Yi, W.; Fischer, J.; Krewer, G.; Akoh, C.C. Phenolic compounds from blueberries can inhibit colon cancer cell proliferation and induce apoptosis. *J. Agric. Food Chem.* **2005**, *53*, 7320–7329. [CrossRef]

48. Martín, M.; Goya, L.; Ramos, S. Preventive effects of cocoa and cocoa antioxidants in colon cancer. *Diseases* **2016**, *4*, 6. [CrossRef]
49. Esposito, D.; Damsud, T.; Wilson, M.; Grace, M.H.; Strauch, R.; Li, X.; Lila, M.A.; Komarnytsky, S. Black currant anthocyanins attenuate weight gain and improve glucose metabolism in diet-induced obese mice with intact, but not disrupted, gut microbiome. *J. Agric. Food Chem.* **2015**, *63*, 6172–6180. [CrossRef]
50. Wu, T.; Tang, Q.; Yu, Z.; Gao, Z.; Hu, H.; Chen, W.; Zheng, X.; Yu, T. Inhibitory effects of sweet cherry anthocyanins on the obesity development in C57BL/6 mice. *Int. J. Food Sci. Nutr.* **2013**, *65*, 351–359. [CrossRef]
51. Remely, M.; Ferk, F.; Sterneder, S.; Setayesh, T.; Roth, S.; Kepcija, T.; Noorizadeh, R.; Rebhan, I.; Greunz, M.; Beckmann, J.; et al. EGCG prevents high fat diet-induced changes in gut microbiota, decreases of dna strand breaks, and changes in expression and dna methylation of *Dnmt1* and *MLH1* in C57BL/6J male mice. *Oxid. Med. Cell. Longev.* **2017**, *2017*, 1–17. [CrossRef] [PubMed]
52. Cheng, M.; Zhang, X.; Miao, Y.; Cao, J.; Wu, Z.; Weng, P. The modulatory effect of (-)-epigallocatechin 3-O-(3-O-methyl) gallate (EGCG3″Me) on intestinal microbiota of high fat diet-induced obesity mice model. *Food Res. Int.* **2017**, *92*, 9–16. [CrossRef] [PubMed]
53. Unno, T.; Sakuma, M.; Mitsuhashi, S. Effect of dietary supplementation of (-)-epigallocatechin gallate on gut microbiota and biomarkers of colonic fermentation in rats. *J. Nutr. Sci. Vitaminol.* **2014**, *60*, 213–219. [CrossRef] [PubMed]
54. Porras, D.; Nistal, E.; Martínez-Flórez, S.; Pisonero-Vaquero, S.; Olcoz, J.L.; Jover, R.; González-Gallego, J.; García-Mediavilla, M.V.; Sánchez-Campos, S. Protective effect of quercetin on high-fat diet-induced non-alcoholic fatty liver disease in mice is mediated by modulating intestinal microbiota imbalance and related gut-liver axis activation. *Free Rad. Biol. Med.* **2017**, *102*, 188–202. [CrossRef]
55. Etxeberria, U.; Arias, N.; Boqué, N.; Macarulla, M.T.; Portillo, M.P.; Martínez, J.A.; Milagro, F.I. Reshaping faecal gut microbiota composition by the intake of trans-resveratrol and quercetin in high-fat sucrose diet-fed rats. *J. Nutr. Biochem.* **2015**, *26*, 651–660. [CrossRef] [PubMed]
56. Roldán-Marín, E.; Krath, B.N.; Poulsen, M.; Binderup, M.-L.; Nielsen, T.H.; Hansen, M.; Barri, T.; Langkilde, S.; Pilar Cano, M.; Sánchez-Moreno, C.; et al. Effects of an onion by-product on bioactivity and safety markers in healthy rats. *Br. J. Nutr.* **2009**, *102*, 1574. [CrossRef]
57. Le Sage, F.; Meilhac, O.; Gonthier, M.-P. Anti-inflammatory and antioxidant effects of polyphenols extracted from Antirhea borbonica medicinal plant on adipocytes exposed to *Porphyromonas gingivalis* and *Escherichia coli* lipopolysaccharides. *Pharmacol. Res.* **2017**, *119*, 303–312. [CrossRef]
58. Pu, P.; Wang, X.-A.; Salim, M.; Zhu, L.-H.; Wang, L.; Chen, K.-J.; Xiao, J.-F.; Deng, W.; Shi, H.-W.; Jiang, H.; et al. Baicalein, a natural product, selectively activating AMPKα2 and ameliorates metabolic disorder in diet-induced mice. *Mol. Cell. Endocrinol.* **2012**, *362*, 128–138. [CrossRef]
59. Lai, Y.-S.; Chen, W.-C.; Ho, C.-T.; Lu, K.-H.; Lin, S.-H.; Tseng, H.-C.; Lin, S.-Y.; Sheen, L.-Y. Garlic essential oil protects against obesity-triggered nonalcoholic fatty liver disease through modulation of lipid metabolism and oxidative stress. *J. Agric. Food Chem.* **2014**, *62*, 5897–5906. [CrossRef]
60. Neyrinck, A.M.; Alligier, M.; Memvanga, P.B.; Névraumont, E.; Larondelle, Y.; Préat, V.; Cani, P.D.; Delzenne, N.M. *Curcuma longa* extract associated with white pepper lessens high fat diet-induced inflammation in subcutaneous adipose tissue. *PLoS ONE* **2013**, *8*, e81252. [CrossRef]
61. Ghosh, S.S.; Bie, J.; Wang, J.; Ghosh, S. Oral supplementation with non-absorbable antibiotics or Curcumin attenuates Western diet-induced atherosclerosis and glucose intolerance in LDLR−/− Mice – role of intestinal permeability and macrophage activation. *PLoS ONE* **2014**, *9*, e108577. [CrossRef] [PubMed]
62. Wang, J.; Ghosh, S.S.; Ghosh, S. Curcumin improves intestinal barrier function: Modulation of intracellular signaling, and organization of tight junctions. *Am. J. Physiol. Cell Physiol.* **2017**, *312*, C438–C445. [CrossRef] [PubMed]
63. Päivärinta, E.; Niku, M.; Maukonen, J.; Storvik, M.; Heiman-Lindh, A.; Saarela, M.; Pajari, A.-M.; Mutanen, M. Changes in intestinal immunity, gut microbiota, and expression of energy metabolism–related genes explain adenoma growth in bilberry and cloudberry-fed Apc Min mice. *Nutr. Res.* **2016**, *36*, 1285–1297. [CrossRef] [PubMed]
64. Qiao, Y.; Sun, J.; Xia, S.; Tang, X.; Shi, Y.; Le, G. Effects of resveratrol on gut microbiota and fat storage in a mouse model with high-fat-induced obesity. *Food Funct.* **2014**, *5*, 1241. [CrossRef] [PubMed]

65. Dao, T.-M.A.; Waget, A.; Klopp, P.; Serino, M.; Vachoux, C.; Pechere, L.; Drucker, D.J.; Champion, S.; Barthélemy, S.; Barra, Y.; et al. Resveratrol increases glucose induced GLP-1 secretion in mice: A mechanism which contributes to the glycemic control. *PLoS ONE* **2011**, *6*, e20700. [CrossRef]
66. Nøhr, M.K.; Kroager, T.P.; Sanggaard, K.W.; Knudsen, A.D.; Stensballe, A.; Enghild, J.J.; Ølholm, J.; Richelsen, B.; Pedersen, S.B. SILAC-MS based characterization of LPS and Resveratrol induced changes in adipocyte proteomics – Resveratrol as ameliorating factor on LPS induced changes. *PLoS ONE* **2016**, *11*, e0159747. [CrossRef]
67. Korsholm, A.; Kjær, T.; Ornstrup, M.; Pedersen, S. Comprehensive metabolomic analysis in blood, urine, fat, and muscle in men with metabolic syndrome: A randomized, placebo-controlled clinical trial on the effects of Resveratrol after four months' treatment. *Int. J. Mol. Sci.* **2017**, *18*, 554. [CrossRef]
68. Tung, Y.-C.; Lin, Y.-H.; Chen, H.-J.; Chou, S.-C.; Cheng, A.-C.; Kalyanam, N.; Ho, C.-T.; Pan, M.-H. Piceatannol exerts anti-obesity effects in C57BL/6 mice through modulating adipogenic proteins and gut microbiota. *Molecules* **2016**, *21*, 1419. [CrossRef]
69. Hijona, E.; Aguirre, L.; Pérez-Matute, P.; Villanueva-Millán, M.J.; Mosqueda-Solis, A.; Hasnaoui, M.; Nepveu, F.; Senard, J.M.; Bujanda, L.; Aldámiz-Echevarría, L.; et al. Limited beneficial effects of piceatannol supplementation on obesity complications in the obese Zucker rat: Gut microbiota, metabolic, endocrine, and cardiac aspects. *J. Physiol. Biochem.* **2016**, *72*, 567–582. [CrossRef]
70. Panche, A.N.; Diwan, A.D.; Chandra, S.R. Flavonoids: An overview. *J. Nutr. Sci.* **2016**, *5*. [CrossRef]
71. Ganesan, K.; Xu, B. Ethnobotanical studies on folkloric medicinal plants in Nainamalai, Namakkal District, Tamil Nadu, India. *Trend. Phytochem. Res.* **2017**, *1*, 153–168.
72. Wang, W.; VanAlstyne, P.C.; Irons, K.A.; Chen, S.; Stewart, J.W.; Birt, D.F. Individual and interactive effects of apigenin analogs on G2/M cell-cycle arrest in human colon carcinoma cell lines. *Nutr. Cancer* **2004**, *48*, 106–114. [CrossRef] [PubMed]
73. Aron, P.M.; Kennedy, J.A. Flavan-3-ols: Nature, occurrence and biological activity. *Mol. Nutr. Food Res.* **2008**, *52*, 79–104. [CrossRef] [PubMed]
74. Zhang, T.; Jayachandran, M.; Ganesan, K.; Xu, B. Black truffle aqueous extract attenuates oxidative stress and inflammation in STZ-induced hyperglycemic rats via Nrf2 and NF-κB pathways. *Front. Pharmacol.* **2018**, *9*, 1257. [CrossRef] [PubMed]
75. Jayachandran, M.; Zhang, T.; Ganesan, K.; Xu, B.; Chung, S.S.M. Isoquercetin ameliorates hyperglycemia and regulates key enzymes of glucose metabolism via insulin signaling pathway in streptozotocin-induced diabetic rats. *Eur. J. Pharmacol.* **2018**, *829*, 112–120. [CrossRef] [PubMed]
76. Jayachandran, M.; Wu, Z.; Ganesan, K.; Khalid, S.; Chung, S.M.; Xu, B. Isoquercetin upregulates antioxidant genes, suppresses inflammatory cytokines and regulates AMPK pathway in streptozotocin-induced diabetic rats. *Chem.Biol. Int.* **2019**, *303*, 62–69. [CrossRef]
77. Hoensch, H.; Groh, B.; Edler, L.; Kirch, W. Prospective cohort comparison of flavonoid treatment in patients with resected colorectal cancer to prevent recurrence. *World J. Gastroenterol.* **2008**, *14*, 2187. [CrossRef]
78. Murota, K.; Nakamura, Y.; Uehara, M. Flavonoid metabolism: The interaction of metabolites and gut microbiota. *Biosci. Biotechnol. Biochem.* **2018**, *82*, 600–610. [CrossRef]
79. Clavel, T.; Henderson, G.; Alpert, C.A.; Philippe, C.; Rigottier-Gois, L.; Dore, J.; Blaut, M. Intestinal bacterial communities that produce active estrogen-like compounds enterodiol and enterolactone in humans. *Appl. Environ. Microbiol.* **2005**, *71*, 6077–6085. [CrossRef]
80. Setchell, K.D.R.; Brown, N.M.; Lydeking-Olsen, E. The clinical importance of the metabolite Equol—A clue to the effectiveness of soy and its isoflavones. *J. Nutr.* **2002**, *132*, 3577–3584. [CrossRef]
81. Claesson, M.J.; Jeffery, I.B.; Conde, S.; Power, S.E.; O'Connor, E.M.; Cusack, S.; Harris, H.M.B.; Coakley, M.; Lakshminarayanan, B.; O'Sullivan, O.; et al. Gut microbiota composition correlates with diet and health in the elderly. *Nature* **2012**, *488*, 178–184. [CrossRef]
82. Spencer, M.D.; Hamp, T.J.; Reid, R.W.; Fischer, L.M.; Zeisel, S.H.; Fodor, A.A. Association between composition of the human gastrointestinal microbiome and development of fatty liver with choline deficiency. *Gastroenterology* **2011**, *140*, 976–986. [CrossRef] [PubMed]
83. De Ferrars, R.M.; Czank, C.; Zhang, Q.; Botting, N.P.; Kroon, P.A.; Cassidy, A.; Kay, C.D. The pharmacokinetics of anthocyanins and their metabolites in humans. *Br. J. Pharmacol.* **2014**, *171*, 3268–3282. [CrossRef]

84. Gil-Cardoso, K.; Ginés, I.; Pinent, M.; Ardévol, A.; Blay, M.; Terra, X. Effects of flavonoids on intestinal inflammation, barrier integrity and changes in gut microbiota during diet-induced obesity. *Nutr. Res. Rev.* **2016**, *29*, 234–248. [CrossRef] [PubMed]
85. Sanders, M.A. Colon cancer stem cells: Implications in carcinogenesis. *Front. Biosci.* **2011**, *16*, 1651. [CrossRef] [PubMed]
86. Van de Wetering, M.; Sancho, E.; Verweij, C.; de Lau, W.; Oving, I.; Hurlstone, A.; van der Horn, K.; Batlle, E.; Coudreuse, D.; Haramis, A.-P.; et al. The β-catenin/TCF-4 complex imposes a crypt progenitor phenotype on colorectal cancer cells. *Cell* **2002**, *111*, 241–250. [CrossRef]
87. Ricci-Vitiani, L.; Fabrizi, E.; Palio, E.; De Maria, R. Colon cancer stem cells. *J. Mol. Med.* **2009**, *87*, 1097–1104. [CrossRef]
88. Zheng, S.; Xin, L.; Liang, A.; Fu, Y. Cancer stem cell hypothesis: A brief summary and two proposals. *Cytotechnology* **2012**, *65*, 505–512. [CrossRef]
89. Liu, S.; Dontu, G.; Wicha, M.S. Mammary stem cells, self-renewal pathways, and carcinogenesis. *Breast Cancer Res.* **2005**, *7*, 86–95. [CrossRef]
90. Zito, G.; Saotome, I.; Liu, Z.; Ferro, E.G.; Sun, T.Y.; Nguyen, D.X.; Bilguvar, K.; Ko, C.J.; Greco, V. Spontaneous tumour regression in keratoacanthomas is driven by Wnt/retinoic acid signalling cross-talk. *Nat. Commun.* **2014**, *5*, 3543. [CrossRef]
91. Munro, M.J.; Wickremesekera, S.K.; Peng, L.; Tan, S.T.; Itinteang, T. Cancer stem cells in colorectal cancer: A review. *J. Clin. Pathol.* **2017**, *71*, 110–116. [CrossRef] [PubMed]
92. Wang, J.-Y.; Chang, C.-C.; Chiang, C.-C.; Chen, W.-M.; Hung, S.-C. Silibinin suppresses the maintenance of colorectal cancer stem-like cells by inhibiting PP2A/AKT/mTOR pathways. *J. Cell. Biochem.* **2011**, *113*, 1733–1743. [CrossRef] [PubMed]
93. Van der Gun, B.T.F.; Melchers, L.J.; Ruiters, M.H.J.; de Leij, L.F.M.H.; McLaughlin, P.M.J.; Rots, M.G. EpCAM in carcinogenesis: The good, the bad or the ugly. *Carcinogenesis* **2010**, *31*, 1913–1921. [CrossRef] [PubMed]
94. Uronis, J.M.; Mühlbauer, M.; Herfarth, H.H.; Rubinas, T.C.; Jones, G.S.; Jobin, C. Modulation of the intestinal microbiota alters colitis-associated colorectal cancer susceptibility. *PLoS ONE* **2009**, *4*, e6026. [CrossRef]
95. Grivennikov, S.I.; Wang, K.; Mucida, D.; Stewart, C.A.; Schnabl, B.; Jauch, D.; Taniguchi, K.; Yu, G.-Y.; Österreicher, C.H.; Hung, K.E.; et al. Adenoma-linked barrier defects and microbial products drive IL-23/IL-17-mediated tumour growth. *Nature* **2012**, *491*, 254–258. [CrossRef]
96. Schwabe, R.F.; Jobin, C. The microbiome and cancer. *Nat. Rev. Cancer* **2013**, *13*, 800–812. [CrossRef]
97. Vannucci, L.; Stepankova, R.; Kozakova, H.; Fiserova, A.; Rossmann, P.; Tlaskalova-Hogenova, H. Colorectal carcinogenesis in germ-free and conventionally reared rats: Different intestinal environments affect the systemic immunity. *Int. J. Oncol.* **2008**, *32*, 609–617. [CrossRef]
98. Li, K.; Xing, S.; Wang, M.; Peng, Y.; Dong, Y.; Li, X. Anticomplement and antimicrobial activities of flavonoids from Entada phaseoloides. *Nat. Prod. Commun.* **2012**, *7*, 1934578X1200700715. [CrossRef]
99. Chen, G.Y.; Shaw, M.H.; Redondo, G.; Nunez, G. The innate immune receptor Nod1 protects the intestine from inflammation-induced tumorigenesis. *Cancer Res.* **2008**, *68*, 10060–10067. [CrossRef]
100. Klimesova, K.; Kverka, M.; Zakostelska, Z.; Hudcovic, T.; Hrncir, T.; Stepankova, R.; Rossmann, P.; Ridl, J.; Kostovcik, M.; Mrazek, J.; et al. Altered gut microbiota promotes colitis-associated cancer in IL-1 receptor–associated Kinase M–deficient mice. *Inflamm. Bowel Dis.* **2013**, *19*, 1266–1277. [CrossRef]
101. Ma, Y.; Ren, Y.; Han, E.Q.; Li, H.; Chen, D.; Jacobs, J.J.; Gitelis, S.; O'Keefe, R.J.; Konttinen, Y.T.; Yin, G.; et al. Inhibition of the Wnt-β-catenin and Notch signaling pathways sensitizes osteosarcoma cells to chemotherapy. *Biochem. Biophys. Res. Commun.* **2013**, *431*, 274–279. [CrossRef] [PubMed]
102. Koury, J.; Zhong, L.; Hao, J. Targeting signaling pathways in cancer stem cells for cancer treatment. *Stem Cells Int.* **2017**, *2017*, 1–10. [CrossRef] [PubMed]
103. Brittan, M.; Wright, N.A. Gastrointestinal stem cells. *J. Pathol.* **2002**, *197*, 492–509. [CrossRef] [PubMed]
104. Deonarain, M.P.; Kousparou, C.A.; Epenetos, A.A. Antibodies targeting cancer stem cells: A new paradigm in immunotherapy? *mAbs* **2009**, *1*, 12–25. [CrossRef]
105. Tan, M.L.; Ooi, J.P.; Ismail, N.; Moad, A.I.H.; Muhammad, T.S.T. Programmed cell death pathways and current antitumor targets. *Pharma. Res.* **2009**, *26*, 1547–1560. [CrossRef] [PubMed]
106. Kahn, M. Can we safely target the WNT pathway? *Nat. Rev. Drug Discover.* **2014**, *13*, 513–532. [CrossRef]
107. Chen, J.K. Inhibition of Hedgehog signaling by direct binding of cyclopamine to Smoothened. *Genes Dev.* **2002**, *16*, 2743–2748. [CrossRef]

108. Bar, E.E.; Chaudhry, A.; Lin, A.; Fan, X.; Schreck, K.; Matsui, W.; Piccirillo, S.; Vescovi, A.L.; DiMeco, F.; Olivi, A.; et al. Cyclopamine-mediated Hedgehog pathway inhibition depletes stem-like cancer cells in glioblastoma. *Stem Cells* **2007**, *25*, 2524–2533. [CrossRef]
109. Oh, S.; Gwak, J.; Park, S.; Yang, C.S. Green tea polyphenol EGCG suppresses Wnt/β-catenin signaling by promoting GSK-3β- and PP2A-independent β-catenin phosphorylation/degradation. *BioFactors* **2014**, *40*, 586–595. [CrossRef]
110. Hong, O.-Y.; Noh, E.-M.; Jang, H.-Y.; Lee, Y.-R.; Lee, B.K.; Jung, S.H.; Kim, J.-S.; Youn, H.J. Epigallocatechin gallate inhibits the growth of MDA-MB-231 breast cancer cells via inactivation of the β-catenin signaling pathway. *Oncol. Lett.* **2017**, *14*, 441–446. [CrossRef]
111. Castillo-Pichardo, L.; Dharmawardhane, S.F. Grape polyphenols inhibit Akt/mammalian target of rapamycin signaling and potentiate the effects of Gefitinib in breast cancer. *Nutr. Cancer* **2012**, *64*, 1058–1069. [CrossRef] [PubMed]
112. Abdal Dayem, A.; Choi, H.; Yang, G.-M.; Kim, K.; Saha, S.; Cho, S.-G. The anti-cancer effect of polyphenols against breast cancer and cancer stem cells: Molecular mechanisms. *Nutrients* **2016**, *8*, 581. [CrossRef] [PubMed]
113. Nautiyal, J.; Banerjee, S.; Kanwar, S.S.; Yu, Y.; Patel, B.B.; Sarkar, F.H.; Majumdar, A.P.N. Curcumin enhances dasatinib-induced inhibition of growth and transformation of colon cancer cells. *Int. J. Cancer* **2010**, *128*, 951–961. [CrossRef] [PubMed]
114. Jaiswal, A.S.; Marlow, B.P.; Gupta, N.; Narayan, S. β-Catenin-mediated transactivation and cell–cell adhesion pathways are important in curcumin (diferuylmethane)-induced growth arrest and apoptosis in colon cancer cells. *Oncogene* **2002**, *21*, 8414–8427. [CrossRef]
115. Min, S.; Lim, J.; Kim, H.; Kim, S.-J.; Kim, Y. *Sasa quelpaertensis* leaf extract inhibits colon cancer by regulating cancer cell stemness in vitro and in vivo. *Int.J. Mol. Sci.* **2015**, *16*, 9976–9997. [CrossRef]
116. Dandawate, P.; Padhye, S.; Ahmad, A.; Sarkar, F.H. Novel strategies targeting cancer stem cells through phytochemicals and their analogs. *Drug Deliv. Translat. Res.* **2012**, *3*, 165–182. [CrossRef] [PubMed]
117. Lee, J.; Kim, Y.-S.; Lee, J.; Heo, S.; Lee, K.; Choi, S.-W.; Kim, Y. Walnut phenolic extract and its bioactive compounds suppress colon cancer cell growth by regulating colon cancer stemness. *Nutrients* **2016**, *8*, 439. [CrossRef]
118. Kumar, S.; Raina, K.; Agarwal, C.; Agarwal, R. Silibinin strongly inhibits the growth kinetics of colon cancer stem cell-enriched spheroids by modulating interleukin 4/6-mediated survival signals. *Oncotarget* **2014**, *5*, 4972–4989. [CrossRef]
119. Soltanian, S.; Riahirad, H.; Pabarja, A.; Jafari, E.; Khandani, B.K. Effect of Cinnamic acid and FOLFOX in diminishing side population and downregulating cancer stem cell markers in colon cancer cell line HT-29. *DARU J. Pharma. Sci.* **2018**, *26*, 19–29. [CrossRef]
120. Schwingshackl, L.; Hoffmann, G. Adherence to Mediterranean diet and risk of cancer: A systematic review and meta-analysis of observational studies. *Int. J. Cancer* **2014**, *135*, 1884–1897. [CrossRef]
121. Epstein, J.; Sanderson, I.R.; MacDonald, T.T. Curcumin as a therapeutic agent: The evidence from in vitro, animal and human studies. *Br. J. Nutr.* **2010**, *103*, 1545–1557. [CrossRef] [PubMed]
122. Banerjee, S.; Kong, D.; Wang, Z.; Bao, B.; Hillman, G.G.; Sarkar, F.H. Attenuation of multi-targeted proliferation-linked signaling by 3,3′-diindolylmethane (DIM): From bench to clinic. *Mutat. Res. Rev. Mutat. Res.* **2011**, *728*, 47–66. [CrossRef] [PubMed]
123. Ezzat, S.; Hegazy, A.; Amer, A.M.; Kamel, G.; El-Alfy, T. Isolation of biologically active constituents from *Moringa peregrina* (Forssk.) Fiori. (family: Moringaceae) growing in Egypt. *Pharmacogn. Mag.* **2011**, *7*, 109. [CrossRef]
124. Sharma, N.; Samarakoon, K.; Gyawali, R.; Park, Y.-H.; Lee, S.-J.; Oh, S.; Lee, T.-H.; Jeong, D. Evaluation of the antioxidant, anti-inflammatory, and anticancer activities of *Euphorbia hirta* ethanolic extract. *Molecules* **2014**, *19*, 14567–14581. [CrossRef] [PubMed]
125. Mongre, R.K.; Sodhi, S.S.; Ghosh, M.; Kim, J.H.; Kim, N.; Park, Y.H.; Kim, S.J.; Heo, Y.J.; Sharma, N.; Jeong, D.K. The novel inhibitor BRM270 downregulates tumorigenesis by suppression of NF-κB signaling cascade in MDR-induced stem like cancer-initiating cells. *Int. J. Oncol.* **2015**, *46*, 2573–2585. [CrossRef]
126. Enkhbat, T.; Nishi, M.; Yoshikawa, K.; Jun, H.; Tokunaga, T.; Takasu, C.; Kashihara, H.; Ishikawa, D.; Tominaga, M.; Shimada, M. Epigallocatechin-3-gallate enhances radiation sensitivity in colorectal cancer cells through Nrf2 activation and autophagy. *Anticancer Res.* **2018**, *38*, 6247–6252. [CrossRef]

127. Tomasello, G.; Mazzola, M.; Bosco, V.; Tomasello, G.; Damiani, P.; Sinagra, E.; Carini, F. Intestinal dysbiosis and hormonal neuroendocrine secretion in the fibromyalgic patient. *Biomed. Papers* **2018**, *162*, 258–262. [CrossRef]
128. Darband, S.G.; Kaviani, M.; Yousefi, B.; Sadighparvar, S.; Pakdel, F.G.; Attari, J.A.; Mohebbi, I.; Naderi, S.; Majidinia, M. Quercetin: A functional dietary flavonoid with potential chemo-preventive properties in colorectal cancer. *J. Cell. Physiol.* **2018**, *233*, 6544–6560. [CrossRef]
129. Bernkopf, D.B.; Daum, G.; Brückner, M.; Behrens, J. Sulforaphane inhibits growth and blocks Wnt/β-catenin signaling of colorectal cancer cells. *Oncotarget* **2018**, *9*, 33982–33994. [CrossRef]
130. O'Gorman, D.M.; Cotter, T.G. Molecular signals in anti-apoptotic survival pathways. *Leukemia* **2001**, *15*, 21–34. [CrossRef]
131. Neergheen, V.S.; Bahorun, T.; Taylor, E.W.; Jen, L.-S.; Aruoma, O.I. Targeting specific cell signaling transduction pathways by dietary and medicinal phytochemicals in cancer chemoprevention. *Toxicology* **2010**, *278*, 229–241. [CrossRef] [PubMed]
132. Khan, N.; Mukhtar, H. Multitargeted therapy of cancer by green tea polyphenols. *Cancer Lett.* **2008**, *269*, 269–280. [CrossRef]
133. Meeran, S.M. Cell cycle control as a basis for cancer chemoprevention through dietary agents. *Front. Biosci.* **2008**, *13*, 2191. [CrossRef] [PubMed]
134. Park, H.Y.; Toume, K.; Arai, M.A.; Sadhu, S.K.; Ahmed, F.; Ishibashi, M. Calotropin: A cardenolide from *Calotropis gigantea* that inhibits Wnt signaling by increasing casein kinase 1α in colon cancer cells. *Chem. BioChem.* **2014**, *15*, 872–878. [CrossRef]
135. Shakibaei, M.; Mobasheri, A.; Lueders, C.; Busch, F.; Shayan, P.; Goel, A. Curcumin enhances the effect of chemotherapy against colorectal cancer cells by inhibition of NF-κB and Src protein kinase signaling pathways. *PLoS ONE* **2013**, *8*, e57218. [CrossRef] [PubMed]
136. Shankar, S.; Ganapathy, S.; Srivastava, R.K. Sulforaphane enhances the therapeutic potential of TRAIL in prostate cancer orthotopic model through regulation of apoptosis, metastasis, and angiogenesis. *Clin. Cancer Res.* **2008**, *14*, 6855–6866. [CrossRef] [PubMed]
137. Suppipat, K.; Park, C.S.; Shen, Y.; Zhu, X.; Lacorazza, H.D. Sulforaphane induces cell cycle arrest and apoptosis in acute lymphoblastic leukemia cells. *PLoS ONE* **2012**, *7*, e51251. [CrossRef]
138. Sung, B.; Prasad, S.; Yadav, V.R.; Aggarwal, B.B. Cancer cell signaling pathways targeted by spice-derived nutraceuticals. *Nutr. Cancer* **2012**, *64*, 173–197. [CrossRef]
139. Shankar, S.; Marsh, L.; Srivastava, R.K. EGCG inhibits growth of human pancreatic tumors orthotopically implanted in Balb C nude mice through modulation of FKHRL1/FOXO3a and neuropilin. *Mol. Cell. Biochem.* **2012**, *372*, 83–94. [CrossRef]
140. Chakraborty, S.; Adhikary, A.; Mazumdar, M.; Mukherjee, S.; Bhattacharjee, P.; Guha, D.; Choudhuri, T.; Chattopadhyay, S.; Sa, G.; Sen, A.; et al. Capsaicin-induced activation of p53-SMAR1 auto-regulatory loop down-regulates VEGF in non-small cell lung cancer to restrain angiogenesis. *PLoS ONE* **2014**, *9*, e99743. [CrossRef]
141. Amado, N.G.; Fonseca, B.F.; Cerqueira, D.M.; Neto, V.M.; Abreu, J.G. Flavonoids: Potential Wnt/beta-catenin signaling modulators in cancer. *Life Sci.* **2011**, *89*, 545–554. [CrossRef] [PubMed]
142. Diehn, M.; Cho, R.W.; Lobo, N.A.; Kalisky, T.; Dorie, M.J.; Kulp, A.N.; Qian, D.; Lam, J.S.; Ailles, L.E.; Wong, M.; et al. Association of reactive oxygen species levels and radioresistance in cancer stem cells. *Nature* **2009**, *458*, 780–783. [CrossRef] [PubMed]
143. Park, W.; Amin, A.R.M.R.; Chen, Z.G.; Shin, D.M. New Perspectives of curcumin in cancer prevention. *Cancer Prev. Res.* **2013**, *6*, 387–400. [CrossRef] [PubMed]
144. Ribeiro, H.S.; Guerrero, J.M.M.; Briviba, K.; Rechkemmer, G.; Schuchmann, H.P.; Schubert, H. Cellular uptake of carotenoid-loaded oil-in-water emulsions in colon carcinoma cells in vitro. *J. Agric. Food Chem.* **2006**, *54*, 9366–9369. [CrossRef]
145. Yang, K.; Pyo, J.; Kim, G.-Y.; Yu, R.; Han, I.; Ju, S.; Kim, W.; Kim, B.-S. Capsaicin induces apoptosis by generating reactive oxygen species and disrupting mitochondrial transmembrane potential in human colon cancer cell lines. *Cell. Mol. Biol. Lett.* **2009**, *14*, 497–510. [CrossRef]
146. Reuter, S.; Gupta, S.C.; Park, B.; Goel, A.; Aggarwal, B.B. Epigenetic changes induced by curcumin and other natural compounds. *Genes Nutr.* **2011**, *6*, 93–108. [CrossRef]

147. Thakur, V.S.; Deb, G.; Babcook, M.A.; Gupta, S. Plant phytochemicals as epigenetic modulators: Role in cancer chemoprevention. *AAPS J.* **2013**, *16*, 151–163. [CrossRef]
148. Den Besten, G.; van Eunen, K.; Groen, A.K.; Venema, K.; Reijngoud, D.-J.; Bakker, B.M. The role of short-chain fatty acids in the interplay between diet, gut microbiota, and host energy metabolism. *J. Lipid Res.* **2013**, *54*, 2325–2340. [CrossRef]
149. Zhang, L.; Mu, C.; He, X.; Su, Y.; Mao, S.; Zhang, J.; Smidt, H.; Zhu, W. Effects of dietary fibre source on microbiota composition in the large intestine of suckling piglets. *FEMS Microbiol. Lett.* **2016**, *363*, fnw138. [CrossRef]
150. Matthews, G.M.; Howarth, G.S.; Butler, R.N. Short-chain fatty acids induce apoptosis in colon cancer cells associated with changes to intracellular redox state and glucose metabolism. *Chemotherapy* **2012**, *58*, 102–109. [CrossRef]
151. Mambet, C.; Chivu-Economescu, M.; Matei, L.; Necula, L.G.; Dragu, D.L.; Bleotu, C.; Diaconu, C.C. Murine models based on acute myeloid leukemia-initiating stem cells xenografting. *World J. Stem Cells* **2018**, *10*, 57–65. [CrossRef] [PubMed]
152. Zeng, H.; Taussig, D.; Cheng, W.-H.; Johnson, L.; Hakkak, R. Butyrate inhibits cancerous HCT116 colon cell proliferation but to a lesser extent in noncancerous NCM460 colon cells. *Nutrients* **2017**, *9*, 25. [CrossRef]
153. Ritchie, L.E.; Taddeo, S.S.; Weeks, B.R.; Lima, F.; Bloomfield, S.A.; Azcarate-Peril, M.A.; Zwart, S.R.; Smith, S.M.; Turner, N.D. Space environmental factor impacts upon murine colon microbiota and mucosal homeostasis. *PLoS ONE* **2015**, *10*, e0125792. [CrossRef] [PubMed]
154. Hooks, K.B.; O'Malley, M.A. Dysbiosis and Its Discontents. *mBio* **2017**, *8*. [CrossRef] [PubMed]
155. Ivanov, I.I.; Atarashi, K.; Manel, N.; Brodie, E.L.; Shima, T.; Karaoz, U.; Wei, D.; Goldfarb, K.C.; Santee, C.A.; Lynch, S.V.; et al. Induction of intestinal Th17 cells by segmented filamentous bacteria. *Cell* **2009**, *139*, 485–498. [CrossRef]
156. Dejea, C.; Wick, E.; Sears, C.L. Bacterial oncogenesis in the colon. *Future Microbiol.* **2013**, *8*, 445–460. [CrossRef]
157. Fukata, M.; Arditi, M. The role of pattern recognition receptors in intestinal inflammation. *Mucosal Immunol.* **2013**, *6*, 451–463. [CrossRef]
158. Garaude, J.; Kent, A.; van Rooijen, N.; Blander, J.M. Simultaneous targeting of Toll- and Nod-like receptors induces effective tumor-specific immune responses. *Sci Translat. Med.* **2012**, *4*, 120ra116. [CrossRef]
159. Clevers, H. The cancer stem cell: Premises, promises and challenges. *Nat. Med.* **2011**, *17*, 313–319. [CrossRef]
160. Blokzijl, F.; de Ligt, J.; Jager, M.; Sasselli, V.; Roerink, S.; Sasaki, N.; Huch, M.; Boymans, S.; Kuijk, E.; Prins, P.; et al. Tissue-specific mutation accumulation in human adult stem cells during life. *Nature* **2016**, *538*, 260–264. [CrossRef]
161. Polakis, P. Wnt signaling in cancer. *Cold Spring Harbor Perspectives in Biology* **2012**, *4*, a008052. [CrossRef] [PubMed]
162. Seidel, D.V.; Azcárate-Peril, M.A.; Chapkin, R.S.; Turner, N.D. Shaping functional gut microbiota using dietary bioactives to reduce colon cancer risk. *Semin. Cancer Biol.* **2017**, *46*, 191–204. [CrossRef] [PubMed]
163. Rycaj, K.; Tang, D.G. Cell-of-origin of cancer versus cancer stem cells: Assays and interpretations. *Cancer Res.* **2015**, *75*, 4003–4011. [CrossRef] [PubMed]
164. Tetteh, P.W.; Farin, H.F.; Clevers, H. Plasticity within stem cell hierarchies in mammalian epithelia. *Trends Cell Biol.* **2015**, *25*, 100–108. [CrossRef]
165. Beyaz, S.; Mana, M.D.; Roper, J.; Kedrin, D.; Saadatpour, A.; Hong, S.-J.; Bauer-Rowe, K.E.; Xifaras, M.E.; Akkad, A.; Arias, E.; et al. High-fat diet enhances stemness and tumorigenicity of intestinal progenitors. *Nature* **2016**, *531*, 53–58. [CrossRef]
166. Haegebarth, A.; Clevers, H. Wnt Signaling, Lgr5, and stem cells in the intestine and skin. *Am. J. Pathol.* **2009**, *174*, 715–721. [CrossRef]
167. Barker, N.; Clevers, H. Leucine-rich repeat-containing G-protein-coupled receptors as markers of adult stem cells. *Gastroenterology* **2010**, *138*, 1681–1696. [CrossRef]
168. Claessen, M.M.H.; Vleggaar, F.P.; Schipper, M.; Oldenburg, B.; Offerhaus, J.; Siersema, P.D. Wnt-pathway activation in early IBD-associated colorectal carcinogenesis: A biomarker for colonic surveillance. *Gastroenterology* **2008**, *134*, A-132. [CrossRef]

169. Sauceda, A.E.Q.; Pacheco-Ordaz, R.; Ayala-Zavala, J.F.; Mendoza, A.H.; González-Córdova, A.F.; Vallejo-Galland, B.; González-Aguilar, G.A. Impact of fruit dietary fibers and polyphenols on modulation of the human gut microbiota. In *Fruit and Vegetable Phytochemicals*; John Wiley & Sons, Ltd.: Hoboken, NJ, USA, 2017; pp. 405–422. [CrossRef]
170. Pyra, K.A.; Saha, D.C.; Reimer, R.A. Prebiotic fiber increases hepatic acetyl CoA carboxylase phosphorylation and suppresses glucose-dependent insulinotropic polypeptide secretion more effectively when used with metformin in obese rats. *J. Nutr.* **2012**, *142*, 213–220. [CrossRef]
171. Laparra, J.M.; Sanz, Y. Interactions of gut microbiota with functional food components and nutraceuticals. *Pharmacol. Res.* **2010**, *61*, 219–225. [CrossRef]
172. Scott, K.P.; Gratz, S.W.; Sheridan, P.O.; Flint, H.J.; Duncan, S.H. The influence of diet on the gut microbiota. *Pharmacol. Res.* **2013**, *69*, 52–60. [CrossRef] [PubMed]
173. Alexander, J.L.; Scott, A.J.; Pouncey, A.L.; Marchesi, J.; Kinross, J.; Teare, J. Colorectal carcinogenesis: An archetype of gut microbiota–host interaction. *ecancermedicalscience* **2018**, *12*. [CrossRef] [PubMed]
174. Morrison, S.J.; Spradling, A.C. Stem cells and niches: Mechanisms that promote stem cell maintenance throughout life. *Cell* **2008**, *132*, 598–611. [CrossRef] [PubMed]
175. Frank, N.Y.; Schatton, T.; Frank, M.H. The therapeutic promise of the cancer stem cell concept. *J. Clin. Investig.* **2010**, *120*, 41–50. [CrossRef]
176. Domingo-Domenech, J.; Vidal, S.J.; Rodriguez-Bravo, V.; Castillo-Martin, M.; Quinn, S.A.; Rodriguez-Barrueco, R.; Bonal, D.M.; Charytonowicz, E.; Gladoun, N.; de la Iglesia-Vicente, J.; et al. Suppression of acquired docetaxel resistance in prostate cancer through depletion of Notch- and Hedgehog-dependent tumor-initiating cells. *Cancer Cell.* **2012**, *22*, 373–388. [CrossRef]

International Journal of
Molecular Sciences

MDPI

Review

Gut Microbiota Metabolism and Interaction with Food Components

Pamela Vernocchi [1,*], Federica Del Chierico [1] and Lorenza Putignani [2]

1 Unit of Human Microbiome, Bambino Gesù Children's Hospital, IRCCS, Viale San Paolo 15, 00146 Rome, Italy; federica.delchierico@opbg.net

2 Unit of Parasitology and Unit of Human Microbiome, Bambino Gesù Children's Hospital, IRCCS, Piazza Sant' Onofrio 4, 00165 Rome, Italy; lorenza.putignani@opbg.net

* Correspondence: pamela.vernocchi@opbg.net; Tel.: +39-0668-594061; Fax: +39-0668-592218

Received: 25 March 2020; Accepted: 14 May 2020; Published: 23 May 2020

check for updates

Abstract: The human gut contains trillions of microbes that play a central role in host biology, including the provision of key nutrients from the diet. Food is a major source of precursors for metabolite production; in fact, diet modulates the gut microbiota (GM) as the nutrients, derived from dietary intake, reach the GM, affecting both the ecosystem and microbial metabolic profile. GM metabolic ability has an impact on human nutritional status from childhood. However, there is a wide variability of dietary patterns that exist among individuals. The study of interactions with the host via GM metabolic pathways is an interesting field of research in medicine, as microbiota members produce myriads of molecules with many bioactive properties. Indeed, much evidence has demonstrated the importance of metabolites produced by the bacterial metabolism from foods at the gut level that dynamically participate in various biochemical mechanisms of a cell as a reaction to environmental stimuli. Hence, the GM modulate homeostasis at the gut level, and the alteration in their composition can concur in disease onset or progression, including immunological, inflammatory, and metabolic disorders, as well as cancer. Understanding the gut microbe–nutrient interactions will increase our knowledge of how diet affects host health and disease, thus enabling personalized therapeutics and nutrition.

Keywords: microbiota metabolism; diet; metabolome; microbiome therapeutics

1. Introduction

The human gastrointestinal (GI) tract contains many mutualistic microbes (over 10 trillion microbial cells), providing numerous specialized metabolites, small and other bioactive molecules that trigger immune and host metabolic pathways. Hence, the gut microbiota (GM) is also called a "metabolic organ" with a metabolic potential, comparable to the same as that which the liver has [1].

The innovation in metabolomics and metagenomics disciplines has allowed the discovery of many microbe-derived small molecules, as well as the genes linked to their production [2]. Hence, the symbiotic relation between GM and host produces a myriad of metabolic signatures, and the technological advances in GM metabolomics are progressively decoding the host–microbes metabolic interaction [3].

GM provides primary metabolites, converting these small molecules into secondary metabolites named "specialized metabolites" [4,5].

Several specialized metabolites show connections with biosynthetic pathways that are unique to an organism or to their class, and by metagenome and metatranscriptome sequencing, combination is possible in order to assign taxonomy and functions [6]. Moreover, the high resolution of mass spectrometry coupled with molecular networking analysis, corroborated with computational approaches, will allow for the understanding of host–microbiome crosstalk mediated by the associated specialized metabolites [5].

Microbial ecosystems may rapidly shift their functionality in response to dietary changes, enhancing human dietary elasticity [7]. Among different environmental variables, short-term dietary interventions coupled with long-term dietary patterns regulate the GM physiology [8].

The processes related to the metabolism of nutrients and xenobiotics affect the following mechanisms: (i) chemical crosstalk between microbial and human metabolic compounds [9,10], (ii) modulation of immune system [11], (iii) protection from pathogens [12], (iv) enteric nervous system regulation [13], (v) colorectal cancer resistance [14–16], (vi) neurological behavior [17], and (vii) reduction of lipid levels in serum and cholesterol balancing [18].

In particular, by deepening the concepts, the effect of the environment (nutrition, social, behavioral, geography) on host genetics and the following GM adaptation, triggers the molecular mechanisms of communication between the microbiome and the host, which is called crosstalk.

Crosstalk could provide an explanation of the development of these diseases, which will help to highlight potential targets and biomarkers able to modulate the abundance of metabolites and relevant functions, as well as operational taxonomic units (OTUs) for health, for further investigations on potential new metabolic functions of the GM [19].

Recently, the interest has grown in terms of understanding of how commensal bacteria metabolites derived from interaction with nutrients could regulate the host immune system. Commensal bacteria are important digestion regulators, and the intestinal content is a mix of microbes that are important for the processing and absorption of several nutrients and metabolites, including bile acids, lipids, amino acids (AAs), vitamins, and short-chain fatty acids (SCFAs) [11]. These nutrients and metabolites, which derive from commensal bacteria, are directly linked to diet and digestion [20], and may modulate immune cells through direct and indirect mechanisms in the context of health and disease.

SCFAs, particularly propionate and butyrate, could downregulate the gene expression of pathogenicity island 1 in *Salmonella typhimurium* required for intestinal epithelial cell invasion [21]. Additionally, the microbial-synthetized molecules and host-derived molecules can be metabolized by commensals, resulting, for example, in the production of primary bile acids following conversion into secondary bile acids, which play a crucial role in defense against pathogens such as the suppression of *Clostridium difficile* growth [22].

Special attention can also be given to the enteric nervous system (ENS), which controls major GI functions independently of the central nervous system. Recent evidence has shown that butyrate, a molecule belonging to SCFAs, can modulate neuronal functions by gene expression of neuromodulators, as well as GI motility. In particular, butyrate increases the proportion of choline acetyltransferase by the Src-kinase signaling pathway and the acetylation of histone H3K9 in enteric neuron, as well as the motility of colon by the activation of cholinergic pathways [13].

Moreover, in cancer, especially in human colon adenocarcinoma cell line (HT-29), acetate and propionate trigger apoptosis at pH 7.5 and necrosis at pH 5.5. These processes are probably induced by mitochondrial depolarization, inner membrane permeabilization, drastic depletion in adenosine triphosphate (ATP) levels, and reactive oxygen species (ROS) accumulation in HT-29 cells [15].

There are several studies on the effect of butyrate on mitochondrial activity, which has also been studied in lymphoblastoid cell lines (LCL) derived from children with autism spectrum disorder (ASD). In this study, authors demonstrated that butyrate in particular has a positive effect on mitochondrial function in LCLs in the context of physiological stress and/or mitochondrial dysfunction, and can help rescue energy metabolism during disease states, having a potential role in host physiology and behavior in ASD [23].

Moreover, non-digestible/fermentable nutrients could also modulate GM activity, determining cholesterol- or triglyceride-lowering effects. In particular, certain bacteria with probiotic characteristics and prebiotics enhanced bile acid deconjugation and subsequent increasing of fecal bile acid excretion that is involved in cholesterol reduction [24].

The production of a high concentration of propionate in rats, through microbial fermentation of resistant starch or fructans, has been identified as a mechanism to explain the reduction in serum and hepatic cholesterol [18]. Moreover, the level of acetate/propionate ratio that reaches the liver by the portal vein is a potential intermediate that could be used to make predictions on the potential lipid-lowering properties belonging to prebiotics and other fermentable carbohydrates [18]. In addition, Saltzman et al. [25,26] may also have demonstrated the mechanisms by which other non-digestible, lipid-lowering carbohydrates produce their effects in humans, in combination with other nutritional factors such as a vegetarian diet or a diet with a high content of cereal fiber, vegetables, and fruits.

The goal of this review is to give a vision on potential future treatments that will be directed to ameliorate host health by modulation of the GM as target, focusing principally on the bacterial metabolic potential, as well as the taxonomy aspect, in order to discover therapeutic and/or diagnostic targets, and also to describe the interaction between the microbiome–host and environmental and dietary changes [19].

2. The Dietary Impact on Gut Microbiota and Metabolic Composition

Foods may be considered to be potential etiopathogenetic factors of GI-related disorders [27]. Long-term dietary intake has an impact on the composition and activity of microbes residing in the gut [28,29], but it remains unclear how fast and how much is the rate of the reproducibility of the human GM to respond to short-range changes in term of macronutrients [7].

Hence, saccharolytic fermentation takes place mainly in the proximal colon, as most bacteria choose to utilize carbohydrates instead of proteins [30]. On the contrary, proteolytic fermentation occurs in the distal colon, producing branched-chain fatty acids (BCFAs) and potentially detrimental metabolites such as ammonia (from amino acid deamination and urea hydrolysis), indoles, and phenols (from amino acid (AAs) carboxylation). Therefore, the intake of foods composed entirely of animal or plant ingredients alters the microbial community [29,31].

An animal-based diet, made up of meat, eggs, and cheese, determines the increase of bile-tolerant microbes such as *Alistipes*, *Bilophila*, and *Bacteroides*, and the decrease of Firmicutes that metabolize dietary plant polysaccharides such as *Roseburia*, *Eubacterium rectale*, and *Ruminococcus bromii* [7].

A high-fat diet, especially in terms of saturated fatty acids, and chronic low-grade tissue inflammation [32] determines a microbial community imbalance, leading to dysbiosis (i.e., increasing at the phylum level of cluster XI of the genus *Clostridium*) and producing an altered metabolic profile in the colon lumen.

Hence, this dysbiosis associated with metabolites (or increased translocation of normal metabolites in the host) is involved in systemic efflux and also in GI disorders [33].

On the contrary, a plant-based diet, which is rich in grains, legumes, fruits, and vegetables, appears to be beneficial for human health by promoting the development of diverse and stable microbial ecosystems. In particular, digestible carbohydrates and fructose have been highlighted to decrease Clostridia and *Bacteroides* [34]. Hence, non-digestible carbohydrates lead to an increase in lactic acid bacteria (LAB), *Ruminococcus*, *Eubacterium rectale*, and *Roseburia*, while reducing *Clostridium* and *Enterococcus* species [34]. Bifidobacteria are also raised up by digestible and non-digestible carbohydrates [35].

The plant-based diet is associated with carbohydrate fermentation products and lower fermentation of AAs [36]. Hence, a plant based diet, may result in an increased Bacteroidetes/Firmicutes ratio and also in a consequent weight loss with the reduction of the energy level extracted from the diet [37].

Moreover, it is possible to divide individuals in two groups according to diet model; in particular, *Bacteroides* are associated with elevated animal protein diet/saturated fat, and *Prevotella* represents the main group associated with an agrarian-type diet, rich in fruit and vegetables, with a high quantity of carbohydrates and simple sugars, and low levels of saturated fats and animal proteins [38–40].

Interestingly, Hjorth and co-workers reported that overweight and obese individuals with a high *Prevotella/Bacteorides* ratio had more success in losing fat by eating fiber and whole grains with respect to individuals with a low *Prevotella/Bacteorides* ratio [41].

The microbial populations in vegan individuals compared to controls showed reduced levels of *Bacteroides* spp., *Bifidobacterium* spp., *Escherichia coli*, and Enterobacteriaceae [42].

Abundance of plant foods produce the increase of *Bifidobacterium* and *Lactobacillus*, which give anti-inflammatory and anti-pathogenic effects and cardiovascular protection [20].

In addition, diet also introduces food microorganisms considered as food ingredients into the distal gut in relation to different type of ingested food [43]; hence, foodborne microbes from both diets, including bacteria, fungi, and viruses, transiently colonize the gut. In particular, some diet-induced changes into bacterial groups are gut-related, including obesity [44] and inflammatory bowel diseases (IBDs), representing chronic disease's epidemic in the Westernized world [45].

Diet can deeply influence disease progression by altering the intestinal microbial composition. It induces pathobiont expansion, with reference to low-abundant species such as *Bilophila wadsworthia*, a sulfite-reducing anaerobe that triggers inflammation through canonical activation of dendritic cells presenting antigen to naive T cells [46].

3. Health Effects Mediated by Food–Microbiota Metabolome

GM is principally related with (i) dietary fiber degradation, (ii) fermentation and anaerobic degradation of proteins and peptides, (iii) glycoconjugates derivation from the host, (iv) bile acid deconjugation and dihydroxylation, (v) vitamins (B and K) and isoprenoids biosynthesis, (vi) cholesterol decrease and AAs and (vii) xenobiotics degradation in order to provide energy through metabolism [3], (viii) development of nervous system, (ix) regulation of appetite, and (x) intestinal and immune system development. In particular, differences in the intake of dietary macro- and micro-nutrients could directly raise the risk of developing chronic diseases.

In addition, consumption of a plant-based versus Western-type dietary pattern results in different gut bacterial communities [29,47] and metabolite production [7], potentially linked to healthy and disease statuses.

Technology innovation and the advent of big data has extended the knowledge on GM, and has fulfilled the description of small molecule identification such as ribosomally post-translationally modified peptides, lipids and glycolipids, AA metabolites, oligosaccharides and polysaccharides, non-ribosomal peptides, terpenoids, and polyketides [48]. We have grouped the dietary patterns into two macro groups: (1) carbohydrates, fibers, and vitamins, and (2) proteins and fats.

3.1. *Dietary Carbohydrates, Fibers, and Vitamins*

Fermentable carbohydrates, such as pectin gums, hemicelluloses, prebiotic ingredients comprising fructose, fructo-oligosaccharides (FOS), and galactose-oligosaccharides (GOS), are different from dietary fiber such as cellulose or wheat bran, which mammalian enzymes are unable to digest [49].

In particular, anaerobic metabolism of non-digestible carbohydrates performed by gut microbes produces SCFAs and gases from different pathways. SCFAs, such as acetate, propionate, and butyrate, mainly in the lumen, are assumed to interact in terms of the production of health benefits, such as antioxidant and anti-cancer activities previously described, avoiding allergic disorders, and also anti-inflammatory effects on the intestinal mucosa [50,51]. In particular, butyrate and propionate can regulate intestinal physiology and immune function, whereas acetate acts as a substrate for lipogenesis and gluconeogenesis [52].

The microbial species, including SCFA producers, play a key role in the degradation of polysaccharides derived from plants [53], cooperating with bacteria specialized in fermentation of FOS and GOS (i.e., *Bifidobacterium* spp.) in order to generate SCFAs and gas, which are also used as carbon and energy sources by other dedicated bacteria (i.e., sulfate-reducing bacteria and methanogens) [54].

Clostridium clusters IV and XIVa (i.e., *Eubacterium*, *Roseburia*, *Faecalibacterium*), and *Lactobacillus* spp. and *Bifidobacterium* spp. represent the principal bacteria that play a key role in SCFAs metabolism [16,55,56].

Recently, important activities have been evidenced for these metabolites in the immune function regulation in the periphery, leading to appropriate immune responses, oral tolerance, and inflammation resolution, as well as the modulation of the inflammatory output of adipose tissue [57].

On the other hand, some water-soluble fibers, such as resistant starch from potatoes, seaweeds [58], and undigested oligosaccharides, are metabolized as SCFAs and lactic acid by colonic bacteria, especially in the distal colon, such as *Faecalibacterium prausnitzii*, *Eubacterium rectale*, *Eubacterium hallii*, and *Ruminococcus bromii* [59]. These can constitute a protective factor in colon carcinogenesis [60]. Regarding colorectal cancer resistance, in in vitro experiments conducted on colon cancer cell lines, propionate, valerate, and especially butyrate caused cellular growth arrest, differentiation, and apoptosis by inhibiting histone deacetylases, consequently leading to the hyperacetylation of selective histone proteins such as histone H4. In the same experimental condition, the previously cited SCFAs altered the expression of the cell cycle regulators p21 and CB1, resulting in growth arrest of colon cancer cells [14].

Additional functions also include reduction of oxidative stress and reinforcement of the colonic defense barrier [61]. Hence, the principal activities of SCFAs are the following [51]: (1) triggering of Foxp3+ T regulatory (Treg) cells and tolerance; (2) induction of IgA secretion from B cells; (3) "competitive exclusion", due to high-fiber diet that spreads commensal bacteria and limits the access of pathogenic bacteria to the gut epithelium; (4) promotion of mucus secretion by gut epithelial cells; (5) contribution to the intestinal barrier integrity, in particular by stimulating the formation of the proteins of tight junctions, such as claudin, occludin, and zonulin [62,63], as well as promotion of tissue repair and wound healing; and (6) inhibition of the proinflammatory transcription factor (NF-κB) and decreasing of oxidative stress (Figure 1).

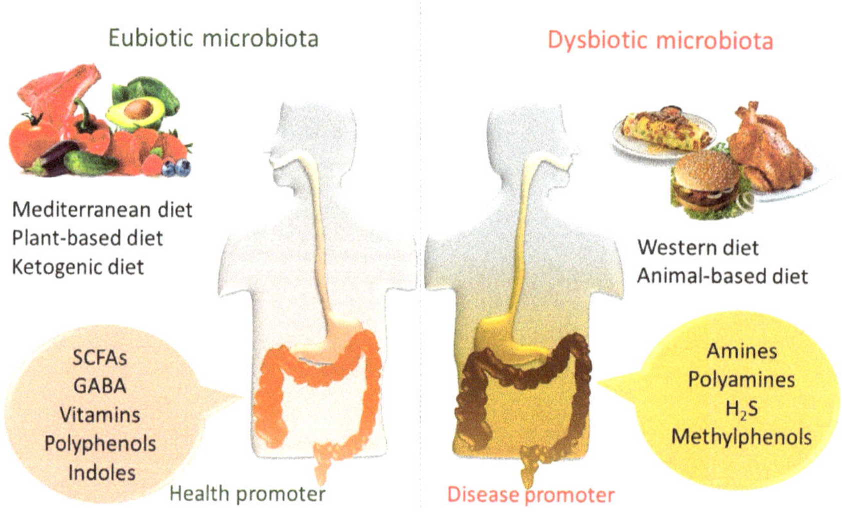

Figure 1. Dietary effect on gut microbiota and health. The gut microbiota takes part in the digestion of food ingredients and in the regulation of host metabolic functions. The nutrient-dependent impact of commensal bacteria to the eubiosis and dysbiosis, thereby the health and disease status, is caused by the production of several microbial metabolites.

The management, with propionate and butyrate, of human Caco-2/TC7 intestinal epithelial cell line has also determined the reduction of the nine important gene expressions related to cholesterol synthesis, including the key gene HMGCR (3-hydroxy-3-methylglutaryl-coenzyme A reductase).

The Mediterranean diet has a very large fiber content and bioavailability, particularly in terms of insoluble fiber, being more than twofold higher than in a typical Western diet (30 vs. 14 g/day) [64,65].

In a recent randomized clinical trial, Haro et al. [17] took into consideration obese individuals randomized to a Mediterranean diet for 2 years in order to describe the reshaping of the GM, showing an increase in *Bacteroides*; *Prevotella*; and most importantly, in *Roseburia*, *Ruminococcus*, *Parabacteroides distasonis*, and *Faecalibacterium prausnitzii*, which are known for their saccharolytic activity and their ability to digest carbohydrates by producing SCFAs.

When the intestinal SCFA (butyrate) level decreases in patients with IBDs, it is clear that GM diversity decreases by reducing the abundance of Firmicutes and Bacteroidetes and increasing the Proteobacteria level [66].

Furthermore, in patients with recurrent Crohn's disease, a decreased level of *F. prausnitzii* has been shown, as well as of butyrate, suggesting that the reduction of *F. prausnitzii* may contribute as a marker of a dysbiotic state that predisposes individuals to the IBD [67].

However, in a recent paper it has been reported that *F. prausnitzii* increased in an obese adolescent population and could be considered as a microbial marker in obesity [68]. Indeed, to date, the increasing interest in finding strategies to modulate the abundance of *F. prausnitzii* in the gut are discussed, as well as its usage as a biomarker for diagnostics and prognostics of intestinal diseases [69].

However, in terms of the consumption of high-carbohydrate food (HCF; i.e., glucose, sucrose, pastas, potatoes, white bread), it is possible that it is correlated with obesity, and in late pregnancy, the GM patterns seem to represent a disruptive microbial composition similar to those of adults with type 2 diabetes [70]. HCF consumption is associated with significant changes in the Firmicutes/Bacteroidetes ratio, with a lowering of butyrate-producing bacteria compared to healthy individuals. HCF is usually very deficient of indigestible carbohydrates such as fibers, which provide important physiological benefits such as stimulating incretin production, serving as an energy source for colonic microbes that promote normal bowel movements [71] and the aforementioned production of SCFAs.

Diet provides vitamins, which are absorbed in the small intestine, although the major part of the microbe-mediated vitamin production takes place in the large intestine. Hence, the vitamins produced can be absorbed by the host through specialized carrier-mediated systems, with the exception of cobalamin. Some microbial species such as *Bacteroides* spp., are able to synthesize vitamins *de novo* (vitamin prototrophs) as water-soluble B-vitamins, whereas on the contrary, other species as *Faecalibacterium* spp. that lack biosynthetic pathways require external sources (vitamin auxotrophs) [72,73]. This is why human gut commensals have been known to be important producers of vitamins, which are necessary as essential coenzymes for a wide class of metabolic reactions. Thus, gut microbes can synthesize vitamin K2, as well as water-soluble B-vitamins such as folic acid, niacin, biotin, pantothenic acid, cobalamin, pyridoxine, riboflavin, and thiamine [74,75].

Long-term consumption of plant-based diets with restriction on caloric intake has been associated with abundant and higher microbial phylogenetic differences [76]. Mainly, an increase of lactic acid bacteria (LAB) has been highlighted, such as *Ruminococcus*, *Enterococcus rectale*, and *Roseburia*, as well as increases in *Bifidobacterium* and *Lactobacillus* perhaps due to polyphenols, also abundant in plant foods, which perform anti-inflammatory, anti-pathogenic effects and cardiovascular protection [35].

Furthermore, polyphenols, which are widely distributed in plants, vegetables, and fruits, are also derived from bacterial metabolism of dietary foods at the gut level, and they are converted into derivatives of aromatic SCFAs as phenylacetate or phenylbutyrate.

These metabolites are produced by gut bacteria such as *Bacteroides*, *Clostridium*, *Eubacterium limosum*, and *Eggerthella lenta*, which have a wide range of activities in the prevention and treatment of several diseases such as diabetes, cancer, neuroinflammation, and aging. Indeed, SCFAs and polyphenols, especially phenylbutyrate, inhibit histone deacetylase (HDAC) activity, which is also involved in impaired intestinal epithelial cell function [77,78].

3.2. Dietary Proteins and Fats

Numerous studies have shown that high-protein diets in humans make a shift from carbohydrates to protein fermentation through the GM [79], whereas they induce changes, particularly leading to a decrease in hypothetically beneficial microbes that produce butyrate [50].

Animal-based diets are usually energy-full and poor in fiber. The microbes that are not able to digest fibers use proteins, fat, and simple sugars for their growth, and are also more appropriate for harvesting the energy taken in excess, as is the case in the Western diet [80].

On the contrary, diets such as the ketogenic diet (KD) that are high in fat and protein and very low in carbohydrates, are normocaloric diets, in which the glucose in the body becomes deficient for both fat oxidation (oxaloacetate derived from tricarboxylic acid cycle, TCA) and energy, which is required for the central nervous system. Hence, the organism is forced to use fats as a primary fuel, in the form of ketone bodies such as 3-hydroxybutyrate (3HB), acetate, and acetoacetate (AcAc), which are produced in the liver by the ketogenesis process. The KD seems to act as an efficient diet therapy for weight reduction, and is also is used for epilepsy and Glucose Transporter 1 Deficiency Syndrome (GLUT1-DS) [81].

In addition, at the gut level, the KD reduces GM diversity, while increasing the relative abundance of *Akkermansia muciniphila*, *Parabacteroides*, and *Lactobacillus*, which produce SCFAs, and also microbiota-dependent seizure protection has been found to be linked to the increase of gamma-aminobutyric acid (GABA) [82] (Figure 1). Hence, a reduction in pro-inflammatory microbes such as *Desulfovibrio* and *Turicibacter* results in being potentially protected against seizures [82].

Metabolic outcome of protein catabolism by the GM is far more diverse [83], including carbohydrate fermentation that mainly produces SCFAs [84], as well as fermentation of AAs, moreover releasing beneficial SCFAs. This produces, in particular, branched-chain fatty acids (BCFAs), hydrogen sulfide (H_2S), ammonia, phenolic and indolic compounds, and also amines and polyamines [85]. BCFAs are developed from branched-chain AAs such as valine, leucine, and isoleucine, which give them biomarkers for bacterial proteolysis [86]. The metabolites such as amines, hydrogen sulfide, *p*-cresol, and ammonia, are harmful for the colonic epithelium at excessive concentrations [38,87]. Some of these may play a role in many diseases such as DNA damage, leaky gut, colon cancer, or IBD [54].

On the contrary, molecules such as indolic compounds (precursors of indoxyl sulfate) take part in the maintenance of the epithelial barrier function [12,88].

BCFAs, indoles, and phenols are not generated by human enzymes and therefore they can only derive from bacterial fermentation. The production of these metabolites is often considered as a marker for evaluating the level of protein fermentation in the colon [89].

The exposure to a Western diet poor in "microbiota-reachable carbohydrates" results in the disappearance of specific bacterial lineages, which could harmfully affect the maturation and function of the immune system, as well as causing the growth in the onset risk of a wide range of inflammatory, metabolic, allergic, and autoimmune disorders [90] (Figure 1).

Animal proteins derived from dairy products (caseins) and red meat stimulate the production of genotoxic endogenous *N*-nitroso compounds in the human gut [91], and the levels of fecal *p*-cresol have been significantly increased and correlated with genetic damage [92].

In addition, animal-based diets also contain a higher quantity of choline and L-carnitine, which have been linked to risk of cardiovascular diseases (CVDs) in humans and in mice, due to their conversion into trimethylamine (TMA) by several gut microbes, such as *Candida*, *Campylobacter*, and *Shigella* species, as well as *Ruminococcus gnavus* [93], following the consequent absorption into portal circulation, and the conversion to trimethylamine *N*-oxides (TMAO) in the liver [94].

It has been shown that TMAO decreases the transport of reverse cholesterol and bile acid synthesis, possibly reducing the normal pathway of intestinal cholesterol elimination, thus explaining the relation between bile acid and TMAO regulation emerging as a possible atherosclerosis mediator [95].

TMAO might also be linked to other undesirable conditions, such as obesity, insulin resistance, and gastrointestinal cancers [51,96].

Recently, some studies have detected choline deficiencies in the development of non-alcoholic fatty liver disease (NAFLD), providing evidence of growth of progression to non-alcoholic steatohepatitis (NASH). Even if NAFLD patients remain asymptomatic, 20% progress to NASH can occur, which emerges as increased mortality through cirrhosis, portal hypertension, and hepatocellular carcinoma [97]. Furthermore, by using metagenomics and metabolomics, it was found to be possible to identify a combination of enterotype–metabotype that appeared to contribute to the definition of a signature that changes during the evolution of the disease [98].

On the contrary, vegans and lacto-ovo vegetarians have highlighted the insignificant postprandial value of TMAO in plasma as response to an L-carnitine meal [99]. Diets rich in saturated dietary lipids are associated with growth of white adipose tissue (WAT) and inflammation mediated by GM molecular mechanisms that induces a macrophage increase in WAT and adverse metabolic consequences [100]. The response of gut microbes differs according to the type of dietary fats; for example, in the presence of lard rich in saturated lipids or fish oil, or rich in polyunsaturated lipids. Genera *Bacteroides*, *Turicibacter*, and *Bilophila* were increased in lard-fed mice; on the contrary, Actinobacteria (*Bifidobacterium* and *Adlercreutzia*), LAB (*Lactobacillus* and *Streptococcus*), Verrucomicrobia (*Akkermansia muciniphila*), Alphaproteobacteria, and Deltaproteobacteria were increased in fish oil-fed mice [100]. In particular, *A. muciniphila*, which has been shown to (i) decrease fat mass gain and WAT macrophage infiltration, (ii) increase gut barrier function, (iii) and increase glucose metabolism when given to mice with diet-induced obesity [101].

For other metabolites such as phenols, a current study highlighted that gut bacterial metabolites such as 4-ethylphenylsulfate (4EPS) and indolepyruvate may derived from AA fermentation, potentially triggered ASD in a mouse model [17]. Moreover, several metabolites (such as *p*-cresol and indolyl-3-acryloylglycine), which are similar to 4EPS and indole pyruvate, have been detected in urine as human autism biomarkers [102].

Furthermore, the oral treatment of maternal immune activation (MIA) mouse model with *Bacteroides fragilis* improves gut permeability; modulates the microbial composition and function; and ameliorates deficiencies in communicative, stereotypic, anxiety-like behaviors, suggesting that gut microbiota and host metabolism affects behavior [17].

Finally, the diet-dependent endogenous GM generate metabolites such as low molecular weight compounds such as vitamins (derived directly from dietary or *de novo* synthetized bacteria), polyamines, and SCFAs, as well as diet-independent products such as lipopolysaccharide (LPS) (Gram-negative cell wall component) and peptidoglycan (Gram-positive cell wall component), which interact within the intestinal microenvironment [103] for normal development and behavior [104].

These metabolites also change the epigenome of host cells, and the epigenetic modifications in turn alter the growth and functions of the cell and modulate gene expression [105] during life of each person [106,107].

4. Microbiome-Based Therapeutics

The understanding of metabolic capabilities of gut microbial inhabitants in all areas of the human body may be the passport to understanding health- and disease-linked mechanisms.

New knowledge on GM modulation mechanisms can provide potential novel strategies to prevent and treat IBD and extraintestinal inflammatory diseases. Dietary interventions with pre-, pro-, and post-biotics or synbiotics showed an increase in saccharolytic fermentation while concomitantly decreasing proteolytic fermentation [108].

Thus far, great progress has been reached through microbial analysis tools and therapeutic strategies through dietary intervention with microbes, as well as the usage of fecal microbiota transplantation (FMT) [109] (Figure 2).

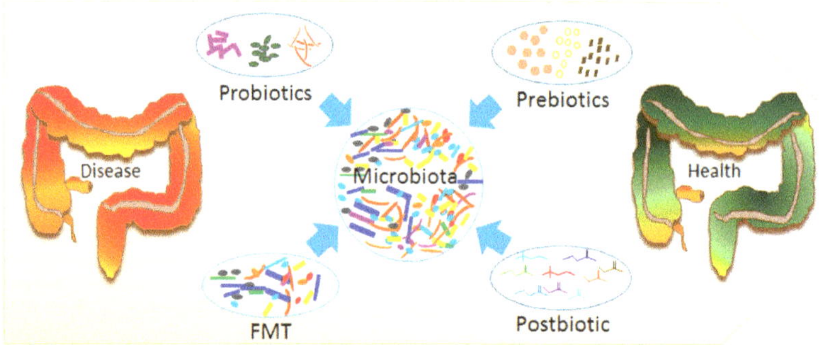

Figure 2. Gut microbiota modulation. Beneficial effect on the intestinal environment through microbial composition modification, by using probiotic, prebiotic, and postbiotic administration and fecal microbiota transplantation (FMT). All these approaches can be optimized by patient-tailored treatments for the better management of the individual's physiology and pathology.

Probiotics are defined as live microorganisms that, when administered in adequate amounts, confer a health benefit to the host [110]. Although this definition has been kept constant for 17 years, the transfer of the concept to the public is yet ongoing. In consideration of the wide spectrum of probiotics, many scientists introduce subgrouping, regarding the system targeted or the potential illness. Some probiotic products may have a target disease (i.e., that which is not addressed to the general population but to patients with a specific disease), suitable to be approached by a particular probiotic product [110].

Therefore, the interpretation of probiotic action and their effects on the host is important to modulate the GM and to treat a broad range of human diseases. The actions are commonly related to probiotic use for their anti-microbial effects, mucosal barrier integrity enhancement, and host immunomodulation [109].

The major extensively commercialized probiotics are composed by *Bifidobacterium* and *Lactobacillus* strains, although other microorganisms, such as the yeast *Saccharomyces*, now are also widely employed [111]. However, different probiotic strains are not similarly powerful, and their effects are mediated by host immune response interactions and within a very intricate gut microbial ecosystem. Through targeted or untargeted metagenomics approaches and predicted pathways, researchers may identify the complex relation between microbiota and host, thus providing the development of probiotics for the next generation. Some positive results have been acquired in animal models and human clinical trials by using recombinant LAB, which expresses beneficial molecules to stimulate the suppression of inflammatory systemic immune response [112,113].

On the other hand, dietary treatment with cow's milk fermented with *Lactobacillus paracasei* CBA L74 (FM-CBAL74) has been associated with high abundance of predicted genes involved in butyrate synthesis in pediatric GI and respiratory infections [114]. Nowadays, as for the use of the genetically modified lactic acid bacteria such as *Lactococcus lactis* and certain species of Lactobacilli, it is possible to create alive recombinant vectors for the progress of new nontoxic mucosal vaccines [115,116].

Personalized probiotic therapy aims to manipulate the GM of the host in order to ameliorate metabolic diseases, asthma, arthritis, and cancer [117–120].

Several plant polysaccharide complexes present in foods, as previously described, cannot be digested by human enzymes due to their insolubility or to a deficiency of human-encoded hydrolytic enzymes—these food substrates are called prebiotics [121]. Prebiotics are defined as dietary supplements and are resistant starch, β-glucans, inulin, pectin, and other GOS and FOS, and are taken in order to increase the level of beneficial bacteria (i.e., Bifidobacteria and LAB). The prebiotics are compounds that selectively exercise an influence on the GM composition or function, exerting a beneficial effect to the host after bacterial metabolism [122]. However, some bacteria acting as probiotics can degrade them, and the consumption of polysaccharides is important for their growth [109].

However, Khalesi et al. [123] collected the literature on the effectiveness of probiotics in healthy adults, evidencing the failure of ability of probiotics to cause persistent changes in GM [123].

Furthermore, guidelines of clinical practice have not reviewed the safety and potential adverse events that may be encountered when using probiotics in the treatment of different pediatric diseases [124]. Therefore, the suitability of probiotics administration to provide real benefits in healthy adults needs further investigation.

Omics-based analyses suggest for prebiotics a promising therapeutic role in metabolic syndrome or inflammatory bowel syndrome (IBS) [125,126], even though their action mode and their effects on different microbiome ecosystems require further studies. The dietary supplement response, however, exhibits an inter-individual variation, partially due to the microbiome composition [127,128].

In addition, a more complete dietary approach, namely, "personalized nutrition" (PN) utilizes a large specific population's metadata (including a rich dataset of microbiome features) and bioinformatic tools, which can allow personalized dietary interventions that are able to modify the microbiome by affecting metabolic homeostasis [129].

The GM may be affected by numerous factors, among which diet may assume a fundamental role. The ability to change the GM composition and activity by tailored dietary interventions is not only attractive but also is an encouragement for future research in disease prevention and wellness management [130].

The advances in human genome sequencing and the field of precision nutrition determines precision lifestyle medicine, which can classify a variety of nutrient metabolisms among subgroups (e.g., ethnicity, health status, lifestyle, cultural preferences, and clinical variables) and inter-individual variability in reactions to dietary interventions [131]. Several studies focused on the co-evolution of humans and their GM to understand to what extent the spread of Western diet and lifestyle have impacted our microbial symbionts and how this has affected human health [46,132].

Personalized strategies to propose optimized nutrition are being consolidated, but more investigations are necessary to improve PN knowledge derived from microbiome studies [133].

Another possibility in GM management is the use of synbiotics, in which prebiotics and probiotics are present in synergistic combinations [134,135].

Previous studies have highlighted the fact that synbiotics have a potential heavy effect on GM metabolic activity modulation, more than probiotics or prebiotics themselves [134,136,137].

It has also been reported that synbiotic administration can improve (i) metabolic status (total cholesterol, low density lipoprotein cholesterol (LDL-c), high density lipoprotein cholesterol (HDL-c), triglycerides etc.) [125], and (ii) serum level of liver function enzymes [123] and inflammatory biomarkers [138] by changing the composition and/or function of the GM.

The non-viable bacterial products or metabolic by-products derived from probiotics are called "postbiotics". Postbiotics are functional products derived from fermentation (i.e., SCFAs, secreted polysaccharides, extracellular polysaccharides (EPS), microbial fractions, functional proteins, cell lysates, teichoic acid, peptidoglycan-derived muropeptides, ethanol, diacetyl, acetaldehydes, and hydrogen peroxide), which could be used in combination with nutritional components to promote health [139]. These metabolic products have a wide range of inhibitory properties towards pathogenic microbes and, therefore, can be used as a substitute of antibiotics [140].

Two common types of postbiotics are represented by paraprobiotics and fermented infant formulas (FIFs). Paraprobiotics, also called ghost probiotics, are non-viable probiotics or inactivated probiotics, generally defined as "inactivated or non-viable microbial cells, when administered in sufficient amounts give benefits to the host" [141]. FIFs are infant formulas fermented by lactic acid-producing or other bacteria, and in many cases do not contain viable bacteria [142].

Therefore, the employment of postbiotic molecules has become a possible strategy for treating many inflammatory diseases through comprehension of microbiota–host metabolism mechanisms.

Indeed, these molecules mime the beneficial and therapeutic effect of probiotics, eluding the living microorganisms' administration to a host with a compromised immune system [109,143].

A potential efficient microbiome intervention is represented by FMT, in which a healthy donor microbiome is transplanted into a patient to correct the individual's own disease-associated microbiome.

The GM modulation by FMT principally follows the probiotic principle, but the patient will not be treated with specific strains, but instead with a fecal suspension infusion from a healthy donor [144]. Currently, this treatment is approved for recurrent *Clostridium difficile* infections (rCDI) in adulthood, and is now also tested in different clinical procedures for treating numerous pathological conditions, ranging from metabolic and neoplastic to autoimmune disorders [145,146], both in adulthood and childhood. However, transplanting an entire community of microbes involves risks, as the pathogen/pathobiont transmission can generate undesirable effects for the transplanted microbiome [147] and incomplete long-term stabilization of an extraneous microbial configuration when introduced into a new host with a unique genetic, immune, metabolic, and nutritional environment [129]. Further studies are necessary to optimize and improve this technique and to increase FMT usage as clinical treatment beyond rCDI. Clinical trials are necessary to standardize adverse events' registration, patients' registers, experimental methods to identify metabolites, and metagenomes associated with different disease treatments by FMT. This is the reason why international consensus and/or recommendation are needed for donor screening for a wide range of diseases [127–129]. Recently, to support the importance of donor screening, it has been suggested that the FMT success depend on the microbial diversity and microbial composition of the donor, generating the hypothesis of the existence of FMT super-donors. The determination and characterization of super-donor gut microbes will help to understand bacterial components of diseases and allow for the creation of more targeted approaches in the future [148–151].

5. Conclusions and Perspective

The research thus far reveals, summarizing the previous results, that the GM can respond to altered diet, potentially facilitating the diversity of human dietary lifestyles.

Diet is one of the most important driving forces able to shape the GM. Dietary interventions and targeted nutritional therapies, such as medical foods, dietary supplements, living microorganisms, nutraceutical food, and FMT, could provide a great promise for the prevention and treatment of microbiota-related diseases. However, much experimental research is needed before these opportunities can be fully realized. Effects of specific nutrients need to be assessed in clinical trials. Finally, the microbiomics field has undergone a massive revolution by the identification of variables that lead to the development of bacterial community structure and functionality, as well as in terms of understanding how these gut microbial ecosystems and their metabolism of dietary ingredients can influence both healthy human status and disorders. It will in future be possible to define new types of therapies that could be developed with the focus on creating "personalized" diets and microbiomes in order to promote health.

Author Contributions: P.V. conceived and wrote the manuscript; F.D.C. participated in the active revision process and produced all figures; L.P. supervised and reviewed the manuscript. All authors have read and agreed to the published version of the manuscript.

Funding: This work was supported by the Ministry of Health, Ricerca Corrente RC 201802G004314, assigned to L.P., Children's Hospital Bambino Gesù, IRCCS.

Conflicts of Interest: The authors declare no conflict of interest.

References

1. O'Hara, A.M.; Shanahan, F. The gut flora as a forgotten organ. *EMBO Rep.* **2006**, *7*, 688–693. [CrossRef] [PubMed]
2. Martinez, K.B.; Leone, V.; Chang, E.B. Microbial metabolites in health and disease: Navigating the unknown in search of function. *J. Biol. Chem.* **2017**, *292*, 8553–8559. [CrossRef] [PubMed]

3. Yadav, M.; Verma, M.K.; Chauhan, N.S. A review of metabolic potential of human gut microbiome in human nutrition. *Arch. Microbiol.* **2018**, *200*, 203–217. [CrossRef] [PubMed]
4. Hooper, L.V.; Littman, D.R.; Macpherson, A.J. Interactions between the microbiota and the immune system. *Science* **2012**, *336*, 1268–1273. [CrossRef] [PubMed]
5. Sharon, G.; Garg, N.; Debelius, J.; Knight, R.; Dorrestein, P.C.; Mazmanian, S.K. Specialized metabolites from the microbiome in health and disease. *Cell Metab.* **2014**, *20*, 719–730. [CrossRef] [PubMed]
6. Quinn, R.A.; Lim, Y.W.; Maughan, H.; Conrad, D.; Rohwer, F.; Whiteson, K.L. Biogeochemical forces shape the composition and physiology of polymicrobial communities in the cystic fibrosis lung. *MBio* **2014**, *5*, e00956-13. [CrossRef]
7. Sheflin, A.M.; Melby, C.L.; Carbonero, F.; Weir, T.L. Linking dietary patterns with gut microbial composition and function. *Gut Microbes* **2017**, *8*, 113–129. [CrossRef]
8. Portune, K.J.; Benítez-Páez, A.; Del Pulgar, E.M.G.; Cerrudo, V.; Sanz, Y. Gut microbiota, diet, and obesity-related disorders-The good, the bad, and the future challenges. *Mol. Nutr. Food Res.* **2017**, *61*. [CrossRef]
9. Maurice, C.F.; Haiser, H.J.; Turnbaugh, P.J. Xenobiotics shape the physiology and gene expression of the active human gut microbiome. *Cell* **2013**, *152*, 39–50. [CrossRef]
10. Sung, J.; Kim, S.; Cabatbat, J.J.T.; Jang, S.; Jin, Y.-S.; Jung, G.Y.; Chia, N.; Kim, P.-J. Global metabolic interaction network of the human gut microbiota for context-specific community-scale analysis. *Nat. Commun.* **2017**, *8*, 15393. [CrossRef]
11. Brestoff, J.R.; Artis, D. Commensal bacteria at the interface of host metabolism and the immune system. *Nat. Immunol.* **2013**, *14*, 676–684. [CrossRef] [PubMed]
12. Clarke, G.; Stilling, R.M.; Kennedy, P.J.; Stanton, C.; Cryan, J.F.; Dinan, T.G. Minireview: Gut microbiota: The neglected endocrine organ. *Mol. Endocrinol.* **2014**, *28*, 1221–1238. [CrossRef] [PubMed]
13. Soret, R.; Chevalier, J.; De Coppet, P.; Poupeau, G.; Derkinderen, P.; Segain, J.P.; Neunlist, M. Short-chain fatty acids regulate the enteric neurons and control gastrointestinal motility in rats. *Gastroenterology* **2010**, *138*, 1772–1782. [CrossRef] [PubMed]
14. Hinnebusch, B.F.; Meng, S.; Wu, J.T.; Archer, S.Y.; Hodin, R.A. The effects of short-chain fatty acids on human colon cancer cell phenotype are associated with histone hyperacetylation. *J. Nutr.* **2002**, *132*, 1012–1017. [CrossRef] [PubMed]
15. Lan, A.; Lagadic-Gossmann, D.; Lemaire, C.; Brenner, C.; Jan, G. Acidic extracellular pH shifts colorectal cancer cell death from apoptosis to necrosis upon exposure to propionate and acetate, major end-products of the human probiotic propionibacteria. *Apoptosis* **2007**, *12*, 573–591. [CrossRef] [PubMed]
16. Nicholson, J.K.; Holmes, E.; Kinross, J.; Burcelin, R.; Gibson, G.; Jia, W.; Pettersson, S. Host-gut microbiota metabolic interactions. *Science* **2012**, *336*, 1262–1267. [CrossRef]
17. Hsiao, E.Y.; McBride, S.W.; Hsien, S.; Sharon, G.; Hyde, E.R.; McCue, T.; Codelli, J.A.; Chow, J.; Reisman, S.E.; Petrosino, J.F.; et al. Microbiota modulate behavioral and physiological abnormalities associated with neurodevelopmental disorders. *Cell* **2013**, *155*, 1451–1463. [CrossRef]
18. Delzenne, N.M.; Williams, C.M. Prebiotics and lipid metabolism. *Curr. Opin. Lipidol.* **2002**, *13*, 61–67. [CrossRef]
19. Visconti, A.; Le Roy, C.I.; Rosa, F.; Rossi, N.; Martin, T.C.; Mohney, R.P.; Li, W.; de Rinaldis, E.; Bell, J.T.; Venter, J.C.; et al. Interplay between the human gut microbiome and host metabolism. *Nat. Commun.* **2019**, *10*, 4505. [CrossRef]
20. Tremaroli, V.; Bäckhed, F. Functional interactions between the gut microbiota and host metabolism. *Nature* **2012**, *489*, 242–249. [CrossRef] [PubMed]
21. Gantois, I.; Ducatelle, R.; Pasmans, F.; Haesebrouck, F.; Hautefort, I.; Thompson, A.; Hinton, J.C.; Van Immerseel, F. Butyrate specifically down-regulates salmonella pathogenicity island 1 gene expression. *Appl. Environ. Microbiol.* **2006**, *72*, 946–949. [CrossRef] [PubMed]
22. Buffie, C.G.; Bucci, V.; Stein, R.R.; McKenney, P.T.; Ling, L.; Gobourne, A.; No, D.; Liu, H.; Kinnebrew, M.; Viale, A.; et al. Precision microbiome reconstitution restores bile acid mediated resistance to *Clostridium difficile*. *Nature* **2015**, *517*, 205–208. [CrossRef] [PubMed]
23. Rose, S.; Bennuri, S.C.; Davis, J.E.; Wynne, R.; Slattery, J.C.; Tippett, M.; Delhey, L.; Melnyk, S.; Kahler, S.G.; MacFabe, D.F.; et al. Butyrate enhances mitochondrial function during oxidative stress in cell lines from boys with autism. *Transl. Psychiatry* **2018**, *8*, 42. [CrossRef] [PubMed]

24. St-Onge, M.P.; Farnworth, E.R.; Jones, P.J. Consumption of fermented and nonfermented dairy products: Effects on cholesterol concentrations and metabolism. *Am. J. Clin. Nutr.* **2000**, *71*, 674–681. [CrossRef]
25. Saltzman, E.; Das, S.K.; Lichtenstein, A.H.; Dallal, G.E.; Corrales, A.; Schaefer, E.J.; Greenberg, A.S.; Roberts, S.B. An oat-containing hypocaloric diet reduces systolic blood pressure and improves lipid profile beyond effects of weight loss in men and women. *J. Nutr.* **2001**, *131*, 1465–1470. [CrossRef]
26. Soliman, G.A. Dietary fiber, atherosclerosis, and cardiovascular disease. *Nutrients* **2019**, *11*, 1155. [CrossRef]
27. Reddel, S.; Putignani, L.; Del Chierico, F. The Impact of Low-FODMAPs, gluten-free, and ketogenic diets on gut microbiota modulation in pathological conditions. *Nutrients* **2019**, *11*, 373. [CrossRef]
28. Walker, A.W.; Ince, J.; Duncan, S.H.; Webster, L.M.; Holtrop, G.; Ze, X.; Brown, D.; Stares, M.D.; Scott, P.; Bergerat, A.; et al. Dominant and diet-responsive groups of bacteria within the human colonic microbiota. *ISME J.* **2011**, *5*, 220–230. [CrossRef]
29. Wu, G.D.; Chen, J.; Hoffmann, C.; Bittinger, K.; Chen, Y.-Y.; Keilbaugh, S.A.; Bewtra, M.; Knights, D.; Walters, W.A.; Knight, R.; et al. Linking long-term dietary patterns with gut microbial enterotypes. *Science* **2011**, *334*, 105–108. [CrossRef]
30. Giuberti, G.; Gallo, A.; Moschini, M.; Masoero, F. New insight into the role of resistant starch in pig nutrition. *Anim. Feed Sci. Technol.* **2015**, *201*, 1–13. [CrossRef]
31. Wang, X.; Gibson, G.R.; Costabile, A.; Sailer, M.; Theis, S.; Rastall, R.A. Prebiotic supplementation of in vitro fecal fermentations inhibits proteolysis by gut bacteria, and host diet shapes gut bacterial metabolism and response to intervention. *Appl. Environ. Microbiol.* **2019**, *85*. [CrossRef] [PubMed]
32. Guilherme, A.; Virbasius, J.V.; Puri, V.; Czech, M.P. Adipocyte dysfunctions linking obesity to insulin resistance and type 2 diabetes. *Nat. Rev. Mol. Cell Biol.* **2008**, *9*, 367–377. [CrossRef] [PubMed]
33. Lee, W.-J.; Hase, K. Gut microbiota-generated metabolites in animal health and disease. *Nat. Chem. Biol.* **2014**, *10*, 416–424. [CrossRef] [PubMed]
34. Singh, R.K.; Chang, H.-W.; Yan, D.; Lee, K.M.; Ucmak, D.; Wong, K.; Abrouk, M.; Farahnik, B.; Nakamura, M.; Zhu, T.H.; et al. Influence of diet on the gut microbiome and implications for human health. *J. Transl. Med.* **2017**, *15*, 73. [CrossRef] [PubMed]
35. Tomova, A.; Bukovsky, I.; Rembert, E.; Yonas, W.; Alwarith, J.; Barnard, N.D.; Kahleova, H. The effects of vegetarian and vegan diets on gut microbiota. *Front. Nutr.* **2019**, *6*, 47. [CrossRef]
36. Qin, J.; Li, R.; Raes, J.; Arumugam, M.; Burgdorf, K.S.; Manichanh, C.; Nielsen, T.; Pons, N.; Levenez, F.; Yamada, T.; et al. A human gut microbial gene catalogue established by metagenomic sequencing. *Nature* **2010**, *464*, 59–65. [CrossRef]
37. Ley, R.E.; Bäckhed, F.; Turnbaugh, P.; Lozupone, C.A.; Knight, R.D.; Gordon, J.I. Obesity alters gut microbial ecology. *Proc. Natl. Acad. Sci. USA* **2005**, *102*, 11070–11075. [CrossRef]
38. Arumugam, M.; Raes, J.; Pelletier, E.; Le Paslier, D.; Yamada, T.; Mende, D.R.; Fernandes, G.R.; Tap, J.; Bruls, T.; Batto, J.-M.; et al. Enterotypes of the human gut microbiome. *Nature* **2011**, *473*, 174–180. [CrossRef]
39. David, L.A.; Maurice, C.F.; Carmody, R.N.; Gootenberg, D.B.; Button, J.E.; Wolfe, B.E.; Ling, A.V.; Devlin, A.S.; Varma, Y.; Fischbach, M.A.; et al. Diet rapidly and reproducibly alters the human gut microbiome. *Nature* **2014**, *505*, 559–563. [CrossRef]
40. Hills, R.D., Jr.; Pontefract, B.A.; Mishcon, H.R.; Black, C.A.; Sutton, S.C.; Theberge, C.R. Gut microbiome: Profound implications for diet and disease. *Nutrients* **2019**, *11*, 1613. [CrossRef]
41. Hjorth, M.F.; Roager, H.M.; Larsen, T.M.; Poulsen, S.K.; Licht, T.R.; Bahl, M.I.; Zohar, Y.; Astrup, A. Pre-treatment microbial Prevotella-to-Bacteroides ratio, determines body fat loss success during a 6-month randomized controlled diet intervention. *Int. J. Obes.* **2018**, *42*, 580–583. [CrossRef] [PubMed]
42. Zimmer, J.; Lange, B.; Frick, J.-S.; Sauer, H.; Zimmermann, K.; Schwiertz, A.; Rusch, K.; Klosterhalfen, S.; Enck, P. A vegan or vegetarian diet substantially alters the human colonic faecal microbiota. *Eur. J. Clin. Nutr.* **2012**, *66*, 53–60. [CrossRef] [PubMed]
43. Laulund, S.; Wind, A.; Derkx, P.M.F.; Zuliani, V. Regulatory and safety requirements for food cultures. *Microorganisms* **2017**, *5*, 28. [CrossRef] [PubMed]
44. Devkota, S.; Wang, Y.; Musch, M.W.; Leone, V.; Fehlner-Peach, H.; Nadimpalli, A.; Antonopoulos, D.A.; Jabri, B.; Chang, E.B. Dietary-fat-induced taurocholic acid promotes pathobiont expansion and colitis in Il10−/− mice. *Nature* **2012**, *487*, 104–108. [CrossRef] [PubMed]
45. Turnbaugh, P.J.; Ley, R.E.; Mahowald, M.A.; Magrini, V.; Mardis, E.R.; Gordon, J.I. An obesity-associated gut microbiome with increased capacity for energy harvest. *Nature* **2006**, *444*, 1027–1031. [CrossRef]

46. Devkota, S.; Chang, E.B. Diet-induced expansion of pathobionts in experimental colitis: Implications for tailored therapies. *Gut Microbes* **2013**, *4*, 172–174. [CrossRef]
47. Cresci, G.A.; Bawden, E. Gut microbiome: What we do and don't know. *Nutr. Clin. Pract.* **2015**, *30*, 734–746. [CrossRef]
48. Donia, M.S.; Fischbach, M.A. HUMAN MICROBIOTA. Small molecules from the human microbiota. *Science* **2015**, *349*, 1254766. [CrossRef]
49. Wong, J.M.W.; de Souza, R.; Kendall, C.W.C.; Emam, A.; Jenkins, D.J.A. Colonic health: Fermentation and short chain fatty acids. *J. Clin. Gastroenterol.* **2006**, *40*, 235–243. [CrossRef]
50. Russell, W.R.; Gratz, S.W.; Duncan, S.H.; Holtrop, G.; Ince, J.; Scobbie, L.; Duncan, G.; Johnstone, A.M.; Lobley, G.E.; Wallace, R.J.; et al. High-protein, reduced-carbohydrate weight-loss diets promote metabolite profiles likely to be detrimental to colonic health. *Am. J. Clin. Nutr.* **2011**, *93*, 1062–1072. [CrossRef]
51. Thorburn, A.N.; Macia, L.; Mackay, C.R. Diet, metabolites, and "western-lifestyle" inflammatory diseases. *Immunity* **2014**, *40*, 833–842. [CrossRef] [PubMed]
52. Macfarlane, G.T.; Macfarlane, S. Fermentation in the human large intestine: Its physiologic consequences and the potential contribution of prebiotics. *J. Clin. Gastroenterol.* **2011**, *45*, S120–S127. [CrossRef] [PubMed]
53. Flint, H.J.; Scott, K.P.; Louis, P.; Duncan, S.H. The role of the gut microbiota in nutrition and health. *Nat. Rev. Gastroenterol. Hepatol.* **2012**, *9*, 577–589. [CrossRef] [PubMed]
54. Marchesi, J.R.; Adams, D.H.; Fava, F.; Hermes, G.D.A.; Hirschfield, G.M.; Hold, G.; Quraishi, M.N.; Kinross, J.; Smidt, H.; Tuohy, K.M.; et al. The gut microbiota and host health: A new clinical frontier. *Gut* **2016**, *65*, 330–339. [CrossRef] [PubMed]
55. Marchix, J.; Goddard, G.; Helmrath, M.A. Host-gut microbiota crosstalk in intestinal adaptation. *Cell. Mol. Gastroenterol. Hepatol.* **2018**, *6*, 149–162. [CrossRef]
56. van Zanten, G.C.; Krych, L.; Röytiö, H.; Forssten, S.; Lahtinen, S.J.; Al-Soud, W.A.; Sørensen, S.; Svensson, B.; Jespersen, L.; Jakobsen, M. Synbiotic *Lactobacillus acidophilus* NCFM and cellobiose does not affect human gut bacterial diversity but increases abundance of lactobacilli, bifidobacteria and branched-chain fatty acids: A randomized, double-blinded cross-over trial. *FEMS Microbiol. Ecol.* **2014**, *90*, 225–236. [CrossRef]
57. Arpaia, N.; Campbell, C.; Fan, X.; Dikiy, S.; van der Veeken, J.; deRoos, P.; Liu, H.; Cross, J.R.; Pfeffer, K.; Coffer, P.J.; et al. Metabolites produced by commensal bacteria promote peripheral regulatory T-cell generation. *Nature* **2013**, *504*, 451–455. [CrossRef]
58. Cheong, K.-L.; Qiu, H.-M.; Du, H.; Liu, Y.; Khan, B.M. Oligosaccharides derived from red seaweed: Production, properties, and potential health and cosmetic applications. *Molecules* **2018**, *23*, 2451. [CrossRef]
59. Louis, P.; Flint, H.J. Formation of propionate and butyrate by the human colonic microbiota. *Environ. Microbiol.* **2017**, *19*, 29–41. [CrossRef]
60. Cherbut, C.; Michel, C.; Lecannu, G. The prebiotic characteristics of fructooligosaccharides are necessary for reduction of TNBS-induced colitis in rats. *J. Nutr.* **2003**, *133*, 21–27. [CrossRef]
61. Hamer, H.M.; Jonkers, D.; Venema, K.; Vanhoutvin, S.; Troost, F.J.; Brummer, R.-J. Review article: The role of butyrate on colonic function. *Aliment. Pharmacol. Ther.* **2008**, *27*, 104–119. [CrossRef] [PubMed]
62. Luissint, A.-C.; Parkos, C.A.; Nusrat, A. Inflammation and the intestinal barrier: Leukocyte-epithelial cell interactions, cell junction remodeling, and mucosal repair. *Gastroenterology* **2016**, *151*, 616–632. [CrossRef] [PubMed]
63. Sturgeon, C.; Fasano, A. Zonulin, a regulator of epithelial and endothelial barrier functions, and its involvement in chronic inflammatory diseases. *Tissue Barriers* **2016**, *4*, e1251384. [CrossRef] [PubMed]
64. Canani, R.B.; Costanzo, M.D.; Leone, L.; Pedata, M.; Meli, R.; Calignano, A. Potential beneficial effects of butyrate in intestinal and extraintestinal diseases. *World J. Gastroenterol.* **2011**, *17*, 1519–1528. [CrossRef]
65. Tosti, V.; Bertozzi, B.; Fontana, L. Health benefits of the mediterranean diet: Metabolic and molecular mechanisms. *J. Gerontol. A Biol. Sci. Med. Sci.* **2018**, *73*, 318–326. [CrossRef]
66. Calder, P.C.; Ahluwalia, N.; Brouns, F.; Buetler, T.; Clement, K.; Cunningham, K.; Esposito, K.; Jönsson, L.S.; Kolb, H.; Lansink, M.; et al. Dietary factors and low-grade inflammation in relation to overweight and obesity. *Br. J. Nutr.* **2011**, *106* (Suppl. S3), S5–S78. [CrossRef]
67. Goldsmith, J.R.; Sartor, R.B. The role of diet on intestinal microbiota metabolism: Downstream impacts on host immune function and health, and therapeutic implications. *J. Gastroenterol.* **2014**, *49*, 785–798. [CrossRef]

68. Del Chierico, F.; Abbatini, F.; Russo, A.; Quagliariello, A.; Reddel, S.; Capoccia, D.; Caccamo, R.; Ginanni Corradini, S.; Nobili, V.; De Peppo, F.; et al. Gut microbiota markers in obese adolescent and adult patients: Age-dependent differential patterns. *Front. Microbiol.* **2018**, *9*, 1210. [CrossRef]
69. Lopez-Siles, M.; Duncan, S.H.; Garcia-Gil, L.J.; Martinez-Medina, M. *Faecalibacterium prausnitzii*: From microbiology to diagnostics and prognostics. *ISME J.* **2017**, *11*, 841–852. [CrossRef]
70. Crusell, M.K.W.; Hansen, T.H.; Nielsen, T.; Allin, K.H.; Rühlemann, M.C.; Damm, P.; Vestergaard, H.; Rørbye, C.; Jørgensen, N.R.; Christiansen, O.B.; et al. Gestational diabetes is associated with change in the gut microbiota composition in third trimester of pregnancy and postpartum. *Microbiome* **2018**, *6*, 89. [CrossRef]
71. Mustad, V.A.; Huynh, D.T.T.; López-Pedrosa, J.M.; Campoy, C.; Rueda, R. The role of dietary carbohydrates in gestational diabetes. *Nutrients* **2020**, *12*, 385. [CrossRef] [PubMed]
72. Das, P.; Babaei, P.; Nielsen, J. Metagenomic analysis of microbe-mediated vitamin metabolism in the human gut microbiome. *BMC Genomics* **2019**, *20*, 208. [CrossRef]
73. Sharma, V.; Rodionov, D.A.; Leyn, S.A.; Tran, D.; Iablokov, S.N.; Ding, H.; Peterson, D.A.; Osterman, A.L.; Peterson, S.N. B-vitamin sharing promotes stability of gut microbial communities. *Front. Microbiol.* **2019**, *10*, 1485. [CrossRef] [PubMed]
74. Rossi, M.; Amaretti, A.; Raimondi, S. Folate production by probiotic bacteria. *Nutrients* **2011**, *3*, 118–134. [CrossRef] [PubMed]
75. Said, H.M.; Nexo, E. Gastrointestinal handling of water-soluble vitamins. *Compr. Physiol.* **2018**, *8*, 1291–1311.
76. Griffin, N.W.; Ahern, P.P.; Cheng, J.; Heath, A.C.; Ilkayeva, O.; Newgard, C.B.; Fontana, L.; Gordon, J.I. Prior dietary practices and connections to a human gut microbial metacommunity alter responses to diet interventions. *Cell Host Microbe* **2017**, *21*, 84–96. [CrossRef]
77. Waldecker, M.; Kautenburger, T.; Daumann, H.; Busch, C.; Schrenk, D. Inhibition of histone-deacetylase activity by short-chain fatty acids and some polyphenol metabolites formed in the colon. *J. Nutr. Biochem.* **2008**, *19*, 587–593. [CrossRef]
78. Alcarraz-Vizán, G.; Boren, J.; Lee, W.-N.P.; Cascante, M. Histone deacetylase inhibition results in a common metabolic profile associated with HT29 differentiation. *Metabolomics* **2010**, *6*, 229–237. [CrossRef]
79. Windey, K.; De Preter, V.; Louat, T.; Schuit, F.; Herman, J.; Vansant, G.; Verbeke, K. Modulation of protein fermentation does not affect fecal water toxicity: A randomized cross-over study in healthy subjects. *PLoS ONE* **2012**, *7*, e52387. [CrossRef]
80. Tilg, H.; Moschen, A.R.; Kaser, A. Obesity and the microbiota. *Gastroenterology* **2009**, *136*, 1476–1483. [CrossRef]
81. Tagliabue, A.; Ferraris, C.; Uggeri, F.; Trentani, C.; Bertoli, S.; de Giorgis, V.; Veggiotti, P.; Elli, M. Short-term impact of a classical ketogenic diet on gut microbiota in GLUT1 Deficiency Syndrome: A 3-month prospective observational study. *Clin. Nutr. ESPEN* **2017**, *17*, 33–37. [CrossRef] [PubMed]
82. Paoli, A.; Mancin, L.; Bianco, A.; Thomas, E.; Mota, J.F.; Piccini, F. Ketogenic diet and microbiota: Friends or enemies? *Genes* **2019**, *10*, 534. [CrossRef] [PubMed]
83. Blachier, F.; Beaumont, M.; Portune, K.J.; Steuer, N.; Lan, A.; Audebert, M.; Khodorova, N.; Andriamihaja, M.; Airinei, G.; Benamouzig, R.; et al. High-protein diets for weight management: Interactions with the intestinal microbiota and consequences for gut health. A position paper by the my new gut study group. *Clin. Nutr.* **2019**, *38*, 1012–1022. [CrossRef] [PubMed]
84. Beaumont, M.; Portune, K.J.; Steuer, N.; Lan, A.; Cerrudo, V.; Audebert, M.; Dumont, F.; Mancano, G.; Khodorova, N.; Andriamihaja, M.; et al. Quantity and source of dietary protein influence metabolite production by gut microbiota and rectal mucosa gene expression: A randomized, parallel, double-blind trial in overweight humans. *Am. J. Clin. Nutr.* **2017**, *106*, 1005–1019. [CrossRef] [PubMed]
85. Blachier, F.; Mariotti, F.; Huneau, J.F.; Tomé, D. Effects of amino acid-derived luminal metabolites on the colonic epithelium and physiopathological consequences. *Amino Acids* **2007**, *33*, 547–562. [CrossRef] [PubMed]
86. Verbeke, K.A.; Boobis, A.R.; Chiodini, A.; Edwards, C.A.; Franck, A.; Kleerebezem, M.; Nauta, A.; Raes, J.; van Tol, E.A.F.; Tuohy, K.M. Towards microbial fermentation metabolites as markers for health benefits of prebiotics. *Nutr. Res. Rev.* **2015**, *28*, 42–66. [CrossRef]
87. Beaumont, M.; Andriamihaja, M.; Lan, A.; Khodorova, N.; Audebert, M.; Blouin, J.-M.; Grauso, M.; Lancha, L.; Benetti, P.-H.; Benamouzig, R.; et al. Detrimental effects for colonocytes of an increased exposure to luminal hydrogen sulfide: The adaptive response. *Free Radic. Biol. Med.* **2016**, *93*, 155–164. [CrossRef]

88. Venkatesh, M.; Mukherjee, S.; Wang, H.; Li, H.; Sun, K.; Benechet, A.P.; Qiu, Z.; Maher, L.; Redinbo, M.R.; Phillips, R.S.; et al. Symbiotic bacterial metabolites regulate gastrointestinal barrier function via the xenobiotic sensor PXR and Toll-like receptor 4. *Immunity* **2014**, *41*, 296–310. [CrossRef]
89. Geypens, B.; Claus, D.; Evenepoel, P.; Hiele, M.; Maes, B.; Peeters, M.; Rutgeerts, P.; Ghoos, Y. Influence of dietary protein supplements on the formation of bacterial metabolites in the colon. *Gut* **1997**, *41*, 70–76. [CrossRef]
90. Sonnenburg, E.D.; Smits, S.A.; Tikhonov, M.; Higginbottom, S.K.; Wingreen, N.S.; Sonnenburg, J.L. Diet-induced extinctions in the gut microbiota compound over generations. *Nature* **2016**, *529*, 212–215. [CrossRef]
91. Kuhnle, G.G.C.; Story, G.W.; Reda, T.; Mani, A.R.; Moore, K.P.; Lunn, J.C.; Bingham, S.A. Diet-induced endogenous formation of nitroso compounds in the GI tract. *Free Radic. Biol. Med.* **2007**, *43*, 1040–1047. [CrossRef] [PubMed]
92. Moore, J.W.; Millard, S.; Babidge, W.; Rowland, R.; Roediger, W.E. Hydrogen sulphide produces diminished fatty acid oxidation in the rat colon in vivo: Implications for ulcerative colitis. *ANZ J. Surg.* **1997**, *67*, 245–249. [CrossRef] [PubMed]
93. Chen, X.; Li, H.-Y.; Hu, X.-M.; Zhang, Y.; Zhang, S.-Y. Current understanding of gut microbiota alterations and related therapeutic intervention strategies in heart failure. *Chin. Med. J.* **2019**, *132*, 1843–1855. [CrossRef] [PubMed]
94. Tang, W.H.W.; Wang, Z.; Levison, B.S.; Koeth, R.A.; Britt, E.B.; Fu, X.; Wu, Y.; Hazen, S.L. Intestinal microbial metabolism of phosphatidylcholine and cardiovascular risk. *N. Engl. J. Med.* **2013**, *368*, 1575–1584. [CrossRef]
95. Wilson, A.; McLean, C.; Kim, R.B. Trimethylamine-N-oxide: A link between the gut microbiome, bile acid metabolism, and atherosclerosis. *Curr. Opin. Lipidol.* **2016**, *27*, 148–154. [CrossRef]
96. Oellgaard, J.; Winther, S.A.; Hansen, T.S.; Rossing, P.; von Scholten, B.J. Trimethylamine N-oxide (TMAO) as a new potential therapeutic target for insulin resistance and cancer. *Curr. Pharm. Des.* **2017**, *23*, 3699–3712. [CrossRef]
97. Henao-Mejia, J.; Elinav, E.; Jin, C.; Hao, L.; Mehal, W.Z.; Strowig, T.; Thaiss, C.A.; Kau, A.L.; Eisenbarth, S.C.; Jurczak, M.J.; et al. Inflammasome-mediated dysbiosis regulates progression of NAFLD and obesity. *Nature* **2012**, *482*, 179–185. [CrossRef]
98. Del Chierico, F.; Nobili, V.; Vernocchi, P.; Russo, A.; Stefanis, C.D.; Gnani, D.; Furlanello, C.; Zandonà, A.; Paci, P.; Capuani, G.; et al. Gut microbiota profiling of pediatric nonalcoholic fatty liver disease and obese patients unveiled by an integrated meta-omics-based approach. *Hepatology* **2017**, *65*, 451–464. [CrossRef]
99. Koeth, R.A.; Wang, Z.; Levison, B.S.; Buffa, J.A.; Org, E.; Sheehy, B.T.; Britt, E.B.; Fu, X.; Wu, Y.; Li, L.; et al. Intestinal microbiota metabolism of L-carnitine, a nutrient in red meat, promotes atherosclerosis. *Nat. Med.* **2013**, *19*, 576–585. [CrossRef]
100. Caesar, R.; Tremaroli, V.; Kovatcheva-Datchary, P.; Cani, P.D.; Bäckhed, F. Crosstalk between gut microbiota and dietary lipids aggravates WAT inflammation through TLR signaling. *Cell Metab.* **2015**, *22*, 658–668. [CrossRef]
101. Everard, A.; Geurts, L.; Caesar, R.; Van Hul, M.; Matamoros, S.; Duparc, T.; Denis, R.G.P.; Cochez, P.; Pierard, F.; Castel, J.; et al. Intestinal epithelial MyD88 is a sensor switching host metabolism towards obesity according to nutritional status. *Nat. Commun.* **2014**, *5*, 5648. [CrossRef] [PubMed]
102. Li, Q.; Han, Y.; Dy, A.B.C.; Hagerman, R.J. The gut microbiota and autism spectrum disorders. *Front. Cell. Neurosci.* **2017**, *11*, 120. [CrossRef]
103. Greer, J.B.; O'Keefe, S.J. Microbial induction of immunity, inflammation, and cancer. *Front. Physiol.* **2011**, *1*, 168. [CrossRef] [PubMed]
104. Bjerregård Madsen, J.; Kaila, A.; Vehviläinen-Julkunen, K.; Miettinen, M. Time allocation and temporal focus in nursing management: An integrative review. *J. Nurs. Manag.* **2016**, *24*, 983–993. [CrossRef] [PubMed]
105. Shenderov, B.A. Probiotic (symbiotic) bacterial languages. *Anaerobe* **2011**, *17*, 490–495. [CrossRef] [PubMed]
106. Murray-Stewart, T.; Woster, P.M.; Casero, R.A. The re-expression of the epigenetically silenced e-cadherin gene by a polyamine analogue lysine-specific demethylase-1 (LSD1) inhibitor in human acute myeloid leukemia cell lines. *Amino Acids* **2014**, *46*, 585–594. [CrossRef] [PubMed]
107. Soliman, M.L.; Combs, C.K.; Rosenberger, T.A. Modulation of inflammatory cytokines and mitogen-activated protein kinases by acetate in primary astrocytes. *J. Neuroimmune Pharmacol.* **2013**, *8*, 287–300. [CrossRef]

108. Macfarlane, G.T.; Gibson, G.R.; Cummings, J.H. Comparison of fermentation reactions in different regions of the human colon. *J. Appl. Bacteriol.* **1992**, *72*, 57–64.
109. Vieira, A.T.; Fukumori, C.; Ferreira, C.M. New insights into therapeutic strategies for gut microbiota modulation in inflammatory diseases. *Clin. Transl. Immunol.* **2016**, *5*, e87. [CrossRef]
110. Martín, R.; Langella, P. Emerging health concepts in the probiotics field: Streamlining the definitions. *Front. Microbiol.* **2019**, *10*, 1047. [CrossRef]
111. Vieira, A.T.; Teixeira, M.M.; Martins, F.S. The role of probiotics and prebiotics in inducing gut immunity. *Front. Immunol.* **2013**, *4*, 445. [CrossRef] [PubMed]
112. Almeida, J.F.; Breyner, N.M.; Mahi, M.; Ahmed, B.; Benbouziane, B.; Boas, P.C.B.V.; Miyoshi, A.; Azevedo, V.; Langella, P.; Bermúdez-Humarán, L.G.; et al. Expression of fibronectin binding protein A (FnBPA) from *Staphylococcus aureus* at the cell surface of *Lactococcus lactis* improves its immunomodulatory properties when used as protein delivery vector. *Vaccine* **2016**, *34*, 1312–1318. [CrossRef] [PubMed]
113. Mays, Z.J.; Nair, N.U. Synthetic biology in probiotic lactic acid bacteria: At the frontier of living therapeutics. *Curr. Opin. Biotechnol.* **2018**, *53*, 224–231. [CrossRef] [PubMed]
114. Berni Canani, R.; De Filippis, F.; Nocerino, R.; Laiola, M.; Paparo, L.; Calignano, A.; De Caro, C.; Coretti, L.; Chiariotti, L.; Gilbert, J.A.; et al. Specific signatures of the gut microbiota and increased levels of butyrate in children treated with fermented cow's milk containing heat-killed lactobacillus paracasei CBA L74. *Appl. Environ. Microbiol.* **2017**, *83*. [CrossRef]
115. LeBlanc, J.G.; Aubry, C.; Cortes-Perez, N.G.; de Moreno de LeBlanc, A.; Vergnolle, N.; Langella, P.; Azevedo, V.; Chatel, J.-M.; Miyoshi, A.; Bermúdez-Humarán, L.G. Mucosal targeting of therapeutic molecules using genetically modified lactic acid bacteria: An update. *FEMS Microbiol. Lett.* **2013**, *344*, 1–9. [CrossRef]
116. LeBlanc, J.G.; Milani, C.; de Giori, G.S.; Sesma, F.; van Sinderen, D.; Ventura, M. Bacteria as vitamin suppliers to their host: A gut microbiota perspective. *Curr. Opin. Biotechnol.* **2013**, *24*, 160–168. [CrossRef]
117. Nakatsu, G.; Li, X.; Zhou, H.; Sheng, J.; Wong, S.H.; Wu, W.K.K.; Ng, S.C.; Tsoi, H.; Dong, Y.; Zhang, N.; et al. Gut mucosal microbiome across stages of colorectal carcinogenesis. *Nat. Commun.* **2015**, *6*, 8727. [CrossRef]
118. Arrieta, M.-C.; Stiemsma, L.T.; Dimitriu, P.A.; Thorson, L.; Russell, S.; Yurist-Doutsch, S.; Kuzeljevic, B.; Gold, M.J.; Britton, H.M.; Lefebvre, D.L.; et al. Early infancy microbial and metabolic alterations affect risk of childhood asthma. *Sci. Transl. Med.* **2015**, *7*, 307ra152. [CrossRef]
119. Ley, R.E.; Turnbaugh, P.J.; Klein, S.; Gordon, J.I. Microbial ecology: Human gut microbes associated with obesity. *Nature* **2006**, *444*, 1022–1023. [CrossRef]
120. Macia, L.; Tan, J.; Vieira, A.T.; Leach, K.; Stanley, D.; Luong, S.; Maruya, M.; Ian McKenzie, C.; Hijikata, A.; Wong, C.; et al. Metabolite-sensing receptors GPR43 and GPR109A facilitate dietary fibre-induced gut homeostasis through regulation of the inflammasome. *Nat. Commun.* **2015**, *6*, 6734. [CrossRef]
121. Holscher, H.D. Dietary fiber and prebiotics and the gastrointestinal microbiota. *Gut Microbes* **2017**, *8*, 172–184. [CrossRef] [PubMed]
122. Bindels, L.B.; Delzenne, N.M.; Cani, P.D.; Walter, J. Towards a more comprehensive concept for prebiotics. *Nat. Rev. Gastroenterol. Hepatol.* **2015**, *12*, 303–310. [CrossRef] [PubMed]
123. Khalesi, S.; Johnson, D.W.; Campbell, K.; Williams, S.; Fenning, A.; Saluja, S.; Irwin, C. Effect of probiotics and synbiotics consumption on serum concentrations of liver function test enzymes: A systematic review and meta-analysis. *Eur. J. Nutr.* **2018**, *57*, 2037–2053. [CrossRef] [PubMed]
124. NASPGHAN Nutrition Report Committee; Michail, S.; Sylvester, F.; Fuchs, G.; Issenman, R. Clinical efficacy of probiotics: Review of the evidence with focus on children. *J. Pediatr. Gastroenterol. Nutr.* **2006**, *43*, 550–557.
125. Beserra, B.T.S.; Fernandes, R.; do Rosario, V.A.; Mocellin, M.C.; Kuntz, M.G.F.; Trindade, E.B.S.M. A systematic review and meta-analysis of the prebiotics and synbiotics effects on glycaemia, insulin concentrations and lipid parameters in adult patients with overweight or obesity. *Clin. Nutr.* **2015**, *34*, 845–858. [CrossRef]
126. Ford, A.C.; Quigley, E.M.M.; Lacy, B.E.; Lembo, A.J.; Saito, Y.A.; Schiller, L.R.; Soffer, E.E.; Spiegel, B.M.R.; Moayyedi, P. Efficacy of prebiotics, probiotics, and synbiotics in irritable bowel syndrome and chronic idiopathic constipation: Systematic review and meta-analysis. *Am. J. Gastroenterol.* **2014**, *109*, 1547–1561. [CrossRef]
127. Healey, G.R.; Murphy, R.; Brough, L.; Butts, C.A.; Coad, J. Interindividual variability in gut microbiota and host response to dietary interventions. *Nutr. Rev.* **2017**, *75*, 1059–1080. [CrossRef]

128. Kovatcheva-Datchary, P.; Nilsson, A.; Akrami, R.; Lee, Y.S.; De Vadder, F.; Arora, T.; Hallen, A.; Martens, E.; Björck, I.; Bäckhed, F. Dietary fiber-induced improvement in glucose metabolism is associated with increased abundance of Prevotella. *Cell Metab.* **2015**, *22*, 971–982. [CrossRef]
129. Suez, J.; Elinav, E. The path towards microbiome-based metabolite treatment. *Nat. Microbiol.* **2017**, *2*, 17075. [CrossRef]
130. De Filippis, F.; Vitaglione, P.; Cuomo, R.; Berni Canani, R.; Ercolini, D. Dietary interventions to modulate the gut microbiome-how far away are we from precision medicine. *Inflamm. Bowel Dis.* **2018**, *24*, 2142–2154. [CrossRef]
131. Hesketh, J. Personalised nutrition: How far has nutrigenomics progressed? *Eur. J. Clin. Nutr.* **2013**, *67*, 430–435. [CrossRef] [PubMed]
132. Mills, S.; Stanton, C.; Lane, J.A.; Smith, G.J.; Ross, R.P. Precision nutrition and the microbiome, part I: Current state of the science. *Nutrients* **2019**, *11*, 923. [CrossRef] [PubMed]
133. Laddu, D.; Hauser, M. Addressing the nutritional phenotype through personalized nutrition for chronic disease prevention and management. *Prog. Cardiovasc Dis.* **2018**. [CrossRef] [PubMed]
134. Hadi, A.; Mohammadi, H.; Miraghajani, M.; Ghaedi, E. Efficacy of synbiotic supplementation in patients with nonalcoholic fatty liver disease: A systematic review and meta-analysis of clinical trials: Synbiotic supplementation and NAFLD. *Crit. Rev. Food Sci. Nutr.* **2019**, *59*, 2494–2505. [CrossRef] [PubMed]
135. Mookiah, S.; Sieo, C.C.; Ramasamy, K.; Abdullah, N.; Ho, Y.W. Effects of dietary prebiotics, probiotic and synbiotics on performance, caecal bacterial populations and caecal fermentation concentrations of broiler chickens. *J. Sci. Food Agric.* **2014**, *94*, 341–348. [CrossRef]
136. Tabrizi, R.; Moosazadeh, M.; Lankarani, K.B.; Akbari, M.; Heydari, S.T.; Kolahdooz, F.; Asemi, Z. The effects of synbiotic supplementation on glucose metabolism and lipid profiles in patients with diabetes: A systematic review and meta-analysis of randomized controlled trials. *Probiotics Antimicrob. Proteins* **2018**, *10*, 329–342. [CrossRef] [PubMed]
137. Vitali, B.; Ndagijimana, M.; Maccaferri, S.; Biagi, E.; Guerzoni, M.E.; Brigidi, P. An in vitro evaluation of the effect of probiotics and prebiotics on the metabolic profile of human microbiota. *Anaerobe* **2012**, *18*, 386–391. [CrossRef]
138. McLoughlin, R.F.; Berthon, B.S.; Jensen, M.E.; Baines, K.J.; Wood, L.G. Short-chain fatty acids, prebiotics, synbiotics, and systemic inflammation: A systematic review and meta-analysis. *Am. J. Clin. Nutr.* **2017**, *106*, 930–945. [CrossRef]
139. Wegh, C.A.M.; Geerlings, S.Y.; Knol, J.; Roeselers, G.; Belzer, C. Postbiotics and their potential applications in early life nutrition and beyond. *Int. J. Mol. Sci.* **2019**, *20*, 4673. [CrossRef]
140. George Kerry, R.; Patra, J.K.; Gouda, S.; Park, Y.; Shin, H.-S.; Das, G. Benefaction of probiotics for human health: A review. *J. Food Drug Anal.* **2018**, *26*, 927–939. [CrossRef]
141. Aguilar-Toalá, J.E.; Garcia-Varela, R.; Garcia, H.S.; Mata-Haro, V.; González-Córdova, A.F.; Vallejo-Cordoba, B.; Hernández-Mendoza, A. Postbiotics: An evolving term within the functional foods field. *Trends Food Sci. Technol.* **2018**, *75*, 105–114. [CrossRef]
142. Szajewska, H.; Skórka, A.; Pieścik-Lech, M. Fermented infant formulas without live bacteria: A systematic review. *Eur. J. Pediatr.* **2015**, *174*, 1413–1420. [CrossRef] [PubMed]
143. Klemashevich, C.; Wu, C.; Howsmon, D.; Alaniz, R.C.; Lee, K.; Jayaraman, A. Rational identification of diet-derived postbiotics for improving intestinal microbiota function. *Curr. Opin. Biotechnol.* **2014**, *26*, 85–90. [CrossRef] [PubMed]
144. Borody, T.J.; Khoruts, A. Fecal microbiota transplantation and emerging applications. *Nat. Rev. Gastroenterol. Hepatol.* **2011**, *9*, 88–96. [CrossRef]
145. Shapiro, H.; Suez, J.; Elinav, E. Personalized microbiome-based approaches to metabolic syndrome management and prevention. *J. Diabetes* **2017**, *9*, 226–236. [CrossRef]
146. Merli, P.; Putignani, L.; Ruggeri, A.; Del Chierico, F.; Gargiullo, L.; Galaverna, F.; Gaspari, S.; Pagliara, D.; Russo, A.; Pane, S.; et al. Decolonization of multi-drug resistant bacteria by fecal microbiota transplantation in five pediatric patients before allogeneic hematopoietic stem cell transplantation: Gut microbiota profiling, infectious and clinical outcomes. *Haematologica* **2020**. [CrossRef]
147. Lee, J.-C.; Lee, H.-Y.; Kim, T.K.; Kim, M.-S.; Park, Y.M.; Kim, J.; Park, K.; Kweon, M.-N.; Kim, S.-H.; Bae, J.-W.; et al. Obesogenic diet-induced gut barrier dysfunction and pathobiont expansion aggravate experimental colitis. *PLoS ONE* **2017**, *12*, e0187515. [CrossRef]

148. Cammarota, G.; Ianiro, G.; Gasbarrini, A.; European FMT Working Group. Faecal microbiota transplantation in clinical practice. *Gut* **2018**, *67*, 196–197. [CrossRef]
149. Cammarota, G.; Ianiro, G.; Kelly, C.R.; Mullish, B.H.; Allegretti, J.R.; Kassam, Z.; Putignani, L.; Fischer, M.; Keller, J.J.; Costello, S.P.; et al. International consensus conference on stool banking for faecal microbiota transplantation in clinical practice. *Gut* **2019**, *68*, 2111–2121. [CrossRef]
150. Mullish, B.H.; McDonald, J.A.K.; Pechlivanis, A.; Allegretti, J.R.; Kao, D.; Barker, G.F.; Kapila, D.; Petrof, E.O.; Joyce, S.A.; Gahan, C.G.M.; et al. Microbial bile salt hydrolases mediate the efficacy of faecal microbiota transplant in the treatment of recurrent *Clostridioides difficile* infection. *Gut* **2019**. [CrossRef]
151. Wilson, B.C.; Vatanen, T.; Cutfield, W.S.; O'Sullivan, J.M. The super-donor phenomenon in fecal microbiota transplantation. *Front. Cell. Infect. Microbiol.* **2019**, *9*, 2. [CrossRef] [PubMed]

International Journal of
Molecular Sciences

MDPI

Article

Unique Gut Microbiome Signatures Depict Diet-Versus Genetically Induced Obesity in Mice

Ravinder Nagpal [1], Sidharth P Mishra [1] and Hariom Yadav [1,2,*]

[1] Department of Internal Medicine-Molecular Medicine, Wake Forest School of Medicine, Winston-Salem, NC 27101, USA; rnagpal@wakehealth.edu (R.N.); spmishra@wakehealth.edu (S.P.M.)

[2] Department of Microbiology and Immunology, Wake Forest School of Medicine, Winston-Salem, NC 27101, USA

* Correspondence: hyadav@wakehealth.edu; Tel.: +1-336-713-5049

Received: 29 March 2020; Accepted: 9 May 2020; Published: 13 May 2020

check for updates

Abstract: The gut microbiome plays an important role in obesity and Type 2 diabetes (T2D); however, it remains unclear whether the gut microbiome could clarify the dietary versus genetic origin of these ailments. Moreover, studies examining the gut microbiome in diet- versus genetically induced obesity/T2D in the same experimental set-up are lacking. We herein characterized the gut microbiomes in three of the most widely used mouse models of obesity/T2D, i.e., genetically induced (leptin-deficient i.e., $Lep^{ob/ob}$; and leptin-receptor-deficient i.e., $Lep^{db/db}$) and high-fat diet (HFD)-induced obese (DIO)/T2D mice, with reference to their normal chow-fed (NC) and low-fat-diet-fed (LF) control counterparts. In terms of β-diversity, $Lep^{ob/ob}$ and $Lep^{db/db}$ mice showed similarity to NC mice, whereas DIO and LF mice appeared as distinct clusters. The phylum- and genus-level compositions were relatively similar in NC, $Lep^{ob/ob}$, and $Lep^{db/db}$ mice, whereas DIO and LF mice demonstrated distinct compositions. Further analyses revealed several unique bacterial taxa, metagenomic functional features, and their correlation patterns in these models. The data revealed that obesity/T2D driven by diet as opposed to genetics presents distinct gut microbiome signatures enriched with distinct functional capacities, and indicated that these signatures can distinguish diet- versus genetically induced obesity/T2D and, if extrapolated to humans, might offer translational potential in devising dietary and/or genetics-based therapies against these maladies.

Keywords: microbiota; obesity; Type 2 diabetes; leptin; diet

1. Introduction

The prevalence of metabolic diseases, including obesity and Type 2 diabetes (T2D), is rapidly increasing worldwide. Since obesity and T2D co-occur and involve multiple physiological functions associated with the intestinal physiology and ecosystem, specific perturbations in the gut microbial ecosystem (gut dysbiosis) are commonly observed in these diseases. Emerging evidence has highlighted the role of multiple factors including host genetics, nutrition, lifestyle, and microbiome in the pathology of obesity/T2D, with a growing body of evidence underpinning the causal and mechanistic role of the gut microbiome in the development and persistence of these diseases [1–4]. However, it remains unclear whether the composition and function of the gut microbiome differ in the pathology of diet- vs. genetically induced obesity/T2D. Among various animal models that have been developed and are routinely used to study the pathogenesis and mechanisms of obesity/T2D, high-fat-diet-induced obese (DIO) mice and genetically induced obese mice, i.e., leptin-deficient ($Lep^{ob/ob}$; hereafter, ObOb) mice and leptin-receptor-deficient ($Lep^{db/db}$; hereafter, DbDb) mice remain the most widely used experimental models. However, it remains unexplored whether and how the magnitude and array of gut dysbiosis varies within these different models of obesity/T2D. Specifically, the relative association of the gut microbiome with diet- versus genetically induced obesity/T2D remains unclear. While several previous studies have separately examined the gut microbiome in these models of obesity/T2D [5–10], the comparison of these models with each other as well as with their normal chow (NC)-fed and low-fat diet (LFD)-fed counterparts in the same experimental set-up remain unstudied. In particular, considering the emerging speculation that an "obese/T2D microbiome" could play a contributing role in obesity/T2D-associated metabolic dysfunctions, the understanding of gut microbiome composition in different mouse models of obesity/T2D is of considerable research interest and importance. In addition, the differentiation of microbiome signatures specific to host diet and genetics remains highly important in translational applications. We herein examined the gut microbiome compositions in DIO, ObOb, and DbDb mice while also comparing these with NC and LFD controls. The composition of the gut microbiome can be influenced by animal facilities, protocols for fecal collection, DNA isolation and sequencing, and bioinformatics analyses; therefore, comparing the relationship and contribution of gut microbiome composition as determined using data from these mouse models analyzed by different labs/facilities and using different sequencing platforms remains a limitation. To our knowledge, this is the first report to compare the gut microbiome signature in these widely used mouse models of obesity/T2D using identical and consistent protocols and analysis tools, and these results should facilitate prospective studies examining the contribution of the gut microbiome in genetically versus diet-induced pathobiology of obesity/T2D.

2. Results

2.1. Diet-Induced and Genetically Induced Obese Mice Harbor Distinct Gut Microbiomes

The analysis of β-diversity demonstrated distinct signatures of the fecal microbiome in all five groups of mouse models (Figure 1a). As evident from the principal coordinate analysis (PCoA) analysis of the Bray–Curtis dissimilarity index (Figure 1a,b), the DbDb and ObOb mice clustered closely together and were relatively closer to the NC mice, whereas the LF and DIO mice stood out as separate clusters. The analysis of unweighted and weighted unifrac distance metrics also demonstrated similar patterns of clustering (Supplementary Figure S1). The α-diversity indices showed similar species richness in terms of phylogenetic diversity (PD whole tree), observed operational taxonomic units (OTUs), and Chao1 (species richness) index in NC, LF, and DIO mice, while these indices were significantly or insignificantly higher in DbDb and ObOb mice ($p < 0.05$ versus NC and LF) (Figure 1c–e). On the other hand, the species evenness (Shannon index) is significantly higher ($p < 0.05$) in DIO and ObOb mice as compared to both NC and LF mice (Figure 1f). At the phylum level, the microbiome composition appeared to be relatively similar in NC, DbDb, and ObOb mice, whereas LF and DIO mice demonstrated distinct composition compared to each other as well as compared to the other three groups (Figure 1g). NC mice had a significantly higher ($p < 0.05$) abundance of Verrucomicrobia than all other groups; LF mice had a significantly higher proportion ($p < 0.05$) of Actinobacteria; DIO mice had a significantly higher ($p < 0.05$) proportion of Firmicutes, Proteobacteria, and Deferribacteres; and DbDb and ObOb mice have similar compositions predominated by Bacteroidetes (Figure 1g). Similar compositional patterns were seen at the bacterial family level as well as at the genus level (Figure 1h,i).

The overall ratio of the predominant phyla *Firmicutes*: *Bacteroidetes* was remarkably higher ($p < 0.01$) in DIO mice compared to all four other groups (Figure 2a). Analysis of organism-level phenotypes showed that the NC and DIO mice had a significantly higher ($p < 0.05$) ratio of aerobes to anaerobes than the other three groups (Figure 2b). The ratio of Gram-positive to -negative bacteria remained similar in NC, DIO, DbDb, and ObOb mice, but was significantly higher ($p < 0.05$) in LF mice (Figure 2c). Interestingly, the proportion of potentially pathogenic bacteria was significantly ($p < 0.05$) higher in DbDb and ObOb mice and lower in DIO mice than in NC and LF mice (Figure 2d). The abundance of bacteria capable of forming biofilms was significantly ($p < 0.05$) lower in DIO, DbDb, and ObOb mice than in NC and LF mice (Figure 2e). Hierarchical clustering analysis of major bacterial families and genera also demonstrated clearly distinct clusters segregated according to the different groups of mice (Figure 2f), wherein ObOb and DbDb were clustered closer to the NC mice whereas LF and DIO mice presented distinct clusters both from each other and from NC, ObOb, and DbDb mice. Correspondingly, hierarchical clustering analysis in terms of the rate of detection of major genera among these mice also demonstrated a similar arrangement of the clusters, wherein ObOb and DbDb mice clustered closer to the NC mice while LF and DIO mice clustered distinctly (Figure 2g).

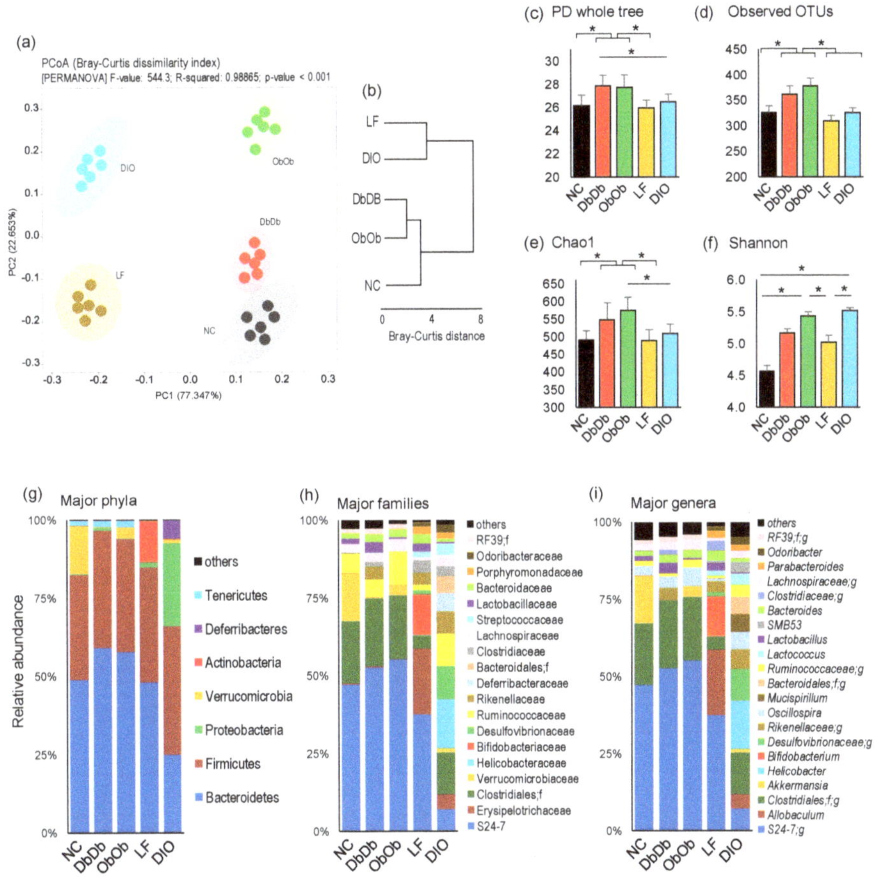

Figure 1. Diet-induced versus genetically induced obese mice exhibit distinct gut microbiome signatures. (**a**) Beta-diversity (principal coordinate analysis; Bray-Curtis dissimilarity index), (**b**) intergroup Bray–Curtis distance, (**c**–**f**) alpha-diversity indices, and (**g**–**i**) microbiome composition at phylum, family, and genus levels in normal-chow-fed leptin-deficient ($Lep^{ob/ob}$, ObOb) and leptin-receptor-deficient ($Lepr^{db/db}$; DbDb) mice and high-fat-diet-fed (diet-induced obese; DIO) mice versus their normal-chow-fed (NC) and low-fat-diet-fed (LF) control counterparts. * $p < 0.05$.

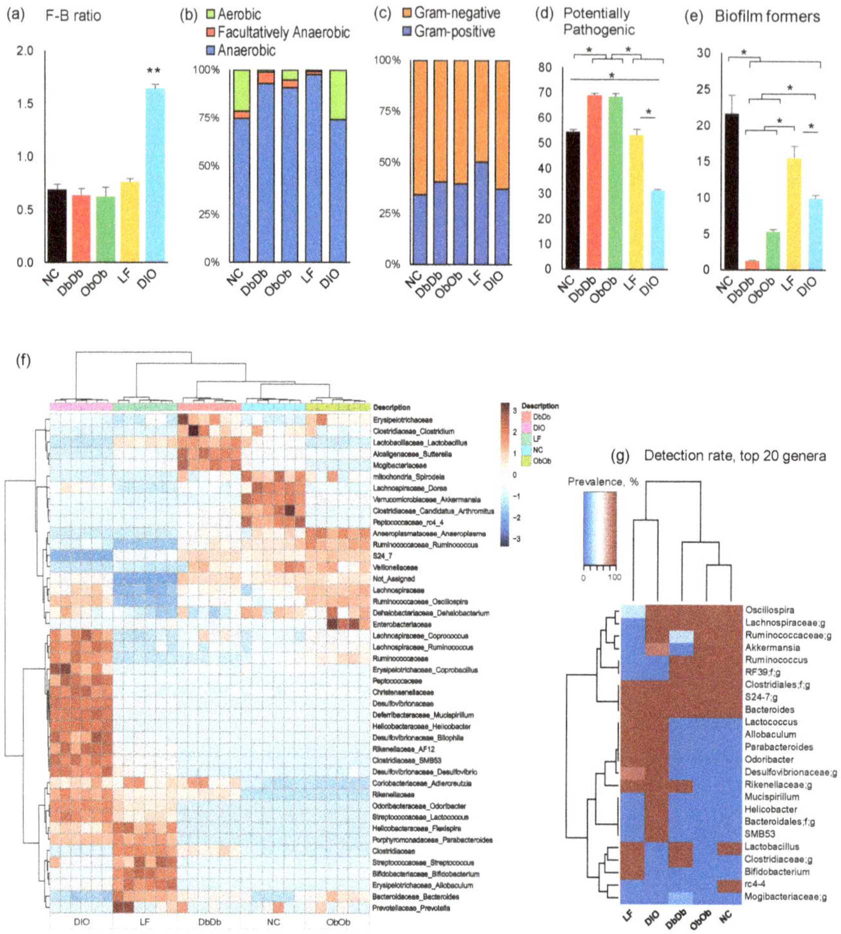

Figure 2. Distinct gut microbiome community structures in diet-induced versus genetically induced obese mice. (**a**) Ratio of major phyla Firmicutes to Bacteroidetes, (**b**) anaerobic to aerobic bacteria, (**c**) Gram-negative to Gram-positive bacteria, (**d**) abundance of bacterial taxa with potential pathogenic phenotypes, (**e**) abundance of bacterial taxa with biofilm-forming traits, (**f**,**g**) heat-map depicting the abundance (**f**) and detection rates (**g**) of major bacterial taxa in normal-chow-fed leptin-deficient (ObOb) and leptin-receptor-deficient (DbDb) mice and high-fat-diet-fed (diet-induced obese; DIO) mice versus their normal-chow-fed (NC) and low-fat-diet-fed (LF) control counterparts. * $p < 0.05$; ** $p < 0.01$.

2.2. Diet-Induced Versus Genetically Induced Obese Mice Demonstrate Unique Signatures of Gut Microbiome Community Structure and Function

We then applied linear discriminatory analysis (LDA) effect size (LEfSe) analysis to the inferred relative abundance data to detect unique taxonomic clades and KEGG (Kyoto Encyclopedia of Genes and Genomes) metagenome orthologs that were significantly ($p < 0.01$) over- or underrepresented (or differentially abundant) and could be considered potential metagenomic biomarkers in these different groups of mice. The statistical analysis of the abundance of these bacterial taxa via LEfSe in the form of a cladogram revealed several unique families and genera that drive differences between these five groups of mice (Figure 3a). Figure 3b further simplifies these data and presents the LDA scores of the genera that were significantly ($p < 0.01$) different among these groups (Figure 3b).

NC mice were distinguished by higher proportions of phylum Verrucomicrobia, genus *Akkermansia*, and family Peptococcaceae, while LF mice presented a higher proportion of the genera *Allobaculum*, *Bifidobacterium*, *Parabacteroides*, and *Bacteroides*. In contrast, ObOb mice were characterized by a higher abundance of families S24-7 and Lachnospiraceae, order Clostridiales and genera *Oscillospira*, *Tenericutes*, and *Ruminococcus*; whereas DbDb mice were unique in terms of higher carriage of phylum Bacteroidetes, genus *Lactobacillus*, and unclassified genera belonging to the orders Clostridiales and RF39 and the family Mogibacteriaceae. On the other hand, DIO mice harbored the highest proportion of many bacterial taxa belonging to Firmicutes, Proteobacteria, *Helicobacter, Desulfovibrio, Rikenella, Deferribacter, Mucispirillum, Lactococcus, Coprococcus, SMB53*, and *Odoribacter* (Figure 3a,b).

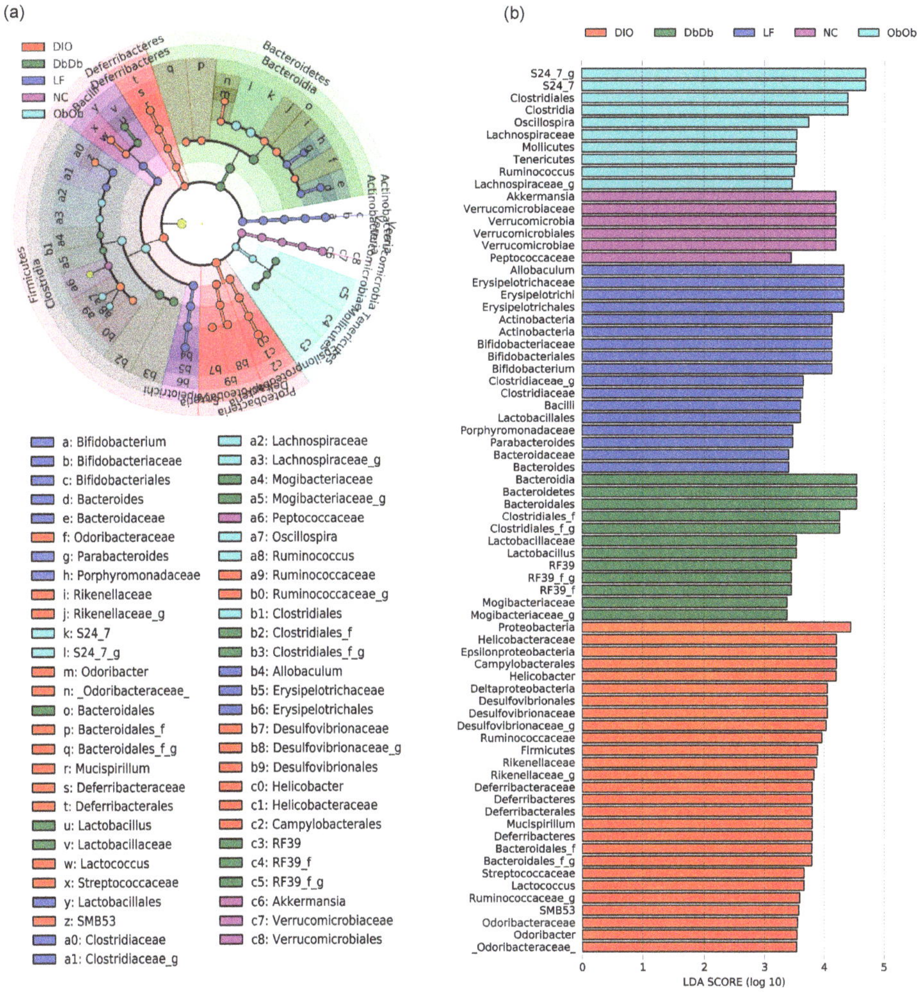

Figure 3. Unique gut microbiome signatures in diet-induced versus genetically induced obese mice. Linear discriminatory analysis (LDA) effect size (LEfSe) analysis cladogram (**a**) and LDA score graph (**b**) illustrating unique bacterial taxa that were differentially and significantly over- or underrepresented (or differentially abundant) and drove differences in normal-chow-fed leptin-deficient ($Lep^{ob/ob}$, ObOb) and leptin-receptor-deficient ($Lepr^{db/db}$; DbDb) mice, high-fat-diet-fed (diet-induced obese; DIO) mice, as well as in normal-chow-fed (NC) and low-fat-diet-fed (LF) control mice.

The PICRUSt (Phylogenetic Investigation of Communities by Reconstruction of Unobserved States) analysis of the KEGG orthologs associated with these bacterial signatures also revealed distinct clusters among these five groups (Figure 4a,b). As evident from the PCoA analysis of the microbiome-associated functional pathways, the DbDb and ObOb mice clustered closely together and were quite similar to the NC mice, whereas the LF and DIO mice stood out as distinct clusters and were highly dissimilar to each other as well as to the other three groups of mice (Supplementary Figure S2a,b). The features were subsequently clustered into functional modules at Levels 1 to 3, as described elsewhere [11]. Figure 4a presents unique features of the Level 1 and 2 modules, while Level 3 features are shown as a bar graph of LDA scores in Figure 4b. The cladogram of LEfSe analysis of the functional Levels 1 and 2 of these annotations clearly reveals distinct arrays of these functions among the different groups (Figure 4a). As further simplified by the LEfSe bar graph, DbDb mice were characterized by a significantly higher ($p < 0.01$) abundance of bacterial taxa associated with the metabolism of cofactors and vitamins, glycan degradation, energy metabolism, and biosynthesis of glycosphingolipids (Figure 4b). In contrast, DIO mice harbored a higher ($p < 0.01$) proportion of bacterial taxa associated with the metabolism of bacterial motility proteins, membrane transporters, metabolism of porphyrin and xenobiotics, biosynthesis of lipopolysaccharides and fatty acids, and the OTUs associated with human diseases including neurodegenerative disease, e.g., Alzheimer's disease (Figure 4b). In contrast, bacteria associated with glycan and lipid biosynthesis, limonene and glycosaminoglycan degradation, carbohydrate and amino acid metabolism, peptidoglycan biosynthesis, etc. were overrepresented ($p < 0.01$) in NC and LF mice, which means that all of these pathways were less represented ($p < 0.01$) in all three experimental groups of mice. The analysis of the metabolism-related functions alone is presented separately in Supplementary Figures S3 and S4.

We then estimated the relationship between taxonomic and functional enrichments in each obese mouse model by computing the correlations between the abundances of taxonomic clades and KEGG orthologs (metagenomic gene families) using the non-parametric test of Spearman's rank correlation. The correlations between bacterial abundance and functions enriched in different mouse models were computed via a statistical approach similar to that described by Segata et al. [12]. However, because of the two independent control groups, i.e., NC as a control group for DbDb and ObOb mice (all chow-fed) and LF mice as control group for DIO mice (low-fat- versus high-fat-fed), we first normalized the DbDb and ObOb datasets against NC dataset, while the DIO dataset was normalized against the LF dataset in terms of the fold-difference in the experimental versus control group. The correlation results as depicted in the form of hierarchal heat map yielded clearly distinct clusters of positive and negative associations between a number of bacterial taxa and metagenomic functions in both diet- and genetically induced obese models (Figure 5a–b). To further simplify these clusters, we extracted the significant correlation subsets (Spearman's rho > 0.7 and p-value < 0.01) and built separate correlation networks for each obese model (Figure 5c–e), which showed several common as well as distinct co-occurrence network arrays among these three obese models. The correlation arrays and networks of the complete un-normalized dataset with all five groups combined are presented in Supplementary Figures S5 and S6.

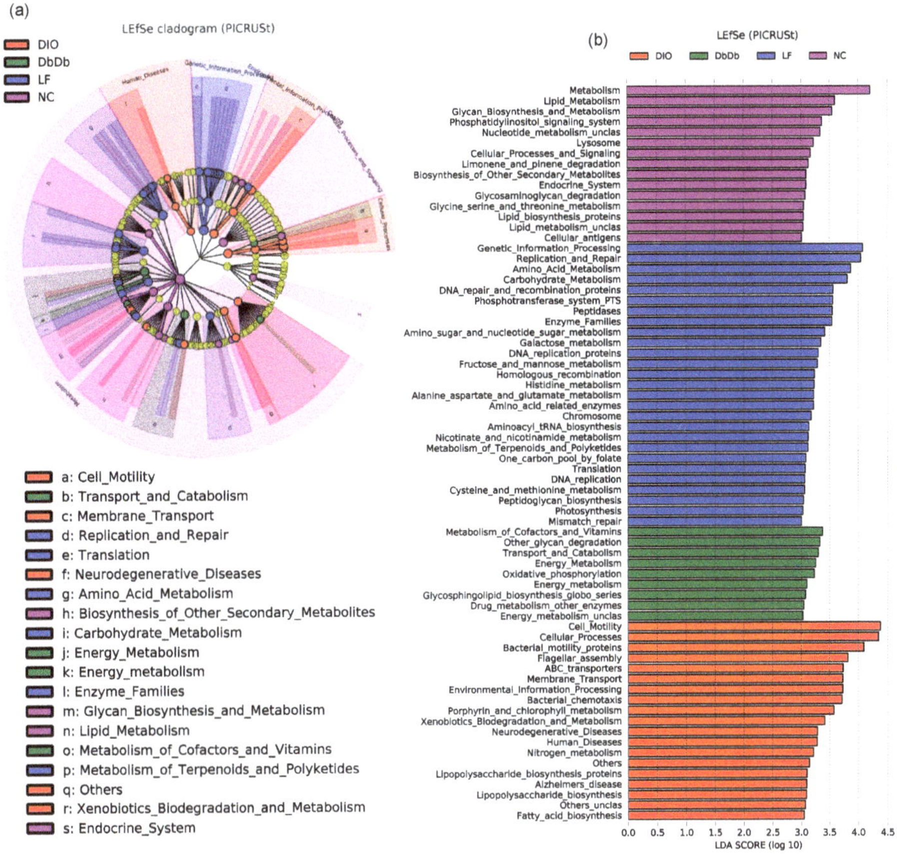

Figure 4. Unique signatures of gut-microbiome-associated metagenomic functions in diet-induced versus genetically induced obese mice. Linear discriminatory analysis (LDA) effect size (LEfSe) analysis cladogram (**a**) and LDA score graph (**b**) illustrating unique metagenomic functional features, as predicted by PICRUSt (Phylogenetic Investigation of Communities by Reconstruction of Unobserved States), that were uniquely and significantly over- or underrepresented (or differentially abundant) and drove differences in normal-chow-fed leptin-deficient ($Lep^{ob/ob}$, ObOb) and leptin-receptor-deficient ($Lepr^{db/db}$; DbDb) mice, high-fat-diet-fed (diet-induced obese; DIO) mice, as well as in normal-chow-fed (NC) and low-fat-diet-fed (LF) control mice.

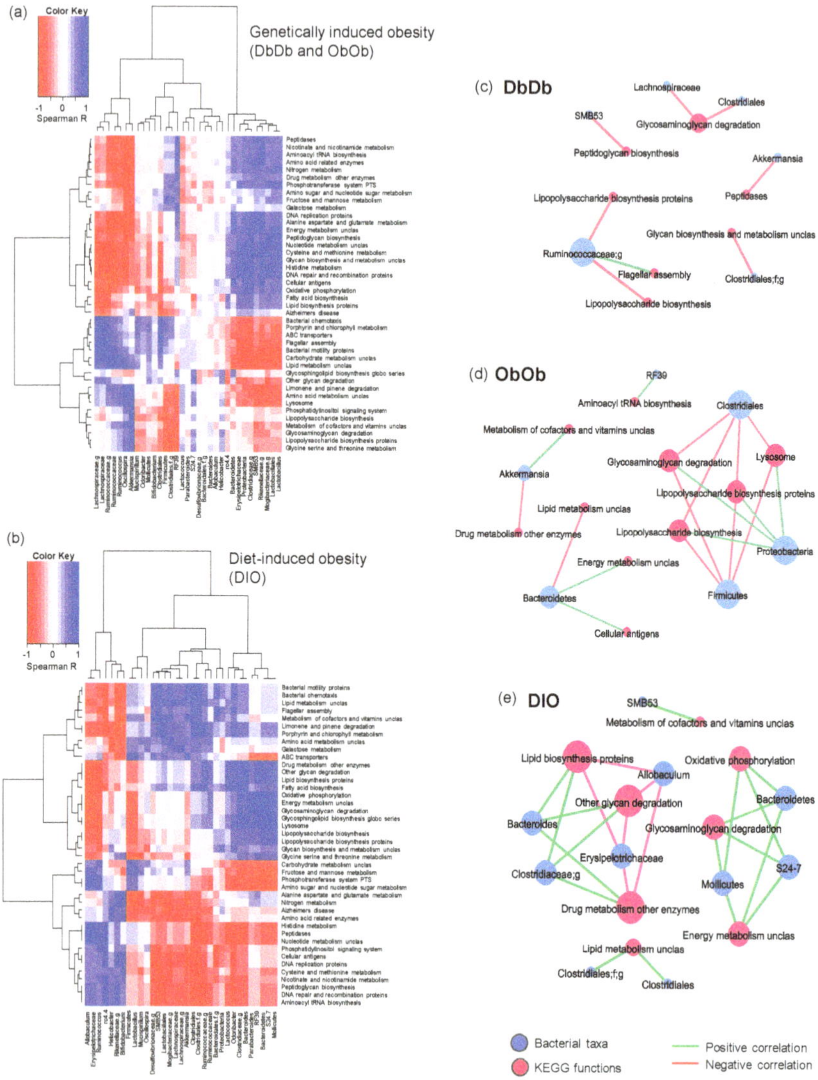

Figure 5. Distinct patterns of correlation between gut bacterial taxa and predicted metagenomic functional features in diet-induced versus genetically induced obese mice. (**a**,**b**) Color heat map depicting the Spearman's correlation of bacterial taxa identified via amplicon sequencing with PICRUSt-curated metagenomic functions in normal-chow-fed leptin-deficient (Lep$^{ob/ob}$, ObOb) and leptin-receptor-deficient (Lepr$^{db/db}$; DbDb) mice (a) and high-fat-diet-induced obese (DIO) mice (b). Correlation networks showing the selected subsets of significant correlations (Spearman's correlation rank > 0.7, p-value < 0.01) between bacterial taxa and predicted metagenomic functions in (**c**) DbDb, (**d**) ObOb, and (**e**) DIO mice.

2.3. Diet-Induced and Genetically Induced Obese Mice Exhibit Higher Intestinal Permeability and Inflammation

The gut permeability, as measured by the assayed serum level of fluorescein isothiocyanate (FITC)-dextran (μg/mL serum) 4 h after oral gavage, revealed significantly higher ($p < 0.05$) gut permeability in all three experimental obese models, i.e., DbDb, ObOb, and DIO mice, as compared to their NC and LF counterparts (Figure 6a). Interestingly, LF mice demonstrated the lowest degree of gut permeability, which was insignificantly but considerably lower than even the NC mice. In line with the gut permeability data, the mRNA expression level of epithelial intercellular-tight-junction-associated proteins, i.e., TJP-1 (tight-junction protein-1; also known as Zonula occludens-1 [ZO-1]) and Occl-1 (Occludin-1) was also found to be significantly lower ($p < 0.05$) in all three obese models versus their respective controls (Figure 6b–c). In addition, compared to the respective controls, all three obese models demonstrated significantly higher ($p < 0.05$) expression of inflammatory markers interleukin (IL)-1b, IL-6, and tumor necrosis factor (TNF)-α while exhibiting significantly ($p < 0.05$) lower expression of IL-10 and significantly ($p < 0.05$; DbDb and ObOb) or insignificantly ($p < 0.1$; DIO) lower expression of transforming growth factor (TGF)-β1 (Figure 6d–h).

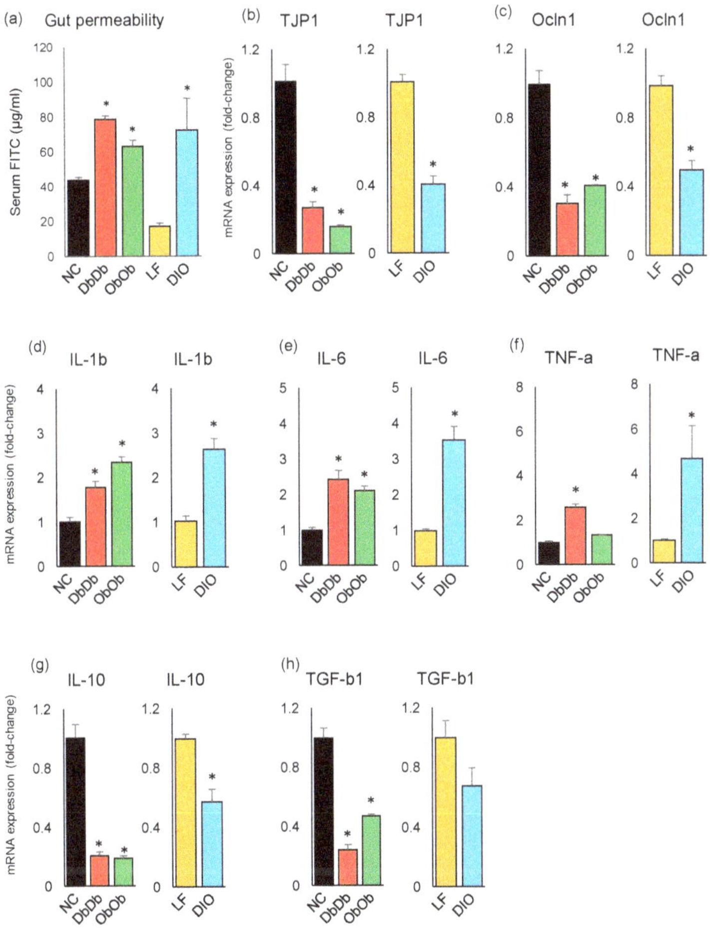

Figure 6. Dysbiotic magnitude of intestinal permeability and inflammatory markers in diet-induced versus genetically induced obese mice. (**a**) Leaky gut (measured by appearance of fluorescein isothiocyanate in blood leaked from gut) (a) and the mRNA expression levels of tight-junction proteins (**b**,**c**) and inflammatory markers (**d**–**h**) in the ileal tissues of chow-fed leptin-deficient (ObOb) and leptin-receptor-deficient (DbDb) obese mice and high-fat-diet-induced obese (DIO) mice, with reference to normal-chow-fed (NC) and low-fat-diet-fed (LF) control mice. * $p < 0.05$.

3. Discussion

Both diet and genetics play an important role in the pathology of obesity and T2D, and all these together are linked with gut microbiome dysbiosis. However, it remains unclear whether the gut microbiome can be used to distinguish the diet-versus genetics-specific pathology of these diseases. We herein used high-fat-diet-induced obese (DIO) mice and the genetically induced (ObOb and DbDb) obese/T2D mouse models to determine whether the gut microbiome can disentangle diet-versus genetically induced pathology. To our knowledge, this is the first study demonstrating the gut microbiome composition in the three most widely used obese/T2D mouse models in a nearly identical experimental set-up while also comparing these with their normal-chow-fed (NC) and low-fat-diet-fed (LF) control counterparts.

As revealed by the β-diversity analysis as well as the hierarchical clustering of major bacterial taxa, the mice within an individual group clustered in very close proximity, demonstrating considerable intragroup conservation of taxonomic composition within the respective groups. However, in terms of intergroup variation, all five groups clustered distinctly, indicating different bacterial community structures when compared with each other. The same pattern was seen in the PCoA analysis of PICRUSt-curated functional data. This indicated that differential availability as well as intake of dietary components might result in favored dependence on certain substrates acquired from different diets, which might have contributed to the characteristically different nature of the microbiomes in these mice. The highest similarity was seen between DbDb, ObOb, and NC mice, whereas DIO and LF mice (plausibly due to their distinct diets) showed the least similarity to each other as well as to DbDb, ObOb, and NC mice (Figure 1a,b; Supplementary Figure S2a). This suggests that the LF and DIO microbiomes were less homogenous for low-fat and high-fat diets, respectively, while indicating relatively inherent similarities in the bacterial community structures of NC, DbDb, and ObOb mice, all of which were maintained on normal chow. Since NC, DbDb, and ObOb groups were all maintained on normal chow while LF and DIO mice were maintained on very different diets, i.e., low-fat diet and high-fat diet, respectively, their clustering patterns corroborated previous reports suggesting that diet may dominate the host genotypic background in shaping the murine gut microbiome [13,14]. However, we found the highest α-diversity (species richness) in DbDb and ObOb mice, as compared to all other groups (Figure 1c–e). A higher α-diversity is generally considered a marker of a healthy gut profile, whereas our data indicated that this may not always be the case, particular in the case of leptin- or leptin-receptor-deficient mice, which are characterized by over-eating of an otherwise normal diet that may eventually lead to a dysbiotic gut microflora with an abnormal overgrowth and overrepresentation of specific bacterial groups that otherwise remain subdominant under normal homeostatic milieus (gut eubiosis). Intriguingly, the species evenness was highest in DIO mice (Figure 1f), which could be explained by the fact that the proportion of certain dominant bacterial taxa, such as those belonging to the phylum *Bacteroidetes*, was lower in these mice, meaning that the overall microbiome composition has a greater number of bacterial taxa than in the other four groups of mice. Likewise, the lower proportion of *Rikenellaceae* and *Lactobacillus* may have accounted for the higher representation of *S24-7*, *Clostridia*, *Oscillospira*, *Terericutes*, and *Rumonococcus*, and hence a higher Shannon index, in ObOb versus NC and LF mice. As expected, the weight gain over the course of 8 week experimental period was highest ($p < 0.01$) in DbDb and ObOb mice followed by DIO mice ($p < 0.05$) as compared to NC and LF mice (Figure S7).

The highest proportion of *Bacteroidetes* being found in DbDb and ObOb mice (Figure 1g) was consistent with separate previous studies reporting higher *Bacteroidetes* abundance in ObOb, DbDb, and NC mice of C57Bl/6J background [6,10,15]. However, we found a remarkably higher abundance of *Firmicutes* and *Proteobacteria* and a lower abundance of *Bacteroidetes* in DIO mice compared to all other groups. A higher *Firmicutes* to *Bacteroidetes* ratio has also previously been associated with obesity and poor glucose tolerance [6,16]. Overall, the control NC and LF mice had similar *Firmicutes* to *Bacteroidetes* proportions and were distinguished mainly in terms of higher abundances of Verrucomicrobia and *Actinobacteria*, respectively. DIO, DbDb, and ObOb obese/T2D mice harbored a very different microbiome composition compared to NC and LF mice, underlining an abnormal (dysbiotic) state of gut microbiome in these obese models. However, unsurprisingly, the microbiome composition differed considerably even among these three models, which may have been associated with their diet (high-fat for DIO vs. normal chow for DbDb and ObOb) and/or genetic background ($Lep^{Ob/Ob}$ and $Lep^{Db/Db}$).

Further examination of the taxonomic apportionment via LEfSe analysis revealed a total of 72 differentially abundant bacterial taxa across all five groups of mice (Figure 3a,b). The largest number of unique taxa was seen for the DIO mice, while the lowest number was recorded for NC mice. This low number of unique taxonomic biomarkers in NC mice may be ascribed to the comparatively higher bacterial community structure similarity between NC, DbDb, and ObOb mice than others, thereby leading to a smaller array of unique and significantly differential taxa. The higher number of unique taxa in DIO followed by LF mice indicated that the preferential abundance of bacterial lineages emanating from the particular higher-level bacterial taxa, probably driven by high-fat and low-fat diets, respectively, led to definitive compositional differences in the gut microbiomes of these mice. Overall, NC mice were characterized by two discriminative features (*Akkermansia* and *Peptococcaceae*) whereas DbDb and ObOb mice presented five and six, respectively, taxonomic features. In contrast, DIO mice presented 11 discriminative features that were uniquely abundant versus all other groups. This clearly suggests that both models (diet- versus genetically induced) of obesity present some degree of microbiome dysbiosis; however, particularly in terms of taxa with altered proportions, a high-fat diet is more perturbing to the microbiome spectrum than genetically induced obesity. In addition, the data from ObOb and DbDb mice might suggest some sort of direct connection between gut epithelial leptin signaling and specific gut microbes. Particularly considering the role of leptin in regulating various processes of metabolism and immunity, it is plausible that leptin activity might influence gut bacterial populations and vice versa, maybe even independently of host dietary intake. Interestingly, in contrast to previous studies reporting a lower proportion of family *Ruminococcaceae* in DIO mice [17], we found a higher abundance of this family in DIO mice This difference may be due to the differences in geographical location of the experimental facilities. In addition, these inconsistencies between different studies reporting the effect of obesity on gut microbiomes suggest the possibility of high model and environmental specificity. Interestingly, compared to NC mice, the abundance of genus *Akkermansia* was significantly lower in DIO, DbDb, and ObOb mice. Notably, *Akkermansia* has been found to play an important role in glucose homeostasis and is being explored as a therapeutically beneficially bacterium to ameliorate insulin resistance and related metabolic disorders [4,18–21]. The PICRUSt-based analysis of differentially abundant functional metagenome features (classified as KEGG orthologs (KOs)) identified 70 functional KOs across all samples. However, no differentially abundant feature was detected in ObOb mice, which may be a reflection of the comparatively higher similarity between NC and ObOb mice in terms of microbial taxa as well as function modules, thus leading to less magnitude of unique and significantly differential features.

We then computed the correlations between bacterial abundance and functions enriched in different mouse models. The abundance of taxa involved in glycosaminoglycan degradation correlated negatively with that of *Clostridiales* in both DbDb and ObOb mice, but associated positively with *Proteobacteria* in ObOb and with *Bacteroidetes* in DIO mice (Figure 5c–e). Interestingly, the abundance of taxa associated with lipopolysaccharide (LPS) biosynthesis correlated negatively with *Firmicutes* and its members, including *Clostridia* and *Ruminococci*, in both DbDb and ObOb mice. This may be because LPS is a major cell-wall component of Gram-negative bacteria, whereas the majority of *Firmicutes*, which were lower in DbDb and ObOb versus DIO mice, are Gram-positive. The abundance of phylum *Bacteroidetes*, which comprises many clades involved in carbohydrate metabolism, correlated positively with the abundance of taxa involved in energy metabolism in both ObOb and DIO mice. In DIO mice, which were fed a high-fat diet and had a lower abundance of *Bacteroidetes*, the abundance of *Bacteroides* correlated positively with taxa involved in lipid biosynthesis, whereas in ObOb mice, taxa involved in lipid metabolism correlated negatively with *Bacteroidetes*. Interestingly, the higher abundance of taxa involved in LPS biosysnthesis in DIO mice corroborated the data of higher gut

levels of phylum, class, order, family, and genus. Alpha-diversity measures were computed within QIIME. Beta-diversity was analyzed using PCoA of the Bray–Curtis dissimilarity index, as described previously [33]. The metabolic and other functional activities of the gut bacterial communities were analyzed using the open source bioinformatics tool PICRUSt against the functional database of KEGG Orthology, as described previously [11].

4.3. Gut Permeability Measurement

Gut permeability was measured as per our previously described methods [34,35]. Briefly, mice were fasted for four hours and were then given an oral gavage of fluorescein isothiocyanate (FITC)-dextran (3–5 kDa; 1 g/kg body weight; Millipore Sigma, Burlington, MA, USA). After four hours of gavage (still on fast), the appearance of FITC fluorescence (excitation at 485 nm and emission at 520 nm) was measured (in duplicate from each mouse) in the serum and the gut permeability was calculated with reference to the FITC standard curve, as described previously [34,35].

4.4. Real-Time PCR Assays

Gene expression assays of tight-junction proteins and inflammation-related markers were conducted as per our previously described methods [34,35]. Briefly, total RNA was extracted from ileum tissues using an RNeasy kit (Qiagen, Germantown, MD, USA) and was reverse-transcribed using a High-Capacity cDNA reverse transcription kit (Applied Biosystems, Foster City, CA, USA). The cDNA was used to quantify the expression of TJP-1, Occl-1, IL-1β, IL-6, IL-10, TNF-α, and TGF-β1 using TaqMan Gene Expression Assays for real-time PCR. 18S rRNA was used as an internal control. Relative gene expression was calculated using the $^{\Delta\Delta CT}$ method and presented as relative fold change. All the assays were performed in triplicate and repeated three times.

4.5. Data Analysis

Alpha-diversity indices, bacterial abundance, gut permeability, and the gene expression data were compared between different groups using Kruskal–Wallis test followed by pair-wise Mann–Whitney U comparison. Resulting *p*-values were corrected by Bonferroni method. LEfSE (linear discriminatory analysis (LDA) effect size) was performed to identify bacterial taxa and metagenome functional modules driving differences in the different groups of mice [36], wherein the alpha parameter significance threshold for the Kruskal–Wallis as well as the Wilcoxon test implemented among classes was set to 0.01, the logarithmic LDA score cut-off was set to 3, and the strategy for multi-class analysis was set to "all-against-all". Bray–Curtis dissimilarity scores inferred from the bacteriome taxonomic data were reduced to a two-dimensional space using principal coordinate analysis (PCoA) to estimate the structural similarity (beta-diversity) of bacteriomes from different groups of mice. Differences in beta-diversity were tested by permutational multivariate analysis of variance (PERMANOVA) using the web-based algorithm tool MicrobiomeAnalyst [37]. Hierarchical clustering and heat maps depicting the patterns of abundance and log values were constructed within the "R" statistical software package (version 3.6.0; https://www.r-project.org/) using the "heatmap.2" and "ggplots" packages. Spearman's correlations of bacterial taxa with KEGG metagenomic functions were calculated in GraphPad Prism software (San Diego, CA, USA; version 6.0; https://www.graphpad.com/scientific-software/prism/). Co-occurrence networks between taxa and functions were calculated by using the open-source software Gephi (https://gephi.org/) to find differential associations caused by similar alterations in the proportion of different taxa and their predicted functions between different groups of mice. Modularity-based co-occurrence networks were analyzed at a Spearman's correlation cutoff of 0.7 and p-value < 0.01; the selected correlation data were imported into the interactive platform, Gephi (version 0.9.2; https://gephi.org), and the following modularity analyses and keystone node identification were conducted within Gephi. Unless otherwise stated, all bar graphs presented herein represent means ± SEM. $p < 0.05$ was considered statistically significant unless otherwise specified.

Supplementary Materials: Supplementary materials can be found at http://www.mdpi.com/1422-0067/21/10/3434/s1. Supplementary figures have been provided along with this manuscript. All the raw sequencing data sets have been submitted to the NCBI Sequence Read Archive database under SRA accession number: SUB7192037 and bio-project number PRJNA615360).

Author Contributions: R.N.: performed mouse experiments, analyzed microbiome, performed data analyses, and wrote the first-draft of manuscript; S.P.M.: performed gene expression assays; H.Y.: conceived the idea, supervised the study, interpreted data, and revised the manuscript. All authors reviewed and approved the final version of the manuscript.

Funding: The authors are highly thankful for grant support from the National Institutes of Health - R01AG018915 and the Pepper Older Americans for Independence Center (P30AG21332), and the Department of Defense-W81XWH-18-1-0118 and W81XWH-19-1-0236, as well funds and services provided from the Center for Diabetes, Obesity and Metabolism, Wake Forest Baptist Medical Center, and the National Center for Advancing Translational Sciences (NCATS), the National Institutes of Health-funded Wake Forest Clinical and Translational Science Institute (WF CTSI) through Grant Award Number UL1TR001420.

Acknowledgments: The authors wish to thank all the colleagues as well as the members of the Wake Forest Animal Resource Program that helped in the study.

Conflicts of Interest: The authors declare no conflict of interest. The funders had no role in the design of the study; in the collection, analyses, or interpretation of data; in the writing of the manuscript, or in the decision to publish the results.

Abbreviations

DIO	Diet-induced obesity
ObOb	Leptin-deficient ($Lep^{ob/ob}$)
DbDb	Leptin-receptor-deficient ($Lepr^{db/db}$)
LF	Low-fat
NC	Normal chow
HFD	High-fat diet
T2D	Type-2 diabetes
LDA	Linear discriminatory analysis
LEfSe	Linear discriminatory analysis effect size
KEGG	Kyoto Encyclopedia of Genes and Genomes
KOs	KEGG orthologs
QIIME	Quantitative Insights into Microbial Ecology
PICRUSt	Phylogenetic Investigation of Communities by Reconstruction of Unobserved States
FITC	Fluorescein isothiocyanate
TJP	Tight-junction proteins
IL	Interleukins
TGF	Transforming growth factor
TNF	Tumor necrosis factor
PCoA	Principal coordinate analysis
EMP	Earth Microbiome Project

References

1. Boulange, C.L.; Neves, A.L.; Chilloux, J.; Nicholson, J.K.; Dumas, M.E. Impact of the gut microbiota on inflammation, obesity, and metabolic disease. *Genome Med.* **2016**, *8*, 42. [CrossRef]
2. Turnbaugh, P.J.; Ley, R.E.; Mahowald, M.A.; Magrini, V.; Mardis, E.R.; Gordon, J.I. An obesity-associated gut microbiome with increased capacity for energy harvest. *Nature* **2006**, *444*, 1027–1031. [CrossRef]
3. Vrieze, A.; Van Nood, E.; Holleman, F.; Salojarvi, J.; Kootte, R.S.; Bartelsman, J.F.; Dallinga-Thie, G.M.; Ackermans, M.T.; Serlie, M.J.; Oozeer, R.; et al. Transfer of intestinal microbiota from lean donors increases insulin sensitivity in individuals with metabolic syndrome. *Gastroenterology* **2012**, *143*, 913–916.e7. [CrossRef] [PubMed]
4. Thingholm, L.B.; Ruhlemann, M.C.; Koch, M.; Fuqua, B.; Laucke, G.; Boehm, R.; Bang, C.; Franzosa, E.A.; Hubenthal, M.; Rahnavard, A.; et al. Obese individuals with and without type 2 diabetes show different gut microbial functional capacity and composition. *Cell Host Microbe* **2019**, *26*, 252–264.e10. [CrossRef] [PubMed]

5. Wang, B.; Chandrasekera, P.C.; Pippin, J.J. Leptin- and leptin receptor-deficient rodent models: Relevance for human type 2 diabetes. *Curr Diabetes Rev.* **2014**, *10*, 131–145. [CrossRef] [PubMed]
6. Everard, A.; Lazarevic, V.; Derrien, M.; Girard, M.; Muccioli, G.G.; Neyrinck, A.M.; Possemiers, S.; Van Holle, A.; Francois, P.; de Vos, W.M.; et al. Responses of gut microbiota and glucose and lipid metabolism to prebiotics in genetic obese and diet-induced leptin-resistant mice. *Diabetes* **2011**, *60*, 2775–2786. [CrossRef] [PubMed]
7. Rajala, M.W.; Patterson, C.M.; Opp, J.S.; Foltin, S.K.; Young, V.B.; Myers, M.G., Jr. Leptin acts independently of food intake to modulate gut microbial composition in male mice. *Endocrinol.* **2014**, *155*, 748–757. [CrossRef]
8. Pfalzer, A.C.; Nesbeth, P.D.; Parnell, L.D.; Iyer, L.K.; Liu, Z.; Kane, A.V.; Chen, C.Y.; Tai, A.K.; Bowman, T.A.; Obin, M.S.; et al. Diet- and genetically-induced obesity differentially affect the fecal microbiome and metabolome in Apc1638N mice. *PLoS ONE* **2015**, *10*, e0135758. [CrossRef]
9. Ellekilde, M.; Krych, L.; Hansen, C.H.; Hufeldt, M.R.; Dahl, K.; Hansen, L.H.; Sorensen, S.J.; Vogensen, F.K.; Nielsen, D.S.; Hansen, A.K. Characterization of the gut microbiota in leptin deficient obese mice - Correlation to inflammatory and diabetic parameters. *Res. Vet. Sci.* **2014**, *96*, 241–250. [CrossRef]
10. Yang, M.; Liu, Y.; Xie, H.; Wen, Z.; Zhang, Y.; Wu, C.; Huang, L.; Wu, J.; Xie, C.; Wang, T.; et al. Gut microbiota composition and structure of the Ob/Ob and Db/Db mice. *Int. J. Endocrinol.* **2019**, *2019*, 1394097. [CrossRef]
11. Langille, M.G.; Zaneveld, J.; Caporaso, J.G.; McDonald, D.; Knights, D.; Reyes, J.A.; Clemente, J.C.; Burkepile, D.E.; Vega Thurber, R.L.; Knight, R.; et al. Predictive functional profiling of microbial communities using 16S rRNA marker gene sequences. *Nat. Biotechnol.* **2013**, *31*, 814–821. [CrossRef] [PubMed]
12. Segata, N.; Haake, S.K.; Mannon, P.; Lemon, K.P.; Waldron, L.; Gevers, D.; Huttenhower, C.; Izard, J. Composition of the adult digestive tract bacterial microbiome based on seven mouth surfaces, tonsils, throat and stool samples. *Genome Biol.* **2012**, *13*, R42. [CrossRef] [PubMed]
13. Carmody, R.N.; Gerber, G.K.; Luevano, J.M., Jr.; Gatti, D.M.; Somes, L.; Svenson, K.L.; Turnbaugh, P.J. Diet dominates host genotype in shaping the murine gut microbiota. *Cell Host Microbe* **2015**, *17*, 72–84. [CrossRef] [PubMed]
14. David, L.A.; Maurice, C.F.; Carmody, R.N.; Gootenberg, D.B.; Button, J.E.; Wolfe, B.E.; Ling, A.V.; Devlin, A.S.; Varma, Y.; Fischbach, M.A.; et al. Diet rapidly and reproducibly alters the human gut microbiome. *Nature* **2014**, *505*, 559–563. [CrossRef]
15. Garcia-Mazcorro, J.F.; Ivanov, I.; Mills, D.A.; Noratto, G. Influence of whole-wheat consumption on fecal microbial community structure of obese diabetic mice. *Peer J.* **2016**, *4*, e1702. [CrossRef]
16. Ley, R.E.; Turnbaugh, P.J.; Klein, S.; Gordon, J.I. Microbial ecology: Human gut microbes associated with obesity. *Nature* **2006**, *444*, 1022–1023. [CrossRef]
17. Daniel, H.; Gholami, A.M.; Berry, D.; Desmarchelier, C.; Hahne, H.; Loh, G.; Mondot, S.; Lepage, P.; Rothballer, M.; Walker, A.; et al. High-fat diet alters gut microbiota physiology in mice. *Isme J.* **2014**, *8*, 295–308. [CrossRef]
18. Naito, Y.; Uchiyama, K.; Takagi, T. A next-generation beneficial microbe: *Akkermansia muciniphila*. *J. Clin. Biochem. Nutr.* **2018**, *63*, 33–35. [CrossRef]
19. Shin, N.R.; Lee, J.C.; Lee, H.Y.; Kim, M.S.; Whon, T.W.; Lee, M.S.; Bae, J.W. An increase in the Akkermansia spp. population induced by metformin treatment improves glucose homeostasis in diet-induced obese mice. *Gut* **2014**, *63*, 727–735. [CrossRef]
20. Roopchand, D.E.; Carmody, R.N.; Kuhn, P.; Moskal, K.; Rojas-Silva, P.; Turnbaugh, P.J.; Raskin, I. Dietary polyphenols promote growth of the gut bacterium *akkermansia muciniphila* and attenuate high-fat diet-induced metabolic syndrome. *Diabetes* **2015**, *64*, 2847–2858. [CrossRef]
21. Zhang, L.; Qin, Q.; Liu, M.; Zhang, X.; He, F.; Wang, G. *Akkermansia muciniphila* can reduce the damage of gluco/lipotoxicity, oxidative stress and inflammation, and normalize intestine microbiota in streptozotocin-induced diabetic rats. *Pathog. Dis.* **2018**, *76*, fty028. [CrossRef] [PubMed]
22. Cani, P.D.; Amar, J.; Iglesias, M.A.; Poggi, M.; Knauf, C.; Bastelica, D.; Neyrinck, A.M.; Fava, F.; Tuohy, K.M.; Chabo, C.; et al. Metabolic endotoxemia initiates obesity and insulin resistance. *Diabetes* **2007**, *56*, 1761–1772. [CrossRef] [PubMed]
23. Nagpal, R.; Yadav, H. Bacterial translocation from the gut to the distant organs: An overview. *Ann. Nutr. Metab.* **2017**, *71*, 11–16. [CrossRef] [PubMed]

24. Bischoff, S.C.; Barbara, G.; Buurman, W.; Ockhuizen, T.; Schulzke, J.D.; Serino, M.; Tilg, H.; Watson, A.; Wells, J.M. Intestinal permeability–a new target for disease prevention and therapy. *BMC Gastroenterol.* **2014**, *14*, 189. [CrossRef] [PubMed]
25. Nagpal, R.; Neth, B.J.; Wang, S.; Craft, S.; Yadav, H. Modified Mediterranean-ketogenic diet modulates gut microbiome and short-chain fatty acids in association with Alzheimer's disease markers in subjects with mild cognitive impairment. *EBioMedicine* **2019**, *47*, 529–542. [CrossRef]
26. Nagpal, R.; Shively, C.A.; Appt, S.A.; Register, T.C.; Michalson, K.T.; Vitolins, M.Z.; Yadav, H. Gut microbiome composition in non-human primates consuming a western or mediterranean diet. *Front Nutr.* **2018**, *5*, 28. [CrossRef]
27. Nagpal, R.; Wang, S.; Ahmadi, S.; Hayes, J.; Gagliano, J.; Subashchandrabose, S.; Kitzman, D.W.; Becton, T.; Read, R.; Yadav, H. Human-origin probiotic cocktail increases short-chain fatty acid production via modulation of mice and human gut microbiome. *Sci. Rep.* **2018**, *8*, 12649. [CrossRef]
28. Ahmadi, S.; Nagpal, R.; Wang, S.; Gagliano, J.; Kitzman, D.W.; Soleimanian-Zad, S.; Sheikh-Zeinoddin, M.; Read, R.; Yadav, H. Prebiotics from acorn and sago prevent high-fat-diet-induced insulin resistance via microbiome-gut-brain axis modulation. *J. Nutr. Biochem.* **2019**, *67*, 1–13. [CrossRef]
29. Thompson, L.R.; Sanders, J.G.; McDonald, D.; Amir, A.; Ladau, J.; Locey, K.J.; Prill, R.J.; Tripathi, A.; Gibbons, S.M.; Ackermann, G.; et al. A communal catalogue reveals Earth's multiscale microbial diversity. *Nature* **2017**, *551*, 457–463. [CrossRef]
30. Caporaso, J.G.; Lauber, C.L.; Walters, W.A.; Berg-Lyons, D.; Huntley, J.; Fierer, N.; Owens, S.M.; Betley, J.; Fraser, L.; Bauer, M.; et al. Ultra-high-throughput microbial community analysis on the Illumina HiSeq and MiSeq platforms. *Isme J.* **2012**, *6*, 1621–1624. [CrossRef]
31. Wang, Q.; Garrity, G.M.; Tiedje, J.M.; Cole, J.R. Naive Bayesian classifier for rapid assignment of rRNA sequences into the new bacterial taxonomy. *Appl. Environ. Microbiol.* **2007**, *73*, 5261–5267. [CrossRef] [PubMed]
32. Caporaso, J.G.; Kuczynski, J.; Stombaugh, J.; Bittinger, K.; Bushman, F.D.; Costello, E.K.; Fierer, N.; Pena, A.G.; Goodrich, J.K.; Gordon, J.I.; et al. QIIME allows analysis of high-throughput community sequencing data. *Nat. Methods* **2010**, *7*, 335–336. [CrossRef] [PubMed]
33. Chong, J.; Liu, P.; Zhou, G.; Xia, J. Using MicrobiomeAnalyst for comprehensive statistical, functional, and meta-analysis of microbiome data. *Nat. Protoc.* **2020**, *15*, 799–821. [CrossRef] [PubMed]
34. Ahmadi, S.; Razazan, A.; Nagpal, R.; Jain, S.; Wang, B.; Mishra, S.P.; Wang, S.; Justice, J.; Ding, J.; McClain, D.A.; et al. Metformin reduces aging-related leaky gut and improves cognitive function by beneficially modulating gut microbiome/goblet cell/mucin axis. *J. Gerontol. A. Biol. Sci. Med. Sci.* **2020**. [CrossRef]
35. Wang, S.; Ahmadi, S.; Nagpal, R.; Jain, S.; Mishra, S.P.; Kavanagh, K.; Zhu, X.; Wang, Z.; McClain, D.A.; Kritchevsky, S.B.; et al. Lipoteichoic acid from the cell wall of a heat killed Lactobacillus paracasei D3-5 ameliorates aging-related leaky gut, inflammation and improves physical and cognitive functions: From C. elegans to mice. *Geroscience* **2020**, *42*, 333–352. [CrossRef]
36. Segata, N.; Izard, J.; Waldron, L.; Gevers, D.; Miropolsky, L.; Garrett, W.S.; Huttenhower, C. Metagenomic biomarker discovery and explanation. *Genome Biol.* **2011**, *12*, R60. [CrossRef]
37. Dhariwal, A.; Chong, J.; Habib, S.; King, I.L.; Agellon, L.B.; Xia, J. MicrobiomeAnalyst: A web-based tool for comprehensive statistical, visual and meta-analysis of microbiome data. *Nucleic. Acids. Res.* **2017**, *45*, W180–W188. [CrossRef]

International Journal of
Molecular Sciences

MDPI

Article

The Influence of Food Intake Specificity in Children with Autism on Gut Microbiota

Aleksandra Tomova [1,*], Katarina Soltys [2,3,4], Petra Kemenyova [1], Miloslav Karhanek [5] and Katarina Babinska [1]

1 Institute of Physiology, Faculty of Medicine, Comenius University in Bratislava, 813 72 Bratislava, Slovakia; petra.kemenyova@fmed.uniba.sk (P.K.); katarina.babinska@fmed.uniba.sk (K.B.)

2 Comenius University, Science Park, Comenius University in Bratislava, 841 04 Bratislava, Slovakia; katarina.soltys@uniba.sk

3 Department of Microbiology and Virology, Faculty of Natural Sciences, Comenius University in Bratislava, 842 15 Bratislava, Slovakia

4 Department of Molecular Biology, Faculty of Natural Sciences, Comenius University in Bratislava, 842 15 Bratislava, Slovakia

5 Biomedical Research Center, Slovak Academy of Sciences, 845 05 Bratislava, Slovakia; miloslav.karhanek@savba.sk

* Correspondence: aleksandra.tomova@fmed.uniba.sk

Received: 31 March 2020; Accepted: 16 April 2020; Published: 17 April 2020

check for updates

Abstract: Autism spectrum disorder (ASD) is a complex of neurodevelopmental conditions with increasing incidence. The microbiota of children with ASD is distinct from neurotypical children, their food habits are also different, and it is known that nutrient intake influences microbiota in a specific way. Thus, this study investigates the food habits of children with ASD and their association with the gut microbiota. Children with ASD had their dietary energy intakes similar to controls, but they more often demonstrated food selectivity, which seemed to result in deficiency of micronutrients such as vitamins K, B6, C, iron, cooper, docosahexaenoic and docosapentanoic acid. Using high-throughput sequencing, a DNA library of intestinal microbiota was performed. Core microbiota was similar in children with and without ASD, but *Dichelobacter*, *Nitriliruptor* and *Constrictibacter* were found to be putative markers of ASD. The changes in gut microbiota that we observed in connection to food selectivity, intake of fats and omega-3 in particular, fermented milk products and animal/plant protein consumption had similar character, independent of diagnosis. However, high fibre intake was connected with a decreased α-diversity only in children with ASD. High carbohydrate and fibre intake influenced β-diversity, changing the abundance of *Bacteroides* and other genera, many of them members of the *Clostidiaceae*. Modulating food habits of ASD children can influence their gut microbiota composition.

Keywords: autism spectrum disorder; intestinal microbiota; food intake

1. Introduction

Autism spectrum disorder (ASD) is a complex of neurodevelopmental conditions with increasing prevalence. A large proportion of children with ASD are reported to have food selectivity. They refuse to eat a diet based on a variety of foods and consume a narrower food repertoire consisting of about two-thirds the number of foods as typically developing children [1,2]. Sensory selectivity, based on tactile/texture, gustatory, and olfactory oversensitivity, is proposed to contribute to the development of food selectivity [1]. Additionally, children with ASD are often subjected to some type of dietary intervention and elimination of some foods or food groups [3]. These feeding habits raise the question of nutritional adequacy [1]. As an example, it was shown that children with ASD consume fewer proteins and more carbohydrates than the general population [4], and their diets may be low in micronutrients [4,5]. The most frequently omitted food group is vegetables, followed by fruits [6]. Evidence shows that feeding problems are associated with gastrointestinal (GI) and intestinal microbiota change in children with ASD. It has been established for the general population and confirmed using animal models that the consumption of particular diets shifts gut microbiota to specific bacterial genera [7]. In particular, a high-fat diet deceases α-diversity, *Firmicutes/Bacteroidetes* ratio, *Blautia* and *Faecalibacterium* abundance and increases *Alistripes* and *Bacteroides* abundance [8]. Supplementation of omega-3 fatty acids has been shown to temporarily shift intestinal microbiota towards bacterial species producing short-chain fatty acids [9]. Dietary intake of non-fermentable fibre is reported to increase such genera as *Helicobacter*, *Enterococcus*, *Desulfovibrio*, *Parabacteroides*, *Pseudoflavonifractor* and *Oscillibacter*, while decreasing genera such as *Lactobacillus*, *Parasutterella*, *Coprobacillus* and *TM7 genera Incertae Sedis*. This leads to changes in the metabolic profile that are beneficial for the prevention of autoimmune diseases [10]. Recent meta-analysis including studies with different kinds of dietary fibre concluded that fibre intake leads to a higher faecal abundance of *Bifidobacterium* and *Lactobacillus* spp. but does not affect α-diversity [11].

The amount of a specific macronutrient in the diet, as well as its source, is important for the microbial configuration of host intestines. Type, composition and quantity of dietary proteins are associated with specific intestinal microbiota, thus they influence microbial metabolites in the intestine, and modulate the function of intestinal barrier and host immune defence [12]. High animal protein intake increases the *Bacteroides/Prevotella* ratio and even changes the microbial enterotype of the host. Furthermore, the nature of food intake affects intestinal microbiota. but in response, the intestinal microbiota itself can regulate eating habits, stimulating the intake of foods beneficial to them and suppressing competitors [13]. Thus, the feeding habits of children with ASD raise several questions including the role of nutritional inadequacy of the diet, the connection of food composition with GI disorders, and the consequences in terms of a change in their intestinal microbiota.

In addition to the above mentioned, it is known that gut microbiota in children with ASD is different compared with control children [14–16]. The ratio of the main phyla *Bacterodetes/Firmicutes* switches, and genera such as *Clostridium*, *Sutterella*, *Lactobacillus*, *Nitriliruptor*, *Youngiibacter*, *Methanomicrobiales*, *Bilophila* and *Desulfovibrio* change in abundance. This change displays an association with behavioural manifestations. This association suggests bidirectional influence of microbiota and brain, explaining the recent interest in the microbiota–gut–brain axis [17]. The different pathways involve the leakage of bacterial metabolites into the blood, and these, through cascades of reactions, alter the tight junctions in the blood–brain barrier, cortex, hippocampus, amygdala and cerebellum [18]. A recent review of nutritional interventions for gut microbiota modulation in ASD [19] has revealed that in this area, the gaps in knowledge are greater than what is known.

Taken together, the specific microbiota reacts to an individual's diet in a personalised way [20]. The interplay between diet, microbiota and host, in children with ASD along with their distinct microbiota, would probably be different from that of neurotypical children.

Thus, this aim of our study to make a step towards elucidating the connection of eating disorders and specificity of food intake, with intestinal microbiota composition in children with ASD, keeping in mind that some microbiota changes might be associated specifically with their neurodevelopmental disorder.

2. Results and Discussion

Participants in the children with ASD group were chosen based on a confirmed diagnosis as described in the Methods section. Children in the control group had similar rates of GI complaints, suggesting that possible differences in microbiota are not associated with GI symptoms. The parameters and criteria used to compare microbiota, such as food selectivity, high-fibre diet etc., within the group of children with ASD were chosen based on the current literature of nutritional and feeding habits' impact on gut microbiota.

2.1. Comparison of Feeding Habits and Intestinal Microbiota of Children with and without ASD

Mealtime behaviours of the individuals are presented in Table 1. Children with ASD displayed mealtime problems more frequently than controls (76.1% vs. 43.8%, $p = 0.017$). Although children with ASD have been reported to have more GI complaints compared to neurotypical children [21], in this study no significant differences either in frequency or in severity of GI disorders were observed between the ASD and control group (Table 1.). Of the most commonly presented GI symptoms, which include bloating, abdominal pain, and constipation, significant differences were observed only in the prevalence of constipation in ASD (28.3% vs. 0%, $p = 0.014$).

Table 1. Gastrointestinal symptoms and nutrition-related characteristics of cohorts of individuals with autism spectrum disorder (ASD) and neurotypical controls.

	ASD	Controls	p
N (all boys)	46	16	
Age range (years)	4.0–8.5	2.8–9.15	
Age (mean ± SD)	6.3 ± 1,5	5.1 ± 1.7	0.017
BMI (kg.m^{-2})	17.1 ± 3.7	16.2 ± 2.0	0.746
SDS BMI	0.12 ± 0.98	0.50 ± 1.56	0.781
Mealtime problems (% of subjects) of that	76.1%	46.8%	0.017
Food selectivity	58.7%	25%	0.020
Aggressive behaviour	34.8%	0.0%	0.006
Stereotyped behaviour	32.6%	18.8%	0.168
Presence of GI symptoms (% of subjects) of that	89.4%	87.5%	0.838
Constipation	28.9%	0.0%	0.014
Diarrhoea	2.2%	6.3%	0.437
Bloating	35.6%	56.3%	0.148
Abdominal pain	35.6%	25.0%	0.439
Pain upon defecation	17.8%	6.3%	0.284
Dietary restrictions initiated by parents	17.4%	0.0%	0.099
Food supplement intake	76.1%	93.8%	0.123

Feeding problems are also common in children with ASD [22]. In our study, the group of children with ASD contained significantly more individuals who displayed food selectivity ("picky eaters") compared to controls (57.7% vs. 25%, $p = 0.02$). Children presenting with selectivity prefer limited food choice, stereotypes during meals, and also significantly more often demonstrate aggressive behaviour or other behavioural disorders during the food intake. Aggressive behaviour, that may also adversely affect their food intake and the variety of food consumed, was observed in more than one third of children with ASD, but in none of the controls ($p = 0.006$). However, energy intake did not differ between the groups, and correspondingly, no significant differences in age-specific values of BMI (SDS BMI) were observed between controls and children with ASD (Table 1). The majority of the study participants fell in the normal range for BMI, as earlier reported [6]. Obesity (SDS BMI > 2 SD) was observed only in one child with ASD, and two control children.

Upon analysing specific nutrient intake, no significant differences in macronutrients were found between the studied groups (Table 2). Similarly, intakes of essential and non-essential amino acids, saturated and unsaturated fatty acids, and total sugars (mono and disaccharides) did not differ between the groups. However, children with ASD had significantly lower intake of docosahexaenoic acid, docosapentanoic acid, iron, cooper, iodine, and vitamins K, B6, and C (data not presented). This suggests a risk of inadequate nutrient intake compared to neurotypical children. Sharp et al. in 2018 identified a decreased daily intake of several nutrients in subjects with ASD when compared to the recommendations [6]. Notably, the abovementioned study found different nutrients to be consumed in lower amounts compared to our study, except for iron, which was lower in both studies.

Table 2. Daily intakes of energy, selected nutrients and foods in cohorts of individuals with ASD and neurotypical controls.

	ASD	Controls	p
N (all boys)	30	16	
Energy and nutrients (mean ± SD)			
Energy, kJ/day	5506 ± 254	5666 ± 344	0.711
Proteins, g/day	45.5 ± 0.4	50.2 ± 0.9	0.273
Animal/plant protein	1.63 ± 0.03	1.76 ± 0.04	0.557
Fats, g/day	53.1 ± 0.6	55.5 ± 1.0	0.656
Carbohydrates, g/day	173.1 ± 1.3	173.9 ± 2.7	0.952
% energy from protein	14.2 ± 0.3	15.0 ± 0.47	0.148
% energy from fat	35.7 ± 1.1	36.8 ± 1.1	0.536
% energy from carbohydrates	50.1 ± 1.2	48.1 ± 1.3	0.306
Dietary fibre, g/ day	9.95 ± 0.56	11.7 ± 1.00	0.114
Omega-3 fatty acids (g/day) of that	0.75 ± 0.01	0.81 ± 0.01	0.473
Linolenic acid (g/day)	0.73 ± 0.01	0.78 ± 0.01	0.548
Eicosapentaenoic acid (mg/day)	10.0 ± 0.60	20.0 ± 1.00	0.325
Docosahexaenoic acid (mg/day)	7.0 ± 0.20	13.0 ± 0.40	0.022
Foods (median, 95% CI)			
Bakery products white (g/day)	62.5 (38.7–90.0)	60.5 (21.4-65.4)	0.037
Bakery products wholegrain (g/day)	4.8 (0.0–21.4)	21.4 (7.1-53.6)	0.102
Fermented milk products (g/day)	125.9 (19.3–244.2)	67.5 (42.9–109.3)	0.393
Fresh fruit (g/day)	84.1 (47.8–133.9)	199.1 (126.0–237.3)	0.001
Fresh vegetables (g/day)	16.6 (2.9–26.8)	61.4 (36.2–95.1)	0.000

Before investigating specific food intake influence on gut microbiota in children with ASD, we compared the gut microbiota of children with and without this disorder. Notably, despite the difference in nutrition, both groups had the same core microbiota (Figure 1), and alpha diversity was not different in children with and without ASD. On the other hand, using linear discriminant analysis effect size(LEfSe), we found univocal biomarkers in the gut microbiota, particularly *Dichelobacter*, *Nitriliruptor* and *Constrictibacter* were typical for ASD group, while *Diaphorobacter* and *Nitratireductor* were typical for the control group (Figure 2). Thus, these bacterial genera were assumed to characterize the differences between children with autism and neurotypical children, rather than to characterize the specificity of the feeding habits.

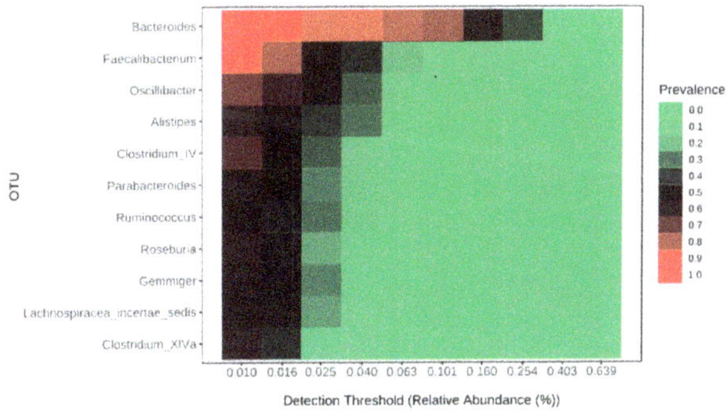

Figure 1. Core intestinal microbiome in children with and without ASD.

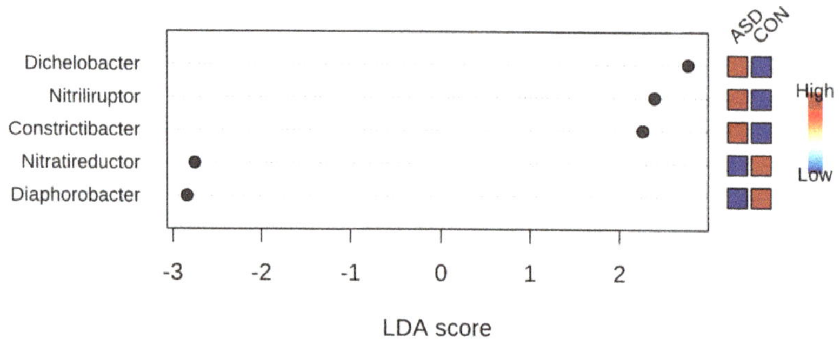

Figure 2. Microbiota biomarkers for ASD children compared to neurotypical, calculated with linear discriminant analysis effect size (LEfSe), MicrobiomeAnalyst.

2.2. Food Selectivity Reflects the Faecal Microbiota Composition

Previous studies showed that children with ASD included significantly more "picky eaters" when compared with controls. The prevalence of food selectivity in our sample of children with ASD corresponded to observations of other authors [2]). Food selectivity correlated with behavioural problems during feeding in ASD children (Pearson correlation 0.64, $p < 0.0001$). Children with autism also had behavioural disorders during feeding more frequently than controls (7.17 vs. 2.68, $p = 0.0006$). Among the children with ASD, "picky eaters" had more GI disorders compared to "non-picky eaters" (total GI score 6.63 vs. 3.6, $p = 0.02$) with belly pain (average score 0.73 vs. 0.25, $p = 0.035$) and constipation (average score 1.19 vs. 0.05, $p = 0.001$) significantly more often. Within the group of children with ASD, "picky eaters" displayed a correlation between GI score and the social interaction scale (A-SUM) of ADI-R (Autism Diagnostic Interview-Revised) ($p = 0.02$, Pearson R = 0.46), which was not seen in non-picky eaters. Thus, these data suggest that "picky eating" is associated with GI and behavioural disorders in children with ASD.

Analysis of the intestinal microbiota at the phylum level in the ASD group displayed more heterogenous composition in "picky eaters" compared to "not picky eaters" (Figure 3). Six out of 58 phyla were significantly differently abundant and another seven tended to be differently abundant in children with and without food selectivity. The ratio of the most represented phyla *Bacteroidetes/Firmicutes* was higher in "picky eaters" compared to "non-picky eaters" (3.34 vs. 1.78), though it was only a trend, based mainly on *Bacteroidetes* abundance.

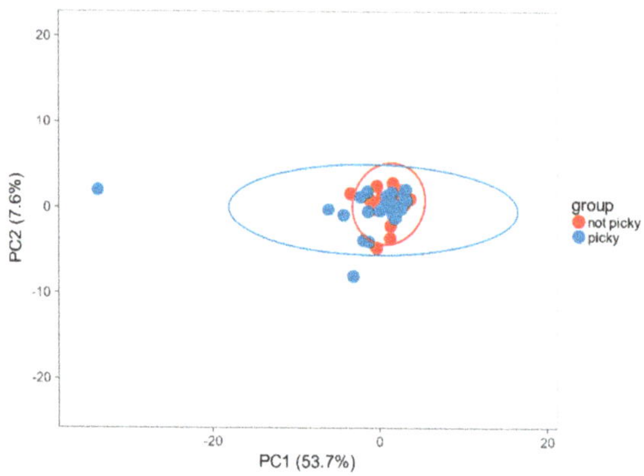

Figure 3. Diversity of intestinal microbiota at the phylum level in "picky eaters" compared to "non-picky eaters". ClustVis PCA.

We found that 58 genera were significantly differently abundant in children with ASD from different subgroups based on the presence or absence of food selectivity. Food selectivity makes faecal microbiota of "picky eaters" distinguishable from "non-picky eaters" within children with ASD (Figure 4). Representatives of *Enterobacteriacaea, Escherichia/Shigella* and *Salmonella*, from *Proteobacteria*, as well as *Clostridium* XIVa, *Anaerofilum* from *Clostridia, Firmicutes*, were characteristic of "picky eaters", according to LEfSe. These genera could also be associated with GI discomfort. Importantly, the most typical genera for ASD "picky eaters" were the same as for neurotypical "picky eaters" (data not shown). *Prevotella, Bacteroides, Parabacteroides* and *Bacteroidetes* characterised "non-picky eaters". *Bacteroides* are often associated with elevated meat intake, while studies connect *Prevotella* with plant-based diets. Both have been described as having pro-inflammatory effects [23]. In our participants, the core microbiota did not depend on food selectivity, dietary fibre, vegetable, fermented milk intake or any other studied criteria.

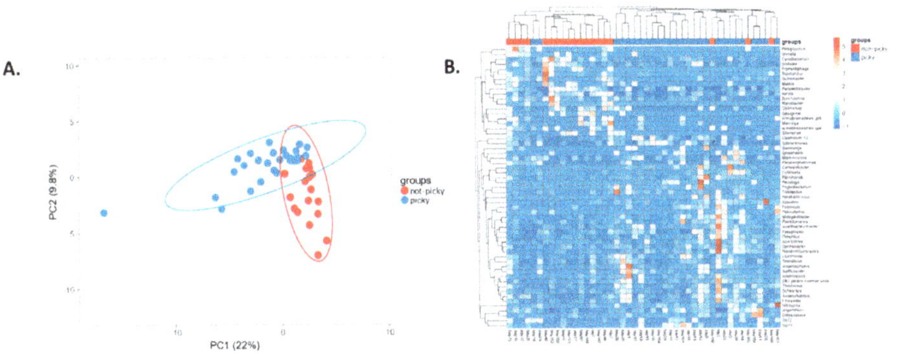

Figure 4. Diversity of intestinal microbiota at the genus level in "picky eaters" compared to "non-picky eaters". **A**. PCA ClustVis. **B**. Heatmap ClustVis, significantly differently abundant genera.

2.3. Carbohydrate Intake and Intestinal Microbiota

Consumption of more than 180 g/day was considered a high carbohydrate intake, and subjects with lower consumption were classified as having low carbohydrate intake. We did not observe an impact on the microbiome alpha diversity of individuals. However, within the cohort of children with ASD, we found a significant difference in microbiota relative abundance between the high and low carbohydrate intake subgroups ($p = 0.01$) (Bray-Curtis index, Permanova) (Figure 5). This was not different in samples of neurotypical children. One of 65 genera that significantly differed in their abundance were *Bacteroides;* their copy number was 2.5 times lower in children with higher intake of carbohydrates, as was previously observed in the neurotypical population. The most abundant genera that significantly differed along with *Bacteroides* included *Oscillibacter, Flavonifractor, Intestinimonas* and *Pseudoflavonifractor* as well. *Lactococcus* was increased with higher carbohydrate intake. Since the difference was found only in the ASD group, we investigated the intake of the dietary fibre as a specific carbohydrate, as well as the consumption of vegetables and fruit.

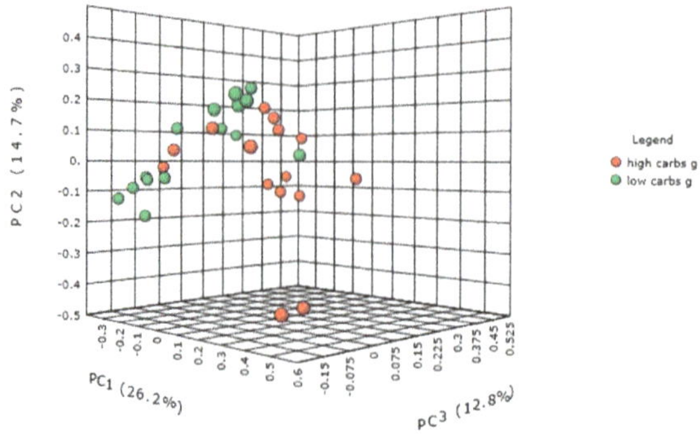

Figure 5. Intestinal microbiota diversity in children with ASD with high and low carbohydrates intake per day, $p = 0.01$.

Higher carbohydrate intake in children with ASD significantly increased the score in social affect, and it displayed a trend with an increased score in reciprocal social interaction and total raw score of ADOS-2 (Autism Diagnostic Observation Schedule, Second Edition).

2.4. Dietary Fibre Intake and Intestinal Microbiota

Dietary fibre comprises edible carbohydrate polymers with three or more monomeric units. It is resistant to endogenous digestive enzymes in the gut and many studies have proven its influence on the intestinal microbiota [24]. In our study, the fibre intake did not differ significantly between the children with and without autism. However, within the ASD group, alpha diversity was significantly lower in the subgroup of children with high fibre intake (Figure 6), unlike the controls, where there was no difference. This observation was not expected, since it opposes the irreversible reduction in microbial diversity in low dietary fibre diet described earlier [25,26]. Nevertheless, a meta-analysis found no change in α-diversity in increased fibre intakes [11]. Children with ASD have increased microbial diversity in the intestine when compared with controls [14], and this can be crucial for the change in microbial diversity under the influence of nutrient intake. Our study observed higher β-diversity in both ASD subgroups compared with both control subgroups. In the ASD group, microbiota in children with higher fibre intake per day notably differed from this in children with lower fibre intake (Figure 7). Seventy-three genera were significantly differently abundant (Figure 8). Low fibre intake significantly increased the abundance of *Hydrogenoanaerobacterium*, *Clostridium* IV, *Anaerotruncus* from *Clostridiaceae* and others. Increased fibre intake decreased the GI score in children with ASD, i.e., it was associated with lower frequency of GI disorders (data not shown), which suggests fibre's importance for healthy GI functioning.

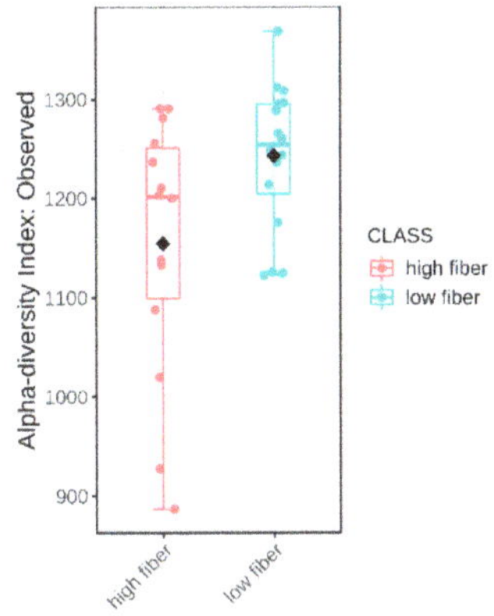

Figure 6. Alpha diversity of intestinal microbiota in children with ASD with high and low daily fibre intake, $p = 0.037$.

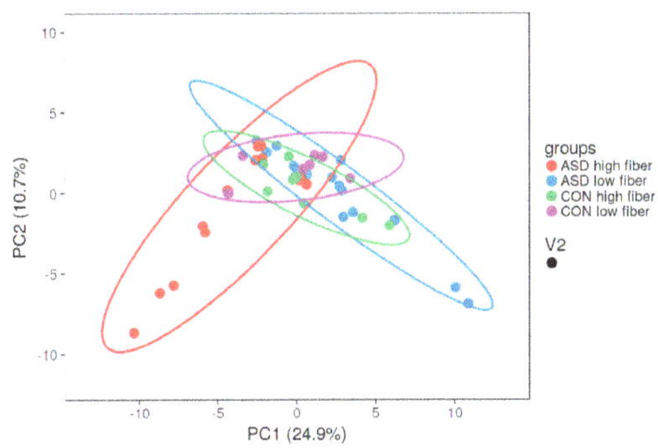

Figure 7. Visualisation of intestinal microbiota diversity at genera level of neurotypical children and children with ASD with high or low daily fibre intake (ClustVis).

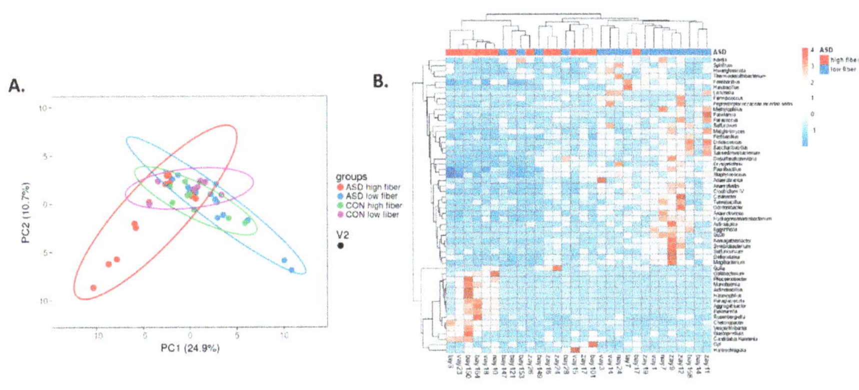

Figure 8. Intestinal microbiota diversity in children with ASD with high or low daily fibre intake, ClustVis. **A**. PCA. **B**. Heatmap, significantly differently abundant genera.

Although the daily fibre intake did not reach a significant difference between children with ASD and neurotypical children, fresh fruit and vegetable intake was significantly higher in neurotypical children compared to children with ASD, and this was expected [6]. Dietary fibres derived from fruits and vegetables, compared with that from cereal, contain a considerably higher proportion of soluble fibres, which exhibit better fermentability, and they have been shown to increase microbiota diversity and change the gut microbiota composition [24]. In our study, α-diversity was higher with higher intake of fresh vegetables than in the control group, but not in the group of children with ASD (data not presented). This, however, may be due to the generally low intake of vegetables in children with ASD. As expected, a high intake of vegetables significantly increased *Bacteroides* and *Hungatella* abundance.

According to the recommendations [27], adequate fibre intake for children is 14 g/1000 kcal (4184 kJ) of energy consumed. In our samples, the average intakes were just above half of the recommended quantity both in children with ASD (8.00 ± 3.26 g, mean, standard deviation) and in the control group (8.57 ± 1.67 g).

This amount of fibre may lead to the disappearance of specific bacterial species and this reduction is insufficient for improving the inflammatory status [25], a status often elevated in children with ASD [28]. Moreover, non-fermentable fibre helps to tune the immune status by the intestinal microbiota and so prevents autoimmune neurological disease [10]. Thus, it is recommended to increase fibre intake in both investigated groups, but especially in children with ASD.

2.5. Fermented Milk Intake and Intestinal Microbiota

An interesting finding of our study is the effect of consumption of fermented milk products on intestinal microbiota. Alpha diversity was significantly higher in children consuming lower quantities of fermented milk products, independently of diagnosis (Figure 9A). Beta-diversity showed differences in microbiota between fermented milk subgroups (higher intake vs. lower intake) of children from both ASD and the control group (Figure 9B). Subgroups of ASD children with low milk intake had significantly increased abundance of *Butyricimonas, Anaerotruncus, Guggenheimella, Acetanaerobacterium, Vallitalea* and other bacteria, most of which belong to class *Clostridia*. In our study, *Lactobacillus, Blautia, Anaerostipes* and *Fusicatenibacter* were typical for children with autism who consumed more fermented milk products (LEfSe), as expected from earlier studies [29,30]. *Fusicatenibacter*, like *Lactobacillus*, produces lactic acid, acetic acid and succinic acid [31]. The control group, unlike ASD, has *Sporomusa* and *Haemophilus*. Our results suggest that fermented milk consumption changes microbial community structure in the gastrointestinal tract, as suggested before [32], but these alterations depend on the type of bacteria in fermented milk [33] and seems to depend on the background intestinal microbiota.

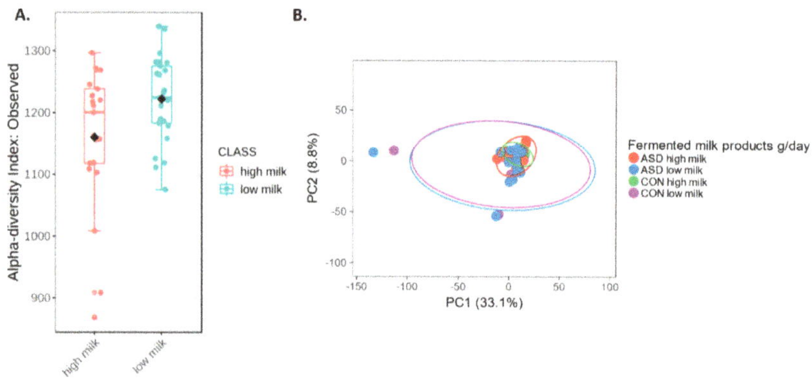

Figure 9. Alpha-diversity (**A**) and beta-diversity (**B**) of intestinal microbiota in children with high and low daily intake of fermented milk products.

2.6. Omega-3 Intake and Intestinal Microbiota

No difference in omega-3 intake in children with and without ASD was found. There were no significant differences in α and β diversity in higher and lower total fat or omega-3 intake in particular in ASD or in control children, as we expected [8]. Children with ASD who had higher omega-3 intakes had significantly higher abundance of *Catonella, Coprobacter, Marvinbryantia* genera, found normally in the faeces of healthy people. We did not observe changes in *Bifidobacterium, Roseburia, Lactobacillus* or others that were identified in previous studies [9,34]. The reason for this could be a natural source for omega-3 fatty acids from the food, reaching the maximum of 1.52g/day, while in the mentioned article, the subjects were supplemented with 4g omega-3 PUFA (polyunsaturated fatty acids)s per day. The microbiota profile in high and low omega-3 intake subgroups of children with autism and controls is shown in the Figure 10.

Docosahexaenoic acid intake was different in children with ASD and controls, but high intake was not associated with significant differences in gut microbiota compared to low intake.

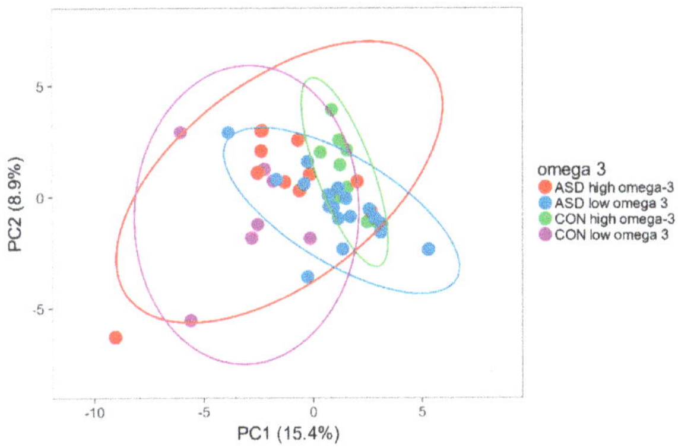

Figure 10. Intestinal microbiota in children with and without ASD with high and low omega-3 daily intake, significantly differently abundant genera.

2.7. Animal vs. Plant Protein Intake

Preferences for plant protein were significantly associated with decreased GI complaints ($p = 0.02$). No differences in diversity were found between the investigated subgroups. *Flavonifractor* from *Clostridia* was the most abundant taxa and significantly increased in children with ASD, who preferred animal proteins. This is rarely isolated from clinical human specimens, and the literature shows an association with the risk of cholecystitis [35] and colorectal cancer [36]. The expected change in the ratio of *Prevotella* to *Bacteroides* [25] was not observed.

The complexity in the interpretation of gut microbiota analysis as well as the difficulties of possible intervention in response to it are based on unique diets, reflected by the unique microbiota of each individual. The limitations facing all investigations of intestinal microbiota, including study, is its dependence on many factors, as age, geography, diet etc. Fortunately, it has been shown that habitual dietary patterns stabilize faecal microbiota in children from 4 to 8 years [37], which makes the findings of this investigation reliable. Limitations of this study imclude the low number of participants and the subjectivity of food intake tests.

Nevertheless, additional studies are needed to confirm the effects discovered in our study.

3. Materials and Methods

The study included 62 boys, of which 46 had ASD and 16 were control non-autistic children. The characteristics of the groups are included in Table 1. Written informed consent was obtained from parents of all participating children.

Children with ASD were recruited from the Academic Research Centre for ASD (ARCA) based at the Institute of Physiology, Faculty of Medicine Comenius University in Bratislava, Slovakia. Psychological evaluation of children with ASD was performed using the ADOS-2 (Autism Diagnostic Observation Schedule, Second Edition) [38] and ADI-R (Autism Diagnostic Interview, Revised) [39] behavioural assessment scales, which are internationally accepted gold standards for the diagnosis of ASD. ADOS-2 evaluated the behaviour in domains of social affect (SA), distinguishing communication (COM) and reciprocal social interaction (RSI), restricted and repetitive behaviour (RRB) scores, and total raw score (Total). ADI-R was evaluated in the areas of qualitative abnormalities in reciprocal social interaction (A) and communication (B), as well as in restricted, repetitive and stereotyped patterns (C) of behaviour. All subjects involved in the study met the criteria for ASD using both diagnostic tools. The diagnosis of ASD met the criteria for DSM-V (The Diagnostic and Statistical Manual of Mental Disorders, Fifth Edition). Control subjects were recruited from local kindergartens and schools and had no psychiatric conditions according to their parent interview. Written informed consent was obtained from parents of all participating children. The protocol was approved by the Ethics Committee of the Comenius University Faculty of Medicine and the University Hospital, APVV 15-0085 approved on 13.06.2016. The study conformed to the code of ethics stated in the Declaration of Helsinki.

Data on gastrointestinal (GI) status and mealtime behaviours were evaluated based on the parental questionnaires. The prevalence and frequency of gastrointestinal symptoms (abdominal pain, bloating, constipation, diarrhoea, hard stools, pain during defecation, voluminous stools) were obtained. Based on their prevalence, the GI score was calculated; the higher the value, the more prevalent the GI problems were. BMI was calculated as body weight in kilograms/square root of height in meters. BMI standard deviation scores (BMI-SDS) were calculated using the reference data of the Slovak population in order to determine the deviation in BMI from the mean BMI of the general population of children of the same age and gender [40]. The prevalence and frequency of adverse mealtime behaviours was recorded, including selectivity in food intake, as well as anger, crying or self-injuries associated with food intake. Child insistence on having the food prepared and served in the same manner was defined as stereotyped behaviour during mealtimes. The score of mealtime problems was calculated based on their prevalence. Higher values indicated more severe mealtime problems. Data on modifications of the child's diet were recorded, as well as information about the use of food supplements in the last 12 months.

Of the total 62 individuals, data from 30 children with ASD and 16 controls were collected about the typical diet of the subjects by a self-administered food frequency questionnaire (FFQ) validated for the Slovak population that included 85 food items. Parents of children were requested to indicate the frequency of consumption of each food item (with options including times per day, times per week, times per month, or almost never) together with the portion size. Average daily consumption of selected food types (g/day) was calculated. Food intake data served for the calculation of the nutrient intake of the individuals, vegetable and fruit intake was adjusted for seasonality. For the conversion of food intake into nutrient intake, the Slovak food composition database was used (Slovak Food Composition Bank. Revision 2004 (2004). Food Research Institute, Bratislava 2004) (http://www.vup.sk/en/index.php?navID=25?start). Calculations were performed using Microsoft Excel standard formulas.

Stool samples were collected by parents at home in sterile flasks. Parents were given a detailed explanation of the procedure and stored the samples at + 4 °C after collection. Samples were delivered to the laboratory within 4 hours and divided into 2 aliquots, one of which was immediately frozen at −80 °C for future DNA purification of intestinal microbiota assessment and the other frozen at −20 °C for calprotectin investigation.

DNA was extracted from frozen stool samples using a commercial extraction system (QIAamp DNA Fast Stool Mini Kit, Qiagen, Hilden, Germany) according to the manufacturer's instructions. The DNA concentration was determined using the NanoDrop 2000 Spectrophotometer (ThermoFisher Scientific, MA, USA).

High-throughput sequencing was performed for the DNA libraries. The PCR amplification of V1-V9 region of 16S rRNA using primer set 27f / 1492r was carried out using the 27f (5′-AGA GTT TGA TCM TGG CTC AG-3′) and the 1492r (5′-CGG TTA CCT TGT TAC GAC TT-3′) primers (Lane, 1991). In this process, 3-50 ng of total input DNA in 20 µL volume PCR reaction was amplified with 4µl of 5x HOT FIREPol Blend Master Mix (Solis BioDyne, Tartu, Estonia), 0.4 µL (10 µM) of each primer (final concentration 0.2 µM) and milli-pore water. PCR conditions were as follows: initial denaturation 95 °C/15 min, cycling 25× (95 °C /20 sec, 60 °C/30 sec, 72 °C/2 min), final polymerization 72 °C/10 min. Amplicons were column-purified (Zymo DNA Clean and Concentrator-5, Zymo Research, Irvine, CA, USA) according to standard protocols and quantified fluorometrically with Qubit™ dsDNA HS Assay Kit (Thermo Fisher Scientific, Waltham, MA, USA). Amplicon sequences were fragmented by a transposon-based approach (Nextera XT, Illumina San Diego, CA, USA) and low-cycle PCR and mutual indexing of the fragments was performed. Fragment size selection and purification with 1.8x AMPure XP beads yielded final DNA libraries that were verified using the Agilent 2100 Bioanalyzer (Agilent Technologies, Waldbronn, Germany) and quantified using the Qubit 2.0 Fluorometer (Thermo Fisher Scientific, Waltham, MA, USA). The 4 nM pool of libraries was further diluted to 10pM and sequenced on Illumina MiSeq platform (Illumina, San Diego, CA, USA) with 2 × 300 bp paired-end sequencing at the Comenius University Science Park (Bratislava, Slovakia). Library sequence data were quality checked using FastQC, Andrews, 2010, a quality control tool for high throughput sequence data available online at: http://www.bioinformatics.babraham.ac.uk/projects/fastqc. Further data processing, including trimming, 16S analysis and visualization, was performed with Geneious (Biomatters Ltd, Auckland, New Zealand). For visualizing and clustering of multivariate data using principal component analysis (PCA) ClustVis (https://biit.cs.ut.ee/clustvis/) [41] and MicrobiomeAnalystR [42,43] (https://github.com/xia-lab/MicrobiomeAnalystR) were applied. For diversity calculations, the Bray-Curtis Index and Permanova test were used.

Data were presented as mean ± SEM values and p values lower than 0.05 were accepted as significant. Since the data distribution matched Gaussian distribution for the correlation analysis, the Pearson correlation coefficient was applied. For all statistical analyses, GraphPad Prism 5 and Microsoft Excel 2016 were used. Data on food intake did not pass the normality test, therefore they are presented as median ±95% CI (confidence interval), and their differences were tested by the Mann–Whitney test. For the testing of categorical variables, the chi-square test and Fisher's exact test were used.

High carbohydrate, protein and fat intake corresponded to the consumption of more than 180, 40 and 50 g/day, respectively. High fibre, fresh vegetables and fresh fruit intake corresponded to the consumption of more than more than 10, 40 and 150 g/day, respectively. High fermented milk or white bakery intake corresponded to the consumption of more than 100 or 40 g/day, respectively. High omega-3 (linolenic, eicosapentaenoic, docosahexaenoic acids) intake corresponded to the consumption of more than 0.75 g/day.

4. Conclusions

Interest in the modulation of intestinal microbiota by specific food intake has increased in recent years. The degree to which eating habits influence gut microbiota in children with ASD is not clear as of yet. Children with ASD frequently have eating and GI disorders, which influence the intestinal microbiota and clinical manifestations and vice versa, with the microbiota influencing eating habits, GI status and behavioural disorders. Our study shows that nutritional inadequacies in children with ASD could be hidden behind normal BMI. Food selectivity, found more frequently in children with ASD, resulted in different micronutrient intake as compared with controls. Although the core microbiota composition was independent of diagnosis or specific nutrient intake, ASD group microbiota was characterised by *Dichelobacter, Nitriliruptor* and *Constrictibacter*, while *Diaphorobacter* and *Nitratireductor* were typical for the control group. For "picky eaters", the gut microbiota was more diverse at the phylum level, distinct at the genera level and characterized by specific genera (*Escherichia/Shigella* and *Salmonella, Clostridium XlVa, Anaerofilum*). Changes in gut microbiota were similar in children with and without ASD. Higher carbohydrate intake changed beta diversity only in children with ASD, for example by decreasing *Bacteroides* abundance. Fibre intake was similar in groups, but increasing it in children with ASD decreased microbial diversity. Fresh fruit and vegetable intake were significantly higher in neurotypical children compared to children with ASD, and their consumption increased microbiota diversity exclusively in the control group. High intake of fermented milk products had strong effects, that were similar in both groups: they increased *Lactobacillus* and *Fusicatenibacter* abundance and decreased microbial diversity. Distinct microbiota were found when omega-3 consumption with food was increased in both groups. Aside from bacteria characterising the microbiota in autism, the reflection of microbiota to nutrients intake was similar in both groups, with the exception of fibre intake.

Food selectivity, as well as the consumption of fermented milk products, total fat, omega-3, animal/plant protein resulted in similar changes in the intestinal microbiota of children with and without autism. However, the effects of carbohydrates, fibre, fruits, vegetables intake were different.

Thus, although food intervention in children with ASD is difficult, such changes could help to alter the intestinal microbiota in such a way as to improve GI and immune status.

Author Contributions: Conceptualization, A.T.; Data curation, K.B.; Methodology, K.S. and P.K.; Resources, A.T., K.S.; Software, M.K.; Visualization, A.T.; Writing – original draft, A.T. and K.B.; Writing – review & editing, K.S. and K.B. All authors have read and agreed to the published version of the manuscript.

Funding: This research was funded by Vedecká Grantová Agentúra MŠVVaŠ SR a SAV (VEGA): 1-0286-18; Agentúra na Podporu Výskumu a Vývoja (APVV): 15-0085; Agentúra na Podporu Výskumu a Vývoja (APVV): 15-0045; Comenius University Science Park in Bratislava: 26240220086; REVOGENE: 26240220067.

Acknowledgments: We would like to thank Davida Smyth, PhD, from *Department of Natural Sciences, The New School, New York* for language editing.

Conflicts of Interest: Authors declare no conflict of interests.

References

1. Cermak, S.A.; Curtin, C.; Bandini, L.G. Food selectivity and sensory sensitivity in children with autism spectrum disorders. *J. Am. Diet. Assoc.* **2010**, *110*, 238–246. [CrossRef] [PubMed]
2. Marí-Bauset, S.; Zazpe, I.; Mari-Sanchis, A.; Llopis-González, A.; Morales-Suárez-Varela, M. Food selectivity in autism spectrum disorders: A systematic review. *J. Child. Neurol.* **2014**, *29*, 1554–1561. [CrossRef] [PubMed]
3. Karhu, E.; Zukerman, R.; Eshraghi, R.S.; Mittal, J.; Deth, R.C.; Castejon, A.M.; Trivedi, M.; Mittal, R.; Eshraghi, A.A. Nutritional interventions for autism spectrum disorder. *Nutr. Rev.* **2019**, nuz092. [CrossRef] [PubMed]
4. Hyman, S.L.; Stewart, P.A.; Schmidt, B.; Cain, U.; Lemcke, N.; Foley, J.T.; Peck, R.; Clemons, T.; Reynolds, A.; Johnson, C.; et al. Nutrient intake from food in children with autism. *Pediatrics* **2012**, *130* (Suppl. 2), S145–S153. [CrossRef]
5. Esteban-Figuerola, P.; Canals, J.; Fernández-Cao, J.C.; Arija Val, V. Differences in food consumption and nutritional intake between children with autism spectrum disorders and typically developing children: A meta-analysis. *Autism* **2019**, *23*, 1079–1095. [CrossRef]
6. Sharp, W.G.; Postorino, V.; McCracken, C.E.; Berry, R.C.; Criado, K.K.; Burrell, T.L.; Scahill, L. Dietary intake, nutrient status, and growth parameters in children with autism spectrum disorder and severe food selectivity: An Electronic Medical Record Review. *J. Acad. Nutr. Diet.* **2018**, *118*, 1943–1950. [CrossRef]
7. Tomova, A.; Bukovsky, I.; Rembert, E.; Yonas, W.; Alwarith, J.; Barnard, N.D.; Kahleova, H. The Effects of Vegetarian and Vegan Diets on Gut Microbiota. *Front. Nutr.* **2019**, *6*, 47. [CrossRef]

8. Wan, Y.; Wang, F.; Yuan, J.; Li, J.; Jiang, D.; Zhang, J.; Li, H.; Wang, R.; Tang, J.; Huang, T.; et al. Effects of dietary fat on gut microbiota and faecal metabolites, and their relationship with cardiometabolic risk factors: A 6-month randomised controlled-feeding trial. *Gut* **2019**, *68*, 1417–1429. [CrossRef]
9. Watson, H.; Mitra, S.; Croden, F.C.; Taylor, M.; Wood, H.M.; Perry, S.L.; Spencer, J.A.; Quirke, P.; Toogood, G.J.; Lawton, C.L.; et al. A randomised trial of the effect of omega-3 polyunsaturated fatty acid supplements on the human intestinal microbiota. *Gut* **2018**, *67*, 1974–1983. [CrossRef]
10. Berer, K.; Martinez, I.; Walker, A.; Kunkel, B.; Schmitt-Kopplin, P.; Walter, J.; Krishnamoorthy, G. Dietary non-fermentable fibre prevents autoimmune neurological disease by changing gut metabolic and immune status. *Sci. Rep.* **2018**, *8*, 10431. [CrossRef]
11. So, D.; Whelan, K.; Rossi, M.; Morrison, M.; Holtmann, G.; Kelly, J.T.; Shanahan, E.R.; Staudacher, H.M.; Campbell, K.L. Dietary fibre intervention on gut microbiota composition in healthy adults: A systematic review and meta-analysis. *Am. J. Clin. Nutr.* **2018**, *107*, 965–983. [CrossRef] [PubMed]
12. Zhao, J.; Zhang, X.; Liu, H.; Brown, M.A.; Qiao, S. Dietary protein and gut microbiota composition and function. *Curr. Protein Pept. Sci.* **2019**, *20*, 145–154. [CrossRef] [PubMed]
13. Alcock, J.; Maley, C.C.; Aktipis, C.A. Is eating behavior manipulated by the gastrointestinal microbiota? Evolutionary pressures and potential mechanisms. *Bioessays* **2014**, *36*, 940–949. [CrossRef] [PubMed]
14. Finegold, S.M.; Dowd, S.E.; Gontcharova, V.; Liu, C.; Henley, K.E.; Wolcott, R.D.; Youn, E.; Summanen, P.H.; Granpeesheh, D.; Dixon, D.; et al. Pyrosequencing study of fecal microflora of autistic and control children. *Anaerobe* **2010**, *16*, 444–453. [CrossRef]
15. Tomova, A.; Soltys, K.; Repiska, G.; Palkova, L.; Filcikova, D.; Minarik, G.; Turna, J.; Prochotska, K.; Babinska, K.; Ostatnikova, D. Specificity of gut microbiota in children with autism spectrum disorder in Slovakia and its correlation with astrocytes activity marker and specific behavioural patterns. *Physiol. Behav.* **2020**, *214*, 112745. [CrossRef]
16. Bezawada, N.; Phang, T.H.; Hold, G.L.; Hansen, R. Autism spectrum disorder and the gut microbiota in children: A systematic review. *Ann. Nutr. Metab.* **2020**, 5363. [CrossRef]
17. Tomova, A.; Pivovarciová, A.; Babinská, K.; Mravec, B. Os črevo-mozog: Význam interakcií medzi črevnou mikrobiotou a mozgom za fyziologických a patologických situácií [Intestinal microbiota and the brain: Multilevel interactions in health and disease]. *Ceskoslovenska Fysiol.* **2015**, *64*, 23–34.
18. Srikantha, P.; Mohajeri, M.H. The possible role of the microbiota-gut-brain-axis in autism spectrum disorder. *Int. J. Mol. Sci.* **2019**, *20*, 2115. [CrossRef]
19. Ristori, M.V.; Quagliariello, A.; Reddel, S.; Ianiro, G.; Vicari, S.; Gasbarrini, A.; Putignani, L. Autism, gastrointestinal symptoms and modulation of gut microbiota by nutritional interventions. *Nutrients* **2019**, *11*, 2812. [CrossRef]
20. Kolodziejczyk, A.A.; Zheng, D.; Elinav, E. Diet-microbiota interactions and personalized nutrition. *Nat. Rev. Microbiol.* **2019**, *17*, 742–753. [CrossRef]
21. Babinska, K.; Pivovarčiová, A.; Filčíková, D.; Tomova, A.; Ostatníková, A. Association of conduct problems and gastrointestinal symptoms in individuals with autism spectrum disorders. *Act. Nerv. Super. Rediviva* **2016**, *58*, 69–72.
22. Bandini, L.G.; Anderson, S.E.; Curtin, C.; Cermak, S.; Evans, E.W.; Scampini, R.; Maslin, M.; Must, A. Food selectivity in children with autism spectrum disorders and typically developing children. *J. Pediatr.* **2010**, *157*, 259–264. [CrossRef] [PubMed]
23. Larsen, J.M. The immune response to Prevotella bacteria in chronic inflammatory disease. *Immunology* **2017**, *151*, 363–374. [CrossRef] [PubMed]
24. Cui, J.; Lian, Y.; Zhao, C.; Du, H.; Han, Y.; Gao, W.; Xiao, H.; Zheng, J. Dietary fibers from fruits and vegetables and their health benefits via modulation of gut microbiota. *Compr. Rev. Food Sci. Food Saf.* **2019**, *18*, 1514–1532. [CrossRef]
25. Makki, K.; Deehan, E.C.; Walter, J.; Bäckhed, F. The impact of dietary fiber on gut microbiota in host health and disease. *Cell Host Microbe* **2018**, *23*, 705–715. [CrossRef]
26. Sonnenburg, E.D.; Smits, S.A.; Tikhonov, M.; Higginbottom, S.K.; Wingreen, N.S.; Sonnenburg, J.L. Diet-induced extinctions in the gut microbiota compound over generations. *Nature* **2016**, *529*, 212–215. [CrossRef]
27. Dahl, W.J.; Stewart, M.L. Position of the academy of nutrition and dietetics: Health implications of dietary fiber. *J. Acad. Nutr. Diet.* **2015**, *115*, 1861–1870. [CrossRef]

28. Saghazadeh, A.; Ataeinia, B.; Keynejad, K.; Abdolalizadeh, A.; Hirbod-Mobarakeh, A.; Rezaei, N. A meta-analysis of pro-inflammatory cytokines in autism spectrum disorders: Effects of age, gender, and latitude. *J. Psychiatr. Res.* **2019**, *115*, 90–102. [CrossRef]
29. Sprague, R. Fusicatenibacter is associated with kefir drinking. *bioRxiv* **2017**, 218313. [CrossRef]
30. Aoyagi, Y.; Amamoto, R.; Park, S.; Honda, Y.; Shimamoto, K.; Kushiro, A.; Tsuji, H.; Matsumoto, H.; Shimizu, K.; Miyazaki, K.; et al. Independent and interactive effects of habitually ingesting fermented milk products containing. *Front. Microbiol.* **2019**, *10*, 1477. [CrossRef]
31. Nierychlo, M.; Andersen, K.S.; Xu, Y.; Green, N.; Albertsen, M.; Dueholm, M.S.; Nielsen, P.H. Species-level microbiome composition of activated sludge–introducing MiDAS 3 ecosystem-specific reference database and taxonomy. *BioRxiv* **2019**, 842393. [CrossRef]
32. Unno, T.; Choi, J.H.; Hur, H.G.; Sadowsky, M.J.; Ahn, Y.T.; Huh, C.S.; Kim, G.B.; Cha, C.J. Changes in human gut microbiota influenced by probiotic fermented milk ingestion. *J. Dairy Sci.* **2015**, *98*, 3568–3576. [CrossRef] [PubMed]
33. González, S.; Fernández-Navarro, T.; Arboleya, S.; de Los Reyes-Gavilán, C.G.; Salazar, N.; Gueimonde, M. Fermented dairy foods: Impact on intestinal microbiota and health-linked biomarkers. *Front. Microbiol.* **2019**, *10*, 1046. [CrossRef] [PubMed]
34. Costantini, L.; Molinari, R.; Farinon, B.; Merendino, N. Impact of omega-3 fatty acids on the gut microbiota. *Int. J. Mol. Sci.* **2017**, *18*, 2645. [CrossRef]
35. Berger, F.K.; Schwab, N.; Glanemann, M.; Bohle, R.M.; Gärtner, B.; Groesdonk, H.V. *Flavonifractor* (*Eubacterium*) *plautii* bloodstream infection following acute cholecystitis. *IDCases* **2018**, *14*, e00461. [CrossRef]
36. Gupta, A.; Dhakan, D.B.; Maji, A.; Saxena, R.; Vishnu Prasoodanan, P.K.; Mahajan, S.; Pulikkan, J.; Kurian, J.; Gomez, A.M.; Scaria, J.; et al. Association of Flavonifractor plautii, a flavonoid-degrading bacterium, with the gut microbiome of colorectal cancer patients in India. *mSystems* **2019**, *4*. [CrossRef]
37. Berding, K.; Holscher, H.D.; Arthur, A.E.; Donovan, S.M. Fecal microbiome composition and stability in 4- to 8-year old children is associated with dietary patterns and nutrient intake. *J. Nutr. Biochem.* **2018**, *56*, 165–174. [CrossRef]
38. Lord, C.; Rutter, M.; Dilavore, P.C.; Risi, S.; Gotham, K.; Bishop, S. Autism Diagnostic Observation Schedule. West. *Psychol. Serv. Torrance CA* **2012**. [CrossRef]
39. Lord, C.; Pickles, A.; McLennan, J.; Rutter, M.; Bregman, J.; Folstein, S.; Fombonne, E.; Leboyer, M.; Minshew, N. Diagnosing autism: Analyses of data from the Autism Diagnostic Interview. *J. Autism Dev. Disord.* **1997**, *27*, 501–517. [CrossRef]
40. Sevcikova, L.; Rovny, I.; Novakova, J.; Hamada, J.; Tatara, M.; Janechova, H. Body development of infants and children in Slovakia. Results of the 6th National Survey in 2001. *Bratisl. Natl. Inst. Public Health Slovak Repub.* **2004**, 124–143.
41. Metsalu, T.; Vilo, J. ClustVis: A web tool for visualizing clustering of multivariate data using principal component analysis and heatmap. *Nucleic Acids Res.* **2015**, *43*, W566–W570. [CrossRef] [PubMed]
42. Chong, J.; Liu, P.; Zhou, G.; Xia, J. Using MicrobiomeAnalyst for comprehensive statistical, functional, and meta-analysis of microbiome data. *Nat. Protoc.* **2020**, *15*, 799–821. [CrossRef] [PubMed]
43. Dhariwal, A.; Chong, J.; Habib, S.; King, I.L.; Agellon, L.B.; Xia, J. MicrobiomeAnalyst: A web-based tool for comprehensive statistical, visual and meta-analysis of microbiome data. *Nucleic Acids Res.* **2017**, *45*, W180–W188. [CrossRef] [PubMed]

International Journal of
Molecular Sciences

MDPI

Article

In Vitro Evaluation of Different Prebiotics on the Modulation of Gut Microbiota Composition and Function in Morbid Obese and Normal-Weight Subjects

Alicja M. Nogacka [1,2,*], Nuria Salazar [1,2], Silvia Arboleya [1,2], Patricia Ruas-Madiedo [1,3], Leonardo Mancabelli [4], Adolfo Suarez [2,5], Ceferino Martinez-Faedo [6,7], Marco Ventura [4], Takumi Tochio [8], Katsuaki Hirano [8], Akihito Endo [9], Clara G. de los Reyes-Gavilán [1,2] and Miguel Gueimonde [1,2]

1. Department of Microbiology and Biochemistry of Dairy Products, Instituto de Productos Lácteos de Asturias (IPLA-CSIC), 33300 Villaviciosa, Asturias, Spain; nuriasg@ipla.csic.es (N.S.); silvia.arboleya@ipla.csic.es (S.A.); ruas-madiedo@ipla.csic.es (P.R.-M.); greyes_gavilan@ipla.csic.es (C.G.d.l.R.-G.); mgueimonde@ipla.csic.es (M.G.)
2. Diet, Human Microbiota and Health Group, Institute of Health Research of the Principality of Asturias (ISPA), 33011 Oviedo, Spain; adolfo.suarez@hcabuenes.es
3. Functionality and Ecology of Beneficial Microorganisms, Institute of Health Research of the Principality of Asturias (ISPA), 33011 Oviedo, Spain
4. Laboratory of Probiogenomics, Department of Life Sciences, University of Parma, 43121 Parma, Italy; leonardo.mancabelli@genprobio.com (L.M.); marco.ventura@unipr.it (M.V.)
5. Digestive Service, Central University Hospital of Asturias (HUCA), 33011 Oviedo, Asturias, Spain
6. Endocrinology and Nutrition Service, Central University Hospital of Asturias (HUCA), 33011 Oviedo, Asturias, Spain; ceferinofaedo@gmail.com
7. Endocrinology, Nutrition, Diabetes and Obesity Group, Institute of Health Research of the Principality of Asturias (ISPA), 33011 Oviedo, Spain
8. β-Food Sciences Co., Chita 478-0046, Japan; t-tochio@bfsci.co.jp (T.T.); k-hirano@bfsci.co.jp (K.H.)
9. Department of Food and Cosmetic Science, Tokyo University of Agriculture, Abashiri 099-2493, Japan; a3endou@nodai.ac.jp

* Correspondence: alicja.nogacka@ipla.csic.es

Received: 20 December 2019; Accepted: 28 January 2020; Published: 30 January 2020

check for updates

Abstract: The gut microbiota remains relatively stable during adulthood; however, certain intrinsic and environmental factors can lead to microbiota dysbiosis. Its restoration towards a healthy condition using best-suited prebiotics requires previous development of in vitro models for evaluating their functionality. Herein, we carried out fecal cultures with microbiota from healthy normal-weight and morbid obese adults. Cultures were supplemented with different inulin-type fructans (1-kestose, Actilight, P95, Synergy1 and Inulin) and a galactooligosaccharide. Their impact on the gut microbiota was assessed by monitoring gas production and evaluating changes in the microbiota composition (qPCR and 16S rRNA gene profiling) and metabolic activity (gas chromatography). Additionally, the effect on the bifidobacterial species was assessed (ITS-sequencing). Moreover, the functionality of the microbiota before and after prebiotic-modulation was determined in an in vitro model of interaction with an intestinal cell line. In general, 1-kestose was the compound showing the largest effects. The modulation with prebiotics led to significant increases in the *Bacteroides* group and *Faecalibacterium* in obese subjects, whereas in normal-weight individuals, substantial rises in *Bifidobacterium* and *Faecalibacterium* were appreciated. Notably, the results obtained showed differences in the responses among the tested compounds but also among the studied human populations, indicating the need for developing population-specific products.

Keywords: in vitro model; microbiota; prebiotics; gas production; obesity; functionality; HT29; RTCA; SCFA; bifidobacterial-ITS

1. Introduction

The human intestinal microbiota represents a very complex and diverse microbial ecosystem that remains relatively stable during adult life [1]. However, several intrinsic and environmental factors can disrupt the microbiota composition, causing a microbiota "dysbiosis" [2]. Given the frequent association of dysbiosis with different disease states, the restoration of the microbiota through dietary modulation strategies could be a suitable approach. Among the different microbiota-modulation strategies, the administration of prebiotic supplements has been associated with health benefits to the gastrointestinal tract, cardiometabolism, mental health and mineral absorption, among others [3]. An international experts group has defined a prebiotic as a "substrate that is selectively utilized by host microorganisms conferring a health benefit" [3]. Most often, these substrates are complex carbohydrate moieties that, due to the presence of β-glycosidic bonds, are resistant to digestion during their passage through the gastrointestinal tract, reaching almost intact the large intestine, where they can be metabolized by the intestinal microbiota [4]. However, it is important to underline that resistance to digestion is not enough, and, by definition, prebiotics substrate must be selectively utilized by the microorganisms, with a subsequent promotion of health.

Traditionally, the selective nature of prebiotics have been specifically associated with the genera *Bifidobacterium* and *Lactobacillus* [3]. Nevertheless, during the last two decades, the development of culture-independent technologies has demonstrated that other intestinal microorganisms could be affected as well. Among these, some butyrate producers, such as members of *Clostridium* cluster XIVa and IV, have been found to be favored by prebiotic supplementation, and negative correlations were also found with some pathogenic bacteria [3,5]. These results indicate more global changes associated with prebiotics consumption than just the effects upon bifidobacteria and lactobacilli and underline the importance of considering the total microbiota when screening compounds for their prebiotic properties.

On the other hand, the metabolism of prebiotics leads to the generation, as main end-products, of bacterial fermentation, of short-chain fatty acids (SCFA), among which are acetate, butyrate, propionate and also branched SCFA (BSCFA: iso-valerate and iso-butyrate) and gases hydrogen, methane and carbon dioxide [6]. These compounds are well-known mediators of the microbiota-host interaction, playing an important role in host health [7,8].

To date, the most widely studied prebiotics include fructooligosaccharides (FOS) of variable chain lengths, commercial preparations often containing a mixture of molecules [9], and galactooligosaccharides (GOS), which are being often used in studies focusing on infants [10]. However, the comparative studies on the impact of different prebiotics upon the microbiota in different population groups are still scarce [11]. In this context, the assessment of the impact of prebiotics in microbiota composition and metabolism using in vitro models as a tool for screening the most effective modulatory strategies prior to accomplish expensive and complex human interventions, is valuable [12]. In vitro models, such as fecal cultures, are broadly used. Moreover, complementing such models with gas production assessment can be used for determining the fermentation profile of prebiotics by the gut microbiota of different population groups [13–18]. Among these, obese subjects constitute an interesting target, since some studies have shown that the use of prebiotics is an effective modulatory strategy in obesity [19], and animal studies provide support for a potential beneficial effect on energy homeostasis and weight loss [20]. In mice, an inverse relationship has been established between the level of bifidobacteria and some features of the metabolic alterations linked to obesity (endotoxemia, fat mass and glucose intolerance) [21]. Some of these were confirmed in human studies, such as the increase of bifidobacterial levels after prebiotic treatment, with beneficial systemic consequences for obese individuals [21–24]. However, there is still limited evidence on the in vitro fermentation profiles of different prebiotic compounds by the microbiota of obese humans [25–28] and its comparison with that of normal-weight individuals. Moreover, the characteristics of the intestinal microbiota in the extreme form of obesity (morbid obesity; MOB) (BMI ≥ 40 kg/m^2) is still not completely known [29,30]. The variability in the response of the obese population to prebiotic and probiotic supplementation in weight loss interventions [31] and the lower microbial richness generally characterizing the microbiota of obese subjects [32] points to the gut microbiota as a target for investigation in this field.

Unfortunately, most often the in vitro screenings of prebiotic substrates have failed to consider the microbiota complexity and the potential differences on the basal microbiota composition among different human groups, with few studies selecting the best-suited compounds for defined population groups [33]. The availability of fast, easy and cheap methods, considering the influence of the basal microbiota, for assessing the fermentability and specificity of potential prebiotics would be of help in the selection of prebiotics for specific applications to human groups. In this context, it is well-known that the microbiota of obese subjects is different from that of normal-weight (NW) individuals [30,34,35], suggesting that the impact of different prebiotic compounds in these human groups may also differ, making advisable selecting the best-suited compounds for each of them. Therefore, in this study, we aimed at the evaluation of fermentative dynamics of different prebiotic substrates and the assessment of their impact on the composition and metabolic activity on the intestinal microbiota of lean and extreme obese individuals.

2. Results

2.1. Gas Production and pH Variations during Fermentation

The check of gas production in real-time allowed us to assess the in vitro fermentative dynamics of the different prebiotics. The decreases in pH and the gas formed by fecal microbiotas of NW and MOB individuals after 24 h of incubation in the presence of different carbon sources are shown in Table 1.

Table 1. Cumulative gas produced (mL) and decreases of pH values (Δ pH) after 24 h of incubation in fecal cultures with normal-weight (NW) and morbid obesity (MOB) microbiota. Kinetic parameters were determined using the modified-Gompertz equation, in which "A" represents the upper asymptote (mL) and "μ" is the rate of gas production (mL/h). The values not sharing the same superscript (a, b, c or d) indicate significant differences (p-value < 0.05) among carbon sources for each population group (NW or MOB).

Group	Condition	Δ pH	Cumulative Gas	A	μ	R^2
MOB	Control	$0.10^{a} \pm 0.06$	$5.10^{a} \pm 0.65$	5.111	0.39	0.979
	Glucose	$-1.22^{b} \pm 0.43$	$18.44^{b} \pm 7.08$	19.306	1.205	0.999
	1-kestose	$-1.34^{b} \pm 0.20$	$21.05^{b} \pm 5.09$	22.299	1.267	0.997
	Actilight	$-1.37^{b} \pm 0.18$	$19.80^{b} \pm 5.40$	20.486	1.163	0.998
	GOS	$-1.37^{b} \pm 0.20$	$19.24^{b} \pm 4.89$	20.394	1.293	0.998
	Inulin	$-0.86^{a} \pm 0.17$	$17.68^{b} \pm 8.54$	21.915	0.822	0.997
	P95	$-1.25^{b} \pm 0.21$	$19.82^{b} \pm 5.82$	21.688	1.294	0.997
	Synergy1	$-1.18^{b} \pm 0.10$	$18.48^{b} \pm 8.32$	18.68	1.189	0.997
NW	Control	$0.07^{a} \pm 0.10$	$5.52^{a} \pm 1.84$	5.421	0.364	0.979
	Glucose	$-1.16^{c} \pm 0.31$	$25.62^{c,d} \pm 6.38$	27.399	1.589	0.999
	1-kestose	$-1.28^{c} \pm 0.16$	$26.57^{d} \pm 5.87$	27.52	1.861	0.999
	Actilight	$-1.25^{c} \pm 0.20$	$19.73^{b,c} \pm 6.36$	20.635	1.384	0.998
	GOS	$-1.28^{c} \pm 0.25$	$22.20^{b,c,d} \pm 5.63$	22.641	1.761	0.998
	Inulin	$-0.77^{a,b} \pm 0.23$	$19.10^{b} \pm 6.51$	21.662	0.962	0.997
	P95	$-1.24^{c} \pm 0.16$	$23.67^{b,c,d} \pm 5.76$	23.952	1.721	0.998
	Synergy1	$-1.08^{b,c} \pm 0.13$	$25.19^{b,c,d} \pm 6.25$	26.454	1.456	0.997

Regarding the comparison among prebiotics in fecal cultures of NW people, the 16S rRNA gene profiling evidenced very low abundances of Fusobacteria and Fusobacteriaceae in the negative control, which were practically undetectable after incubation with prebiotics (Supplementary Table S1). qPCR analyses showed that all carbon sources (prebiotics and glucose) caused a significant increase of the absolute levels of *Faecalibacterium* and *Bifidobacterium* at 24 h of incubation but did not significantly affect the population of *Bacteroides* (Figure 2). In addition, the absolute levels of the species *B. longum* increased significantly in fecal cultures of NW people with all substrates tested, and the same was true for *Bifidobacterium adolescentis*, with the exception of the prebiotic inulin (Figure 2B). Moreover, ITS-sequencing evidenced a reduction of *Bifidobacterium breve* relative abundances after incubation of NW fecal cultures with all substrates assayed (Supplementary Table S1).

Focusing on the comparison among fecal cultures of NW and MOB subjects, the absolute quantification (qPCR) of the main bacterial groups evidenced significant differences in the basal microbiota composition (time 0), for most microorganisms analyzed, between MOB and NW individuals (Supplementary Table S2). Moreover, the alpha-diversity (Chao-1 index) determined with the 16S rRNA gene profiling data demonstrated a reduced diversity ($p < 0.05$) in MOB subjects, as compared to NW (132.85 ± 43.39 *vs.* 169.19 ± 21.29, respectively). In the cultures with glucose, the differences in favor of the NW cultures for the microbial groups *Akkermansia*, *Faecalibacterium*, *B. adolescentis* and *Clostridium* cluster XIVa were maintained along incubation, giving rise to significantly higher counts of these microorganisms in NW cultures, as compared to MOB. Among prebiotics, inulin contributed to maintain differences already existing in the basal population for *Bifidobacterium catenulatum* in favor of the fecal cultures of NW individuals. The genus *Akkermansia* was significantly higher at 24 h of incubation in the cultures of NW individuals with respect to MOB in all conditions, these differences being not evident in cultures of the basal microbiota with no carbohydrates added (negative control). However, the most noticeable effect among prebiotics was that promoted by GOS on the *Bacteroides* group in MOB. Interestingly, ITS analysis showed a clear differential pattern of abundances of several bifidobacterial species between fecal cultures of NW and MOB subjects after incubation with different carbon sources (Supplementary Figure S1). The comparison between fecal samples of both human population groups in basal conditions (time 0, before incubation) reflected higher abundances of *Bifidobacterium mongoliense* and *B.crudilactis* in NW individuals (2.78 ± 4.13 % and 14.98 ± 17.04 %, respectively) than in MOB subjects (0.16 ± 0.16 % and 1.96 ± 1.46 %, respectively) and a lower species richness (number of species of bifidobacteria detected by ITS-sequencing) in the NW fecal cultures at time 0 and after 24 h incubation in all conditions assayed (Supplementary Table S3). Moreover, the qPCR quantification of bifidobacterial species confirmed higher ($p < 0.05$) levels of *B. longum* (NW: 5.85 ± 0.48; MOB: 5.60 ± 0.32) and *B. catenulatum* (NW: 6.80 ± 0.70; MOB: 5.34 ± 1.10) in NW subjects. All these results point to substantial differences at the species level in the microbiota of NW and MOB subjects that are conditioning differences among fecal cultures from groups NW and MOB subjects after the incubation with prebiotics.

2.3. Production of Short-chain Fatty Acids

In a similar way as for microbiota composition, differences on the levels of SCFA were determined in fecal cultures depending on the prebiotic tested and the population group considered (Figure 3). Focusing on the comparison among prebiotics, GOS and 1-kestose were the substrates promoting the highest increase of total SCFA at 24 h of incubation in NW and MOB fecal cultures, respectively. 1-kestose gave rise to the highest increase of acetate and propionate among prebiotics tested in fecal cultures of MOB subjects, whereas GOS was the main promoter of acetate production in cultures from NW individuals. Inulin and Synergy1 were the prebiotics with a lower impact on the production of acetate and propionate in fecal cultures of both MOB and NW people. All prebiotics enhanced butyrate production in fecal cultures of NW individuals, with no clear differences among the different compounds (Figure 3B). In contrast, not statistically significant increases of butyrate were evidenced in fecal cultures of MOB subjects (Figure 3A), which could be due to the high variability in the production of this compound by the fecal cultures analyzed. In spite of that indicated above, no significant differences were obtained in the increments of acetic, propionic, butyric, BSCFA and total SCFA between fecal cultures of MOB and NW individuals at 24 h of incubation (Mann Whitney U test, p-value > 0.05).

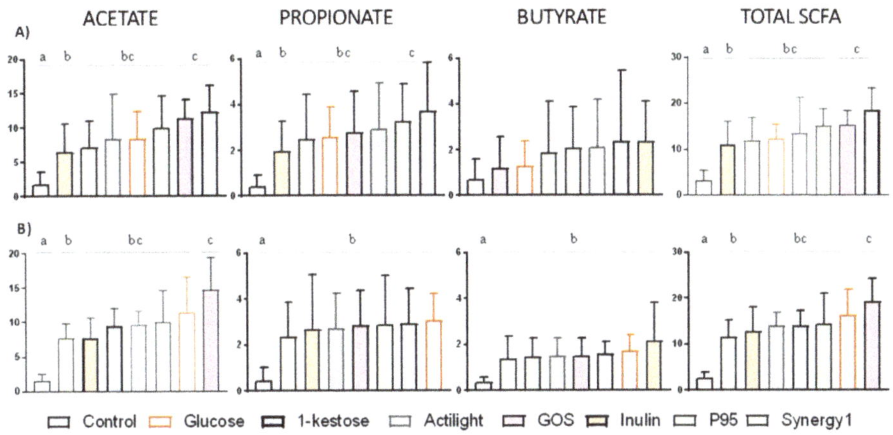

Figure 3. Increments in ascending order, with respect to time 0, in the concentration (mM) of the major short-chain fatty acids (acetic, propionic and butyric) after 24 h of incubation with different carbon sources in fecal cultures from MOB (**A**) and NW (**B**) groups. Differences are shown for each short-chain fatty acid (SCFA); columns that do not share the same letter are significantly different ($p < 0.05$).

2.4. Interaction of the Isolated Microbiotas and Supernatants from the Fecal Cultures with HT29 Cells

Functional differences of fecal supernatants (FS) and isolated microbiotas (IM) collected before and after incubation with representative prebiotics (1-kestose, Actilight, inulin and GOS) were evaluated through an in vitro model using the HT29 intestinal cell line (Figure 4). Regarding FS, a significant decrease of the Area Under the Curve (AUC) values was evidenced after incubation of the HT29 cell line with samples from fecal MOB cultures added with prebiotics (AUC values ranging between −0.07 and 0.04), with respect to the value before the addition of substrates (0.63 ± 0.42). The only exception to this was inulin, for which no significant variations were obtained (Figure 4A). It is interesting to note that AUC determined with FS from cultures of MOB subjects added with prebiotics resemble those obtained with FS of NW subjects. Notably, in these last samples, the AUC before incubation with HT29 were lower than in MOB cultures. These data suggest that the functionality of the FS from MOB subjects could be restored (becoming similar to that of NW subjects) after culturing with some of the prebiotics studied (1-kestose, Actilight and GOS). In co-cultures of HT29 with IM from fecal cultures of MOB and NW groups, all prebiotics promoted significant increases of AUC values, with the exception of inulin in both population cohorts.

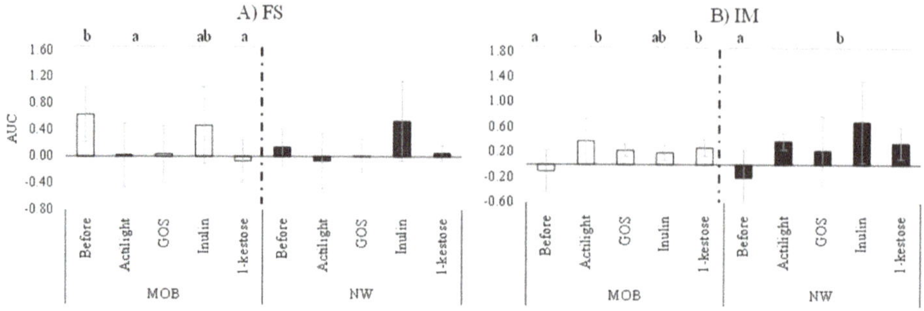

Figure 4. Real-time monitoring the interaction between (**A**) fecal supernatants and (**B**) isolated microbiota obtained before and after incubation with prebiotics and HT29 intestinal epithelial cells. Values (media ± SD) correspond to the Area Under the Curve (AUC) resulting from monitoring the cell index (CI) during 10 h. Significant differences (p-value < 0.05) represent the comparison of results before and after prebiotics addition in each condition.

3. Discussion

The definitive way to prove the impact of prebiotics on gut microbiota and health is through human intervention studies. However, these studies are expensive and time-consuming, and it is advisable to perform an initial screening of the different candidate substrates by using affordable in vitro models that could help to predict in vivo functionality [12]. In a previous work, we described an in vitro model that allowed us to predict the functionality of IM and FS of different human populations groups [37]. Its application in the present work for assessing the impact of several prebiotics in NW and MOB microbiota highlighted potential modulatory benefits on the gut microbiota of some FOS and GOS. Particularly, real-time monitoring the interaction with the HT29 intestinal cell line of FS from fecal cultures of MOB subjects added with different prebiotics evidenced a response that approaches values obtained of FS from NW cultures in the same conditions. It suggests a possible restoration of the unbalanced functionality of the microbiota by some of the prebiotic substrates tested. Nevertheless, inulin preserved a behavior more similar to the initial situation in NW and MOB groups, probably due to the fact that inulin was the prebiotic with less marked effects on microbiota composition and activity, which is consistent with previous reports by other authors [14,16].

Another functional approach tested in this study was the application of a gas monitoring profiling system during fecal cultures of NW and MOB microbiotas with different prebiotics. Although this system has been recently applied to human fecal cultures [18,38], this is the first report on their use to prebiotics evaluation. Differences in gas production ability among prebiotic substrates can be partly explained by possible differences in the fermentability among compounds but also by differential effects of these substrates on the intestinal microbiota. In fact, prebiotics may differ in their ability to modulate the growth and activity of those microorganisms with limited gas production and/or to up-regulating microbial gas-consuming reactions (methanogenesis, sulfate reduction and acetogenesis) [39]. In addition to the potential direct impact on gas-producing bacteria, prebiotics are also known to affect other microorganisms of the intestinal microbiota, such as bifidobacteria, not releasing gas but producing acetate and lactate; these compounds could be involved in cross-feeding mechanisms with gas-producing microorganisms such as *Clostridium* spp. or sulphate-reducing bacteria [40]. Therefore, the final gas formed will depend, not only on the chemical and physical structure of the prebiotic, but on several other factors related with the composition and metabolic activity of the intestinal microbiota. In this context, different fermentation dynamics of fecal cultures from MOB and NW people were demonstrated in this study. A lower rate of gas production was appreciated with the MOB microbiota in all tested carbon sources. This feature suggests a metabolically less active microbiota.

In the context of obesity, an inverse association between body mass index and H_2 and CH_4 gas detection in breath tests has been reported [41,42], which is in good agreement with the lower cumulative gas in MOB cultures obtained herein. Although emphasis has been given to the potential inflammatory or carcinogenic properties of colonic gases, emerging evidence suggests that these gases might have a beneficial effect in colonic health [43]. One of the main gases produced by anaerobic fermentation is H_2 [44], and an imbalance in its metabolism (H_2-producing and H_2-consuming bacteria) might facilitate inflammation [43]. It is intriguing to consider whether the promotion of gas production in MOB subjects could improve antioxidant and antiapoptotic status than contributing to decrease inflammation [45].

The prebiotics used in the present work have previously proven efficacy for modulating the microbiota of the general or specific population groups, both in in vivo and in vitro models [21]. In our case, the most pronounced effects in NW and MOB fecal cultures were obtained with 1-kestose and GOS. In this way, 1-kestose has been found to be metabolized by different intestinal microorganisms [46,47]. Moreover, a recent study reported that the administration to healthy volunteers of 1-kestose (5 g/day) during eight weeks promoted an increase of the intestinal populations of *Faecalibacterium prausnitzii* and *Bifidobacterium* spp. [48]. The potential beneficial changes promoted by GOS and FOS on the intestinal microbiota found in the present work are in good agreement with the widely reported effects of these substrates on the microbiota composition of the general population, assessed in clinical trials and used in vitro models [21,49].

It is important to emphasize that in the present work we have focused on morbid obese subjects, which could make the comparison with other studies difficult since, in the literature available, obese individuals are not often subcategorized. Moreover, the obesity-associated microbiota shifts are still not completely known, and several confounding factors often make this task difficult [32,50]. In this regard, the modulation of the microbiota of obese subjects by prebiotics has produced contradictory results on the genus *Bacteroides*, with increases reported after the administration of α-glucooligosaccharide and arabinogalactan [25,26] and decreases with FOS [23]. Discrepancies on experimental results are likely due to a differential effect depending on the type of prebiotics, the experimental design and the analytical techniques used to determine the composition of the intestinal microbiota by different authors. It is also necessary to point out that in the present work all prebiotics tested were able to up-modulate the levels of the genus *Faecalibacterium*, a microorganism with well-known anti-inflammatory properties [51] and, therefore, interesting in the context of obesity, which is generally accompanied by a low-grade inflammation [52,53]. A bifidogenic effect was observed only for the species *B. longum* after the addition of the prebiotics 1-kestose and GOS to fecal cultures. These observations and the inverse association of this species with serum lipopolysaccharides and endotoxemia [22,54] suggests that some of these prebiotics, specially 1-kestose and GOS, could be good candidates to modulate the microbiota in the context of obesity.

A broad prebiotic enhancement of absolute levels of the genus *Bifidobacterium* was seen in fecal cultures of the NW group, in contrast to the absence of effect in cultures from MOB subjects. In order to expand the study of the prebiotic impact on bifidobacteria, we performed an ITS-region profiling. To our knowledge, this is the first study going in detail on bifidobacteria species variation in NW and MOB fecal cultures. Even though, regarding the influence of prebiotics, the ITS-profiling only stood out as a decrease after the supplementation with carbon sources of *B. animalis* subsp. *lactis* and *B. crudilactis* in NW cultures and of *B. breve* in cultures of MOB subjects, a clear distinction between MOB and NW microbiotas was still evident. Firstly, we noticed a greater richness of bifidobacteria species in MOB microbiota. Additionally, differentially higher levels of *B. monogoliense* and *B. crudilactis* were present in the microbiota of the NW group, which may be explained by a higher consumption of dairy fermented foods [55,56].

The production of SCFA in fecal cultures is in good agreement with variations in pH, gas production and impact caused on the microbiota composition by the different substrates. Thus, differences in the production of propionate between fecal cultures of MOB and NW promoted by

1-kestose could be directly related with its higher capacity (together with GOS) to differentially promote the increase of *Bacteroides* (the main propionate producer in the human colon and a producer of acetate) in fecal cultures of MOB individuals. Interestingly the intestinal microbiota, mainly that from MOB subjects, showed a better ability to produce SCFA with some of the prebiotics tested than from glucose. This could be related with the enrichment of the microbiome in some metabolic pathways involved in the initial steps of breaking down indigestible dietary polysaccharides, including pathways for starch/sucrose metabolism, galactose metabolism and butanoate metabolism, previously reported in the obese population [57].

It is also worth mentioning that the basal differences in the metabolic activity and microbiota composition, added to the specific effects of prebiotics found in the present work depending on the donor population, highlights the importance of selecting the best-suited compounds for the desired target population and the potential limitations of extrapolating conclusions from one population group to another.

To summarize, the present study provides evidence about the in vitro fermentation profiles of different prebiotics by microbiotas from NW and MOB individuals. In our study, we did not perform total metagenome analyses, but instead, we performed a microbiota characterization by 16S rRNA gene profiling and complemented it with two functional tests, gas production and interaction with an intestinal line, allowing the assessment of both microbiota compositions and some functional properties of these microbiotas. By monitoring gas production along fermentation, we found a higher capacity of gas production by fecal cultures of NW subjects than from MOB individuals. 1-kestose was the fructan showing the highest gas accumulation and largest microbiota modulation activity in MOB subjects, together with GOS, underlining the utilization of this compound by the intestinal microbiota of these individuals. The fecal cultures incubated with some of the prebiotics tested also showed differences at the functional level when assessed upon epithelial cell lines. Even though the in vitro models present inherent limitations and a difficult interpretation with respect to physiological conditions, the application of in vitro models to the analysis of microbiota composition and functionality could allow the selection of the most suitable prebiotics for different populations prior to their assessment in human intervention studies. Moreover, our results underline the interest of further exploring the prebiotic role of 1-kestose due to their modulatory capacity of the microbiota composition and activity in MOB subjects.

4. Materials and Methods

4.1. Prebiotics and Carbon Sources

Two types of prebiotics, based on their monosaccharide's composition (fructose or galactose), were evaluated. Among FOS, the trisaccharide 1-kestose (>99%; β Food Science Co. Ltd., Chita, Japan); Actilight® (DP = 3–5, enzymatically produced, 95% purity; Beghin Meiji, Lila, France); P95 (DP = 2–8, obtained by hydrolysis, 95% purity; Beneo-Orafti, Oreye, Belgium); Synergy1 (FOS plus inulin in proportion 1:1, 92% purity; Beneo-Orafti, Oreye, Belgium) and the long-chain fructan–inulin (DP>36; Sigma-Aldrich, Madrid, Spain) extracted from dahlia tubers were included in the study. A GOS from the brand Bimuno Daily (Clasado, Shinfield, England) with 79.70% (*w/w*) of purity was also evaluated. Glucose (Fluka Analytical, Madrid, Spain) was also used as a nonprebiotic universal carbon source. Sterilization of all substrates was carried out by filtration through a pore size of 0.45 μm, except for the inulin, which was autoclaved.

4.2. Volunteers and Fecal Sample Collection

Fecal samples were obtained from nine healthy NW adults (77.78% women; BMI <25 kg/m^2) and nine MOB volunteers (75% women; BMI ≥40 kg/m^2) recruited at the Digestive and Endocrinology Services, respectively, of Asturias Central University Hospital (HUCA, Asturias, Spain). The mean age of the volunteers was 38 ± 9 and 45 ± 10 for NW and MOB subjects, respectively. All participants have followed an unrestricted diet and have not taken antibiotics during the previous six months. The study was approved by the Regional Ethical Committee of Asturias Public Health Service (Ref. N° 120/13, 20 November 2013), and an informed written consent was obtained from each volunteer. Samples were collected and immediately introduced into anaerobic jars (Anaerocult A System; Merck, Darmstadt, Germany) for transportation to the laboratory within 1 h after collection. A 1/10 (*w*/*v*) dilution was made in prereduced PBS solution and homogenized in a Lab-Blender 400 stomacher (Seward Medical, London, UK) for 5 min.

4.3. Fecal Batch Culture Fermentation

Independent batch fermentations were performed at pH-uncontrolled in a carbohydrate-free basal medium (CFBM) [58], with feces from different human donors and different carbohydrates added. Briefly, CFBM was prepared and reduced overnight in an anaerobic chamber MG500 (Don Whitley Scientific, West Yorkshire, UK) one day before the sample processing. On the day of the assay, fresh fecal samples, collected and homogenized as stated above, were added (10% *v*/*v*) to the reduced CFBM and then were distributed into 100 mL bottles of the ANKOMRF system (ANKOM Technology, USA). An overnight incubation in anaerobic conditions was performed at 37 °C prior to the addition of carbon sources in order to allow microbiota to stabilize in the culture medium.

A set of independent fermentations were performed with feces from each donor, using as carbon sources either inulin-type fructans (1-kestose, Actilight, P95, Synergy1 and inulin), GOS or glucose (nonprebiotic positive control) at a final concentration of 0.3% (*v*/*v*). A bottle with no carbon source added was used as a control. Fermentations were carried out under anaerobic conditions at 37 °C during 24 h. The pH of cultures was determined with a pHmeter (SensION + PH3, Hach; Barcelona, Spain) and was considered as an indicator of the progression of fermentation. Samples (1 mL) were taken in duplicate before incubation (time 0) and considered as basal conditions and at 24 h of incubation. Samples were centrifuged at full speed for 10 min, and supernatants and pellets were stored separately at −20 °C until their use for microbiota and metabolite analyses.

4.4. Gas Monitorization

The cumulative gas produced along the different fermentations was monitored in real-time by using the ANKOM RF system. The system provides increases in pressure (psi), which can be converted to mL of gas produced using the ideal gas equation.

$$V = Vj \cdot Ppsi \cdot 0.068004084 \quad (1)$$

where V = gas volume at 39 °C in mL, Vj = headspace of digestion jar (glass bottle) in mL and Ppsi = cumulative pressure recorded by Gas Monitor System software.

The data of gas production were fitted to a modified-Gompertz equation, a model frequently used to fit data of bacterial, plant growth, tumor proliferation and gas production [59], by using the formula:

$$y = A \times \exp\left\{-\exp\left[\frac{\mu \times e}{A}(\lambda - t) + 1\right]\right\} \quad (2)$$

in which variable "A" represents the upper asymptote (mL), "μ" is the rate of gas production (mL/h) and "λ" is the time lag before the exponential phase (h).

4.5. Microbiota Composition and SCFA Quantification

DNA was extracted from the pellets harvested using the QIAamp DNA Stool Mini kit (Qiagen GmbH; Hilden, Germany), as previously described [60], and the isolated DNA was stored at −20 °C until use for qPCR analyses and 16S ribosomal and intergenic ribosomal transcriber spaces (ITS).

4.5.1. qPCR Analyses

Absolute levels of some relevant intestinal microbial groups (*Bacteroides–Prevotella–Porphyromonas* group, *Lactobacillus* group, *Akkermansia*, *Clostridium* cluster XIVa, *Bifidobacterium* and *Faecalibacterium* genus), as well as total bacteria, were determined at 0 and 24 h of fermentation by qPCR using previously described primers and conditions [31,61]. Variations in the levels of the species *B. longum*, *B. catenulatum* and *B. adolescentis* were assessed as described elsewhere [22].

4.5.2. 16S rRNA Gene Based Microbiota Profiling

Purified DNA was used as a template for amplification of partial 16S rRNA gene sequences by PCR using the primers and conditions described by Milani and coworkers [62]. The obtained amplicons were then sequenced by using the MiSeq (Illumina) platform at GenProbio srl (Italy). The individual reads obtained were filtered, trimmed and processed [63]. 16S rRNA operational taxonomic units were defined at $\geq$ 97% sequence homology using the UCLUST tool developed by Edgar [64]. All reads were classified to the lowest possible taxonomic rank using QIIME and a reference dataset from the SILVA database [65].

4.5.3. ITS Region-Based Profiling of Bifidobacterial Microbiota

To gain further insight into the fecal bifidobacterial populations present in the samples and how the different prebiotics affected them, the 16S–23S internal transcriber spaces of the ribosomal DNA (ITS region) was amplified by PCR using the primer pair Probio_bif_uni/Probio_bif_rev. and further sequenced as indicated in the previous section. An improved bifidobacterial ITS database, containing all publicly available bifidobacterial genomes and a custom bioinformatics script [66], were used. Relative abundance of bifidobacterial composition filtered by a minimum presence ($\geq$1%) of each species in all databases were represented by a heatmap following instructions described elsewhere [67], centered and scaled by the "scale" function in RStudio version 1.2.5001.

4.5.4. SCFA Analyses

The analysis of SCFA was performed by gas chromatography (GC) in the fecal culture supernatants in order to determine the molar concentrations of three main compounds: acetate, propionate and butyrate. The remaining BSFA, namely isobutyrate and isovalerate, were also quantified and summed up for further analysis. Briefly, culture supernatants collected during the fecal fermentation (0.250 mL) were mixed with 0.3 mL methanol, 0.05 mL internal standard solution (2-ethylbutyric 1.05 mg/mL) and 0.05 mL of 20% formic acid. This mixture was centrifuged, and the supernatant was used for quantification of SCFA by GC, as described previously [37]. Samples were analyzed in triplicate. Increments in molar concentrations of the main SCFA and BSCFA with respect to the time 0 were calculated for each fermentation batch with the different carbon sources.

4.6. Monitoring the Interaction of Isolated Microbiotas (IM) and Fecal Supernatants (FS) with HT29 Cells

Briefly, the behavior of HT29 cells monolayers in a confluent state was assessed upon exposure to IM and FS collected after incubation of fecal samples with different carbon sources by using a real-time cell analyzer (RTCA-DP) xCelligence apparatus (ACEA Bioscience Inc., San Diego, CA, USA). The culture conditions and the maintenance of the intestinal epithelial cell line HT29 (ECACC 91072201) are detailed in a previous work where the functional model was developed [37]. IM were purified from 10-fold concentrated fecal cultures by using a density gradient method previously described [68]. Purified microbiotas were inactivated by UV light exposure (15 W; Selecta, Barcelona, Spain) and adjusted to 1×10^8 bacteria/mL using a Neubauer-improved camera (Blau Brand, Germany).

For the functional assessment of IM and FS, HT29 monolayers in a confluent state were coincubated with 6.5×10^7 bacteria/mL of UV-inactivated purified microbiotas in McCoy´s medium (MM) (bacteria to cell ratio 10:1). In the case of fecal supernatants, the behavior of HT29 cells monolayers was assessed with filtered fecal supernatants (pH adjusted to 7.55 ± 0.05) and diluted 40% with MM. Additionally, a negative control consisting of MM without bacteria or fecal supernatants was included in each experiment. Each sample was tested in duplicate using two independent E-plates. The monitoring was followed for every 10 min under standard incubation conditions. CI values recorded were normalized by the time of the sample addition and by the control sample, as previously described [69]. For statistical comparison purposes, the "Area Under the Curve" (AUC), representing the CI values along 10 h of incubation for each sample, was calculated as explained in [37].

4.7. Statistics Analyses

Unless otherwise specified, all experimental data are reported as mean ± standard deviation. Statistical analysis of results was performed using the software SPSS v.25 (SPSS Inc., Chicago, USA). Data were compared for the effect caused on the parameters analyzed by the addition of different carbon sources in fecal cultures from each population cohort (NW and MOB) at the end of fermentation (24 h). For variables with a normal distribution (Shapiro-Wilk test) and homoscedasticity (Levene test), one-way ANOVA followed by post hoc LSD comparison were conducted (Supplementary Table S4). In the remaining cases (variables showing non-normal distribution), a Kruskal-Wallis test followed by a post hoc Dunn's test of pairwise comparisons were applied when necessary (Supplementary Table S4). A significant p-value of 0.05 was used for the interpretation of results. For two-group comparisons between MOB and NW (at time 0 and after incubation with all conditions), a two-tailed Student's t-test or Mann-Whitney U test was conducted for the evaluation of data by parametric or nonparametric contrast, respectively.

Supplementary Materials: Supplementary materials can be found at http://www.mdpi.com/1422-0067/21/3/906/s1.

Author Contributions: Conceptualization, A.E., T.T., K.H., C.G.R.-G. and M.G.; volunteers recruitment, A.S. and C.M.-F.; methodology, A.N., N.S., S.A., P.R-M., L.M., A.S. and C.M.-F.; formal analysis, A.N., N.S., S.A., L.M., M.V., C.G.R.-G. and M.G.; writing—original draft preparation, A.N., C.G.R.-G. and M.G.; writing—review and editing, A.N., N.S., S.A., P.R.-M., L.M., A.S., C.M.-F., M.V., T.T., K.H., A.E., C.G.R.-G. and M.G. All authors have read and agreed to the published version of the manuscript.

Funding: The present work was financed by the project AGL2013-43770-R (receiving funds from the Ministry of Economy and Competitiveness—MINECO and the European Union FEDER funds) and by a contract with the company β- Food Sciences (Chita, Japan). Alicja M Nogacka was the recipient of a predoctoral contract granted by MINECO (reference BES-2014-068796), and S.A. received a postdoctoral Juan de la Cierva contract (Ministry of Science, Innovation and Universities, Spain, Ref. IJCI-2017-32156). Nuria Salazar is the recipient of a postdoctoral contract awarded by the Biosanitary Research Foundation in Asturias (FINBA, Spain).

Acknowledgments: We wish to thank the volunteers participating in the study and the Scientific-Technical Service of IPLA-CSIC for carrying out the SCFA determination.

Conflicts of Interest: Takumi Tochio and Katsuaki Hirano are employees at the company β- Food Sciences.

Abbreviations

AUC	Area Under the Curve
CI	Cell Index
FS	Fecal Supernatant
GC	Gas Chromatography
IM	Isolated Microbiota
ITS	16S-23S internal transcriber spaces
LSD	Least Significant Difference post-hoc
MM	McCoy´s medium
MOB	Morbid Obesity
NW	Normal-Weight Adults
qPCR	Real-time polymerase chain reaction
rRNA	Ribosomal ribonucleic acid
RTCA	Real-time Cell Analyzer

References

1. Lloyd-Price, J.; Abu-Ali, G.; Huttenhower, C. The healthy human microbiome. *Genome Med.* **2016**, *8*, 51. [CrossRef]
2. Levy, M.; Kolodziejczyk, A.A.; Thaiss, C.A.; Elinav, E. Dysbiosis and the immune system. *Nat. Rev. Immunol.* **2017**, *17*, 219–232. [CrossRef] [PubMed]
3. Gibson, G.R.; Hutkins, R.; Sanders, M.E.; Prescott, S.L.; Reimer, R.A.; Salminen, S.J.; Scott, K.; Stanton, C.; Swanson, K.S.; Cani, P.D.; et al. Expert consensus document: The International Scientific Association for Probiotics and Prebiotics (ISAPP) consensus statement on the definition and scope of prebiotics. *Nat. Rev. Gastroenterol. Hepatol.* **2017**, *14*, 491–502. [CrossRef] [PubMed]
4. Vernazza, C.L.; Rabiu, A.R.; Gibson, G.R. Human Colonic Microbiology and the Role of Dietary Intervention: Introduction to Prebiotics. In *Prebiotics: Development & Application*; Gibson, G.R., Rastall, R.A., Eds.; John Wiley & Sons, Ltd.: Chichester, UK, 2006; ISBN 13-978-0-470-02313-6.
5. DeVuyst, L.; Leroy, F. Cross-feeding between bifidobacteria and butyrate-producing colon bacteria explains bifdobacterial competitiveness, butyrate production, and gas production. *Int. J. Food Microbiol.* **2011**, *149*, 73–80. [CrossRef] [PubMed]
6. Cummings, J.H.; Macfarlane, G.T. Role of intestinal bacteria in nutrient metabolism. *Clin. Nutr.* **1997**, *16*, 3–11. [CrossRef]
7. Ríos-Covián, D.; Ruas-Madiedo, P.; Margolles, A.; Gueimonde, M.; de los Reyes-Gavilán, C.G.; Salazar, N. Intestinal Short Chain Fatty Acids and their Link with Diet and Human Health. *Front. Microbiol.* **2016**, *7*, 185. [CrossRef]
8. Rowland, I.; Gibson, G.; Heinken, A.; Scott, K.; Swann, J.; Thiele, I.; Tuohy, K. Gut microbiota functions: Metabolism of nutrients and other food components. *Eur. J. Nutr.* **2018**, *57*, 1–24. [CrossRef]
9. Dominguez, A.L.; Rodrigues, L.R.; Lima, N.M.; Teixeira, J.A. An Overview of the Recent Developments on Fructooligosaccharide Production and Applications. *Food Bioprocess Technol.* **2014**, *7*, 324–337. [CrossRef]
10. Wilson, B.; Whelan, K. Prebiotic inulin-type fructans and galacto-oligosaccharides: Definition, specificity, function, and application in gastrointestinal disorders. *J. Gastroenterol. Hepatol.* **2017**, *32*, 64–68. [CrossRef]
11. Sanders, M.E.; Merenstein, D.J.; Reid, G.; Gibson, G.R.; Rastall, R.A. Probiotics and prebiotics in intestinal health and disease: From biology to the clinic. *Nat. Rev. Gastroenterol. Hepatol.* **2019**, *16*, 605–616. [CrossRef]
12. Williams, C.F.; Walton, G.E.; Jiang, L.; Plummer, S.; Garaiova, I.; Gibson, G.R. Comparative Analysis of Intestinal Tract Models. *Annu. Rev. Food Sci. Technol.* **2015**, *6*, 329–350. [CrossRef] [PubMed]
13. Ghoddusi, H.B.; Grandison, M.A.; Grandison, A.S.; Tuohy, K.M. In vitro study on gas generation and prebiotic effects of some carbohydrates and their mixtures. *Anaerobe* **2007**, *13*, 193–199. [CrossRef] [PubMed]
14. Beards, E.; Tuohy, K.; Gibson, G. Bacterial, SCFA and gas profiles of a range of food ingredients following in vitro fermentation by human colonic microbiota. *Anaerobe* **2010**, *16*, 420–425. [CrossRef] [PubMed]
15. Koecher, K.J.; Noack, J.A.; Timm, D.A.; Klosterbuer, A.S.; Thomas, W.; Slavin, J.L. Estimation and Interpretation of Fermentation in the Gut: Coupling Results from a 24 h Batch in Vitro System with Fecal Measurements from a Human Intervention Feeding Study Using Fructo-oligosaccharides, Inulin, Gum Acacia, and Pea Fiber. *J. Agric. Food Chem.* **2014**, *62*, 1332–1337. [CrossRef] [PubMed]
16. Hernot, D.C.; Boileau, T.W.; Bauer, L.L.; Middelbos, I.S.; Murphy, M.R.; Swanson, K.S.; Fahey, G.C. In Vitro Fermentation Profiles, Gas Production Rates, and Microbiota Modulation as Affected by Certain Fructans, Galactooligosaccharides, and Polydextrose. *J. Agric. Food Chem.* **2009**, *57*, 1354–1361. [CrossRef] [PubMed]
17. VandenAbbeele, P.; Venema, K.; Van de Wiele, T.; Verstraete, W.; Possemiers, S. Different Human Gut Models Reveal the Distinct Fermentation Patterns of Arabinoxylan versus Inulin. *J. Agric. Food Chem.* **2013**, *61*, 9819–9827. [CrossRef]
18. Yao, C.K.; Rotbart, A.; Ou, J.Z.; Kalantar-Zadeh, K.; Muir, J.G.; Gibson, P.R. Modulation of colonic hydrogen sulfide production by diet and mesalazine utilizing a novel gas-profiling technology. *Gut Microbes* **2018**, *9*, 1–13. [CrossRef]
19. Dahiya, D.K.; Puniya, M.; Shandilya, U.K.; Dhewa, T.; Kumar, N.; Kumar, S.; Puniya, A.K.; Shukla, P. Gut Microbiota Modulation and Its Relationship with Obesity Using Prebiotic Fibers and Probiotics: A Review. *Front. Microbiol.* **2017**, *8*, 563. [CrossRef]
20. Delgado, G.T.C.; Tamashiro, W.M.D.S.C. Role of prebiotics in regulation of microbiota and prevention of obesity. *Food Res. Int.* **2018**, *113*, 183–188. [CrossRef]

21. Roberfroid, M.; Gibson, G.R.; Hoyles, L.; McCartney, A.L.; Rastall, R.; Rowland, I.; Wolvers, D.; Watzl, B.; Szajewska, H.; Stahl, B.; et al. Prebiotic effects: Metabolic and health benefits. *Br. J. Nutr.* **2010**, *104*, S1–S63. [CrossRef]
22. Salazar, N.; Dewulf, E.M.; Neyrinck, A.M.; Bindels, L.B.; Cani, P.D.; Mahillon, J.; de Vos, W.M.; Thissen, J.-P.; Gueimonde, M.; de los Reyes-Gavilán, C.G.; et al. Inulin-type fructans modulate intestinal *Bifidobacterium* species populations and decrease fecal short-chain fatty acids in obese women. *Clin. Nutr.* **2015**, *34*, 501–507. [CrossRef] [PubMed]
23. Dewulf, E.M.; Cani, P.D.; Claus, S.P.; Fuentes, S.; Puylaert, P.G.; Neyrinck, A.M.; Bindels, L.B.; de Vos, W.M.; Gibson, G.R.; Thissen, J.-P.; et al. Insight into the prebiotic concept: Lessons from an exploratory, double blind intervention study with inulin-type fructans in obese women. *Gut* **2013**, *62*, 1112–1121. [CrossRef]
24. Parnell, J.A.; Klancic, T.; Reimer, R.A. Oligofructose decreases serum lipopolysaccharide and plasminogen activator inhibitor-1 in adults with overweight/obesity. *Obesity* **2017**, *25*, 510–513. [CrossRef] [PubMed]
25. Aguirre, M.; Bussolo de Souza, C.; Venema, K. The Gut Microbiota from Lean and Obese Subjects Contribute Differently to the Fermentation of Arabinogalactan and Inulin. *PLoS ONE* **2016**, *11*, e0159236. [CrossRef]
26. Sarbini, S.R.; Kolida, S.; Gibson, G.R.; Rastall, R.A. In vitro fermentation of commercial α-gluco-oligosaccharide by faecal microbiota from lean and obese human subjects. *Br. J. Nutr.* **2013**, *109*, 1980–1989. [CrossRef]
27. Sarbini, S.R.; Kolida, S.; Deaville, E.R.; Gibson, G.R.; Rastall, R.A. Potential of novel dextran oligosaccharides as prebiotics for obesity management through in vitro experimentation. *Br. J. Nutr.* **2014**, *112*, 1303–1314. [CrossRef]
28. Aguirre, M.; Jonkers, D.M.A.E.; Troost, F.J.; Roeselers, G.; Venema, K. In Vitro Characterization of the Impact of Different Substrates on Metabolite Production, Energy Extraction and Composition of Gut Microbiota from Lean and Obese Subjects. *PLoS ONE* **2014**, *9*, e113864. [CrossRef]
29. Cani, P.D. Severe obesity and gut microbiota: does bariatric surgery really reset the system? *Gut* **2019**, *68*, 5–6. [CrossRef]
30. Castaner, O.; Goday, A.; Park, Y.-M.; Lee, S.-H.; Magkos, F.; Shiow, S.-A.T.E.; Schröder, H. The Gut Microbiome Profile in Obesity: A Systematic Review. *Int. J. Endocrinol.* **2018**, *2018*, 4095789. [CrossRef]
31. Dao, M.C.; Clément, K. Gut microbiota and obesity: Concepts relevant to clinical care. *Eur. J. Intern. Med.* **2018**, *48*, 18–24. [CrossRef]
32. Aron-Wisnewsky, J.; Prifti, E.; Belda, E.; Ichou, F.; Kayser, B.D.; Dao, M.C.; Verger, E.O.; Hedjazi, L.; Bouillot, J.-L.; Chevallier, J.-M.; et al. Major microbiota dysbiosis in severe obesity: Fate after bariatric surgery. *Gut* **2019**, *68*, 70–82. [CrossRef] [PubMed]
33. Valdés, L.; Salazar, N.; González, S.; Arboleya, S.; Ríos-Covián, D.; Genovés, S.; Ramón, D.; de los Reyes-Gavilán, C.G.; Ruas-Madiedo, P.; Gueimonde, M. Selection of potential probiotic bifidobacteria and prebiotics for elderly by using in vitro faecal batch cultures. *Eur. Food Res. Technol.* **2017**, *243*, 157–165. [CrossRef]
34. Ley, R.E.; Backhed, F.; Turnbaugh, P.; Lozupone, C.A.; Knight, R.D.; Gordon, J.I. Obesity alters gut microbial ecology. *Proc. Natl. Acad. Sci. USA* **2005**, *102*, 11070–11075. [CrossRef] [PubMed]
35. Gomes, A.C.; Hoffmann, C.; Mota, J.F. The human gut microbiota: Metabolism and perspective in obesity. *Gut Microbes* **2018**, *9*, 308–325. [CrossRef] [PubMed]
36. McDonald, D.; Hyde, E.; Debelius, J.W.; Morton, J.T.; Gonzalez, A.; Ackermann, G.; Aksenov, A.A.; Behsaz, B.; Brennan, C.; Chen, Y.; et al. American Gut: An Open Platform for Citizen Science Microbiome Research. *mSystems* **2018**, *3*, e00031-18. [CrossRef]
37. Nogacka, A.M.; Ruas-Madiedo, P.; Gómez, E.; Solís, G.; Fernández, N.; Suárez, M.; Suárez, A.; Salazar, N.; de los Reyes-Gavilán, C.G.; Gueimonde, M. Real-time monitoring of HT29 epithelial cells as an in vitro model for assessing functional differences among intestinal microbiotas from different human population groups. *J. Microbiol. Methods* **2018**, *152*, 210–216. [CrossRef]
38. Rotbart, A.; Moate, P.J.; Yao, C.K.; Ou, J.Z.; Kalantar-Zadeh, K. A novel mathematical model for the dynamic assessment of gas composition and production in closed or vented fermentation systems. *Sensors Actuators B Chem.* **2018**, *254*, 354–362. [CrossRef]
39. Mego, M.; Manichanh, C.; Accarino, A.; Campos, D.; Pozuelo, M.; Varela, E.; Vulevic, J.; Tzortzis, G.; Gibson, G.; Guarner, F.; et al. Metabolic adaptation of colonic microbiota to galactooligosaccharides: A proof-of-concept-study. *Aliment. Pharmacol. Ther.* **2017**, *45*, 670–680. [CrossRef]

40. Sarbini, S.R.; Rastall, R.A. Prebiotics: Metabolism, Structure, and Function. *Funct. Food Rev.* **2011**, *3*, 93–106. [CrossRef]
41. Jung, S.E.; Joo, N.S.; Han, K.S.; Kim, K.N. Obesity is inversely related to hydrogen-producing small intestinal bacterial overgrowth in non-constipation irritable bowel syndrome. *J. Korean Med. Sci.* **2017**, *32*, 948–953. [CrossRef]
42. Wilder-Smith, C.H.; Olesen, S.S.; Materna, A.; Drewes, A.M. Breath methane concentrations and markers of obesity in patients with functional gastrointestinal disorders. *United Eur. Gastroenterol. J.* **2018**, *6*, 595–603. [CrossRef] [PubMed]
43. Carbonero, F.; Benefiel, A.C.; Gaskins, H.R. Contributions of the microbial hydrogen economy to colonic homeostasis. *Nat. Rev. Gastroenterol. Hepatol.* **2012**, *9*, 504–518. [CrossRef] [PubMed]
44. Macfarlane, G.T.; Gibson, G.R. Carbohydrate Fermentation, Energy Transduction and Gas Metabolism in the Human Large Intestine. In *Gastrointestinal Microbiology*; Mackie, R.I., White, B.A., Eds.; Chapman & Hall: New York, NY, USA, 1997; pp. 269–318. [CrossRef]
45. Monteiro-Sepulveda, M.; Touch, S.; Mendes-Sá, C.; André, S.; Poitou, C.; Allatif, O.; Cotillard, A.; Fohrer-Ting, H.; Hubert, E.-L.; Remark, R.; et al. Jejunal T Cell Inflammation in Human Obesity Correlates with Decreased Enterocyte Insulin Signaling. *Cell Metab.* **2015**, *22*, 113–124. [CrossRef] [PubMed]
46. Endo, A.; Nakamura, S.; Konishi, K.; Nakagawa, J.; Tochio, T. Variations in prebiotic oligosaccharide fermentation by intestinal lactic acid bacteria. *Int. J. Food Sci. Nutr.* **2016**, *67*, 125–132. [CrossRef]
47. Ose, R.; Hirano, K.; Maeno, S.; Nakagawa, J.; Salminen, S.; Tochio, T.; Endo, A. The ability of human intestinal anaerobes to metabolize different oligosaccharides: Novel means for microbiota modulation? *Anaerobe* **2018**, *51*, 110–119. [CrossRef]
48. Tochio, T.; Kadota, Y.; Tanaka, T.; Koga, Y. 1-Kestose, the Smallest Fructooligosaccharide Component, Which Efficiently Stimulates *Faecalibacterium prausnitzii* as Well as Bifidobacteria in Humans. *Foods* **2018**, *7*, 140. [CrossRef]
49. Rajendran, C.K.S.R.; Okolie, C.L.; Udenigwe, C.C.; Mason, B. Structural features underlying prebiotic activity of conventional and potential prebiotic oligosaccharides in food and health. *J. Food Biochem.* **2017**, *41*, e12389. [CrossRef]
50. Duvallet, C.; Gibbons, S.M.; Gurry, T.; Irizarry, R.A.; Alm, E.J. Meta-analysis of gut microbiome studies identifies disease-specific and shared responses. *Nat. Commun.* **2017**, *8*, 1784. [CrossRef]
51. Hiippala, K.; Jouhten, H.; Ronkainen, A.; Hartikainen, A.; Kainulainen, V.; Jalanka, J.; Satokari, R. The Potential of Gut Commensals in Reinforcing Intestinal Barrier Function and Alleviating Inflammation. *Nutrients* **2018**, *10*, 988. [CrossRef]
52. Sokol, H.; Pigneur, B.; Watterlot, L.; Lakhdari, O.; Bermúdez-Humarán, L.G.; Gratadoux, J.J.; Blugeon, S.; Bridonneau, C.; Furet, J.P.; Corthier, G.; et al. *Faecalibacterium prausnitzii* is an anti-inflammatory commensal bacterium identified by gut microbiota analysis of Crohn disease patients. *Proc. Natl. Acad. Sci. USA* **2008**, *105*, 16731–16736. [CrossRef]
53. Machiels, K.; Joossens, M.; Sabino, J.; De Preter, V.; Arijs, I.; Eeckhaut, V.; Ballet, V.; Claes, K.; Van Immerseel, F.; Verbeke, K.; et al. A decrease of the butyrate-producing species *Roseburia hominis* and *Faecalibacterium prausnitzii* defines dysbiosis in patients with ulcerative colitis. *Gut* **2014**, *63*, 1275–1283. [CrossRef] [PubMed]
54. Cani, P.D.; Neyrinck, A.M.; Fava, F.; Knauf, C.; Burcelin, R.G.; Tuohy, K.M.; Gibson, G.R.; Delzenne, N.M. Selective increases of bifidobacteria in gut microflora improve high-fat-diet-induced diabetes in mice through a mechanism associated with endotoxaemia. *Diabetologia* **2007**, *50*, 2374–2383. [CrossRef] [PubMed]
55. Delcenserie, V.; Taminiau, B.; Gavini, F.; de Schaetzen, M.-A.; Cleenwerck, I.; Theves, M.; Mahieu, M.; Daube, G. Detection and characterization of *Bifidobacterium crudilactis* and *B. mongoliense* able to grow during the manufacturing process of French raw milk cheeses. *BMC Microbiol.* **2013**, *13*, 239. [CrossRef] [PubMed]
56. Milani, C.; Duranti, S.; Napoli, S.; Alessandri, G.; Mancabelli, L.; Anzalone, R.; Longhi, G.; Viappiani, A.; Mangifesta, M.; Lugli, G.A.; et al. Colonization of the human gut by bovine bacteria present in Parmesan cheese. *Nat. Commun.* **2019**, *10*, 1286. [CrossRef]
57. Turnbaugh, P.J.; Ley, R.E.; Mahowald, M.A.; Magrini, V.; Mardis, E.R.; Gordon, J.I. An obesity-associated gut microbiome with increased capacity for energy harvest. *Nature* **2006**, *444*, 1027–1031. [CrossRef] [PubMed]
58. Al-Tamimi, M.A.H.M.; Palframan, R.J.; Cooper, J.M.; Gibson, G.R.; Rastall, R.A. In vitro fermentation of sugar beet arabinan and arabino-oligosaccharides by the human gut microflora. *J. Appl. Microbiol.* **2006**, *100*, 407–414. [CrossRef] [PubMed]

59. Ware, A.; Power, N. Modelling methane production kinetics of complex poultry slaughterhouse wastes using sigmoidal growth functions. *Renew. Energy* **2017**, *104*, 50–59. [CrossRef]
60. Arboleya, S.; Binetti, A.; Salazar, N.; Fernández, N.; Solís, G.; Hernández-Barranco, A.; Margolles, A.; de los Reyes-Gavilán, C.G.; Gueimonde, M. Establishment and development of intestinal microbiota in preterm neonates. *FEMS Microbiol. Ecol.* **2012**, *79*, 763–772. [CrossRef]
61. Bartosch, S.; Fite, A.; Macfarlane, G.T.; McMurdo, M.E.T. Characterization of Bacterial Communities in Feces from Healthy Elderly Volunteers and Hospitalized Elderly Patients by Using Real-Time PCR and Effects of Antibiotic Treatment on the Fecal Microbiota. *Appl. Environ. Microbiol.* **2004**, *70*, 3575–3581. [CrossRef]
62. Milani, C.; Hevia, A.; Foroni, E.; Duranti, S.; Turroni, F.; Gueimonde, M.; van Sinderen, D.; Lugli, A.; Sanchez, B.; Martı, R.; et al. Assessing the fecal microbiota: an optimized ion torrent 16S rRNA gene-based analysis protocol. *PLoS ONE* **2013**, *8*, e68739. [CrossRef]
63. Nogacka, A.; Salazar, N.; Suárez, M.; Milani, C.; Arboleya, S.; Solís, G.; Fernández, N.; Alaez, L.; Hernández-Barranco, A.M.; de los Reyes-Gavilán, C.G.; et al. Impact of intrapartum antimicrobial prophylaxis upon the intestinal microbiota and the prevalence of antibiotic resistance genes in vaginally delivered full-term neonates. *Microbiome* **2017**, *5*, 93. [CrossRef] [PubMed]
64. Edgar, R.C. Search and clustering orders of magnitude faster than BLAST. *Bioinforma. Appl. Note* **2010**, *26*, 2460–2461. [CrossRef] [PubMed]
65. Quast, C.; Pruesse, E.; Yilmaz, P.; Gerken, J.; Schweer, T.; Glo, F.O.; Yarza, P. The SILVA ribosomal RNA gene database project: Improved data processing and web-based tools. *Nucleic Acids Res.* **2013**, *41*, 590–596. [CrossRef] [PubMed]
66. Milani, C.; Lugli, G.A.; Turroni, F.; Mancabelli, L.; Duranti, S.; Viappiani, A.; Mangifesta, M.; Segata, N.; van Sinderen, D.; Ventura, M. Evaluation of bifidobacterial community composition in the human gut by means of a targeted amplicon sequencing (ITS) protocol. *FEMS Microbiol. Ecol.* **2014**, *90*, 493–503. [CrossRef] [PubMed]
67. Metsalu, T.; Vilo, J. ClustVis: A web tool for visualizing clustering of multivariate data using Principal Component Analysis and heatmap. *Nucleic Acids Res.* **2015**, *43*, W566–W570. [CrossRef]
68. Hevia, A.; Delgado, S.; Margolles, A.; Sánchez, B. Application of density gradient for the isolation of the fecal microbial stool component and the potential use thereof. *Sci. Rep.* **2015**, *5*, 16807. [CrossRef]
69. Valdés, L.; Gueimonde, M.; Ruas-Madiedo, P. Monitoring in real time the cytotoxic effect of *Clostridium difficile* upon the intestinal epithelial cell line HT29. *J. Microbiol. Methods* **2015**, *119*, 66–73. [CrossRef]

International Journal of
Molecular Sciences

MDPI

Article

Pistachio Consumption Alleviates Inflammation and Improves Gut Microbiota Composition in Mice Fed a High-Fat Diet

Simona Terzo [1,2], Flavia Mulè [2], Gaetano Felice Caldara [2], Sara Baldassano [2], Roberto Puleio [3], Maria Vitale [3], Giovanni Cassata [3], Vincenzo Ferrantelli [3] and Antonella Amato [2,*]

1 Department of Experimental Biomedicine and Clinical Neuroscience (BioNec), University of Palermo, Via del Vespro 129, 90127 Palermo, Italy; simona.terzo01@unipa.it

2 Department of Biological- Chemical- Pharmaceutical Science and Technology (STEBICEF), University of Palermo- Viale delle Scienze, Edificio 16, 90128 Palermo, Italy; flavia.mule@unipa.it (F.M.); opba@unipa.it (G.F.C.); sara.baldassano@unipa.it (S.B.)

3 Istituto Zooprofilattico Sperimentale della Sicilia "A. Mirri", Via Gino Marinuzzi 3, 90129 Palermo, Italy; roberto.puleioizs@gmail.com (R.P.); maria.vitale@izssicilia.it (M.V.); giovanni.cassata@izssicilia.it (G.C.); vincenzo.ferrantelli@izssicilia.it (V.F.)

* Correspondence: antonella.amato@unipa.it; Tel.: +39-091-23897506

Received: 11 December 2019; Accepted: 4 January 2020; Published: 6 January 2020

check for updates

Abstract: High-fat diet (HFD) induces inflammation and microbial dysbiosis, which are components of the metabolic syndrome. Nutritional strategies can be a valid tool to prevent metabolic and inflammatory diseases. The aim of the present study was to evaluate if the chronic intake of pistachio prevents obesity-associated inflammation and dysbiosis in HFD-fed mice. Three groups of male mice (four weeks old; n = 8 per group) were fed for 16 weeks with a standard diet (STD), HFD, or HFD supplemented with pistachios (HFD-P; 180 g/kg of HFD). Serum, hepatic and adipose tissue inflammation markers were analyzed in HFD-P animals and compared to HFD and STD groups. Measures of inflammation, obesity, and intestinal integrity were assessed. Fecal samples were collected for gut microbiota analysis. Serum TNF-α and IL-1β levels were significantly reduced in HFD-P compared to HFD. Number and area of adipocytes, crown-like structure density, IL-1β, TNF-α, F4-80, and CCL-2 mRNA expression levels were significantly reduced in HFD-P subcutaneous and visceral adipose tissues, compared to HFD. A significant reduction in the number of inflammatory foci and IL-1β and CCL-2 gene expression was observed in the liver of HFD-P mice compared with HFD. *Firmicutes*/*Bacteroidetes* ratio was reduced in HFD-P mice in comparison to the HFD group. A pistachio diet significantly increased abundance of healthy bacteria genera such as *Parabacteroides*, *Dorea*, *Allobaculum*, *Turicibacter*, *Lactobacillus*, and *Anaeroplasma*, and greatly reduced bacteria associated with inflammation, such as *Oscillospira*, *Desulfovibrio*, *Coprobacillus*, and *Bilophila*. The intestinal conductance was lower in HFD-P mice than in the HFD mice, suggesting an improvement in the gut barrier function. The results of the present study showed that regular pistachio consumption improved inflammation in obese mice. The positive effects could be related to positive modulation of the microbiota composition.

Keywords: obesity-related inflammation; pistachio intake; gut microbiota; HFD mice; adipose tissue

1. Introduction

Obesity and overweight in western societies and developing countries has become one of the most important public health problems. These, in part, result from the consumption of unbalanced hypercaloric diets that cause excessive visceral fat accumulation [1]. Obesity is associated with chronic low-grade inflammation, which can impair glucose and fatty acid metabolism, leading to insulin resistance and metabolic syndrome [2]. Most studies have focused on adipocytes as the source of inflammatory mediators in this pathology. Storage of excess of triacylglycerol induces hyperplasia and hypertrophy of the adipocytes with altered release of adipokines and pro-inflammatory cytokines, which in turn enhance the recruitment of immune cells, especially macrophages [3]. Therefore, the macrophages in obese adipose tissues are considered to be a major source of pro-inflammatory cytokines, such as TNF-α and IL-6, which are involved in abnormal metabolisms [4].

However, recent studies have suggested that changes in the composition of the gut microbiota might be associated with the development of metabolic disorders related to obesity [5–7]. Indeed, a diet that is rich in saturated fat and poor in fiber is responsible for weight gain, changes in gut microbiota [8], and increased intestinal permeability [9]. The intestinal barrier dysfunction causes an increased circulation of lipopolysaccharides (LPS) derived from gram-negative bacteria [10,11]. In turns, LPS spread participates in metabolic endotoxemia development, adipose tissue dysfunction, and systemic inflammation, triggering obesity-related complications [12].

Nutritional strategies can represent a valid support to prevent metabolic and inflammatory diseases. Increased consumption of fruit and vegetables could prevent chronic diseases such as cardiovascular disease and could prevent body weight gain [13]. Additionally, plant-based foods reduce metabolic syndrome risk [14]. Functional food, i.e., food that can modulate the richness and biodiversity of the gut microbiota and consequently induce a healthier metabolic status, has received increased attention from researchers worldwide [15,16]. It is widely accepted that the consumption of nuts such as almonds, walnuts, and pistachios, as a part of the daily diet provides beneficial effects on human health [17]. Among nuts, Pistachio (*Pistacia vera* L.) is the healthiest due to its fatty acid composition and bioactive compound content (such as lutein and anthocyanin) [18,19]. In recent years, the anti-inflammatory effects of pistachios and the anti-inflammatory activity of its components have been the object of numerous studies. In particular, the anti-inflammatory effects have been reported in both in vitro models [20,21] and in various animal models [22–24]. The antimicrobial properties of polyphenolic fractions obtained from roasted pistachios have also been demonstrated [25,26].

Moreover, we have already shown that the daily pistachio intake prevents and improves some obesity-related metabolic dysfunctions such as dyslipidemia and hepatic steatosis in mice with diet-induced obesity, through a positive modulation of lipid-metabolizing gene expression [27]. Nevertheless, no study has characterized the links between pistachio supplementation, adiposity-related inflammation, and gut microbiota alterations. High-fat diet (HFD) mice are considered a good obese model to characterize the beneficial potential of various treatments on obesity-related disorders since they develop dyslipidemia, hyperglycemia [28,29], type 2 diabetes mellitus [30], hepatic steatosis [31], atherosclerosis [32], and neurodegeneration [33].

Therefore, the purpose of the present study was to investigate whether chronic pistachio consumption is able to prevent the associated visceral–obesity inflammation, the altered composition of gut microbiota, and the intestinal barrier integrity in HFD-obese mice.

2. Results

2.1. Impact of Pistachio Consumption on Body Weight and Metabolic Parameters

As previously reported [27,31], after 16 weeks on HFD, mice showed a significant increase in body weight, triglyceride, and cholesterol plasma concentration in comparison with the standard diet (STD)-fed lean animals. In HFD supplemented with pistachio (HFD-P)-fed mice, triglyceride and cholesterol concentrations were significantly reduced, in comparison with untreated obese mice, whereas the body weight and food intake were similar (Table 1).

Table 1. Effects of pistachio consumption on high-fat diet (HFD)-related dysmetabolisms.

	Lean	HFD	HFD-P
Body weight (g)	32.3 ± 0.9 g	46.2 ± 1.1 g *	46 ± 1.2 g *
Food Intake (g)	4.05 ± 0.2 g	3.4 ± 0.08 g	3.3 ± 0.07 g
Triglycerides (mg/dL)	82 ± 4.5 mg/dL	119 ± 5.5 mg/dL *	93.1 ± 5.1 mg/dL #
Cholesterol (mg/dL)	100 ± 5 mg/dL	192 ± 4 mg/dL *	150 ± 4 mg/dL #

Body weight, food intake, triglyceride and cholesterol plasma concentrations of lean, HFD, and HFD supplemented with pistachios (HFD-P) animals at the end of the experimental period. Data are expressed as mean ± SEM (n = 8/group). * $p < 0.05$ compared with lean; # $p < 0.05$ compared with HFD.

2.2. Impact of Pistachio Consumption on TNF-α and IL-1β Expression

To examine whether pistachio consumption prevents the systemic inflammation induced by HFD, the serum levels of the pro-inflammatory cytokines TNF-α and IL-1β were evaluated by ELISA. As shown in Figure 1, intake of pistachios significantly decreased the HFD-induced high levels of IL-1β and TNF-α.

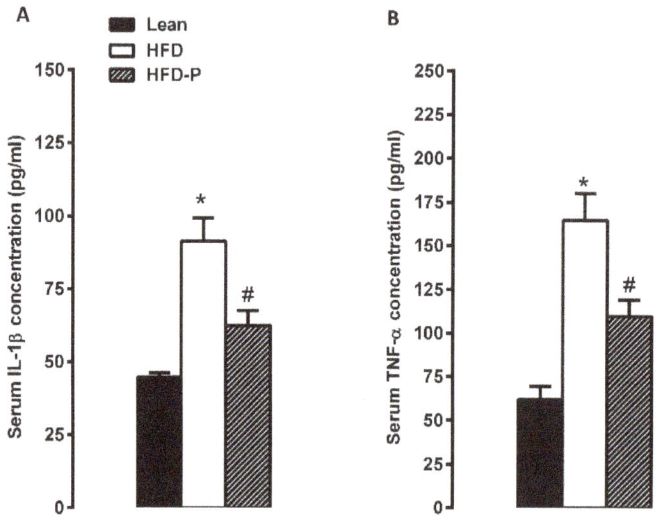

Figure 1. Effects of pistachio consumption on pro-inflammatory cytokines. Serum circulating levels of IL-1β (**A**) and TNF-α (**B**) in the lean, HFD, and HFD-P groups. Data are expressed as mean ± SEM; (n = 8/group). * $p < 0.05$ compared with lean; # $p < 0.05$ compared with HFD.

2.3. Impact of Pistachio Consumption on Adipocytes Hypertrophy

Adipocyte area (μm^2) and adipocyte size distribution (%) were analyzed in visceral adipose tissues (VAT) and subcutaneous adipose tissues (SAT). The adipocytes area in the HFD was significantly higher than that in the lean group; however, the degree of increase was significantly suppressed by HFD-P suggesting that pistachio chronic intake reduces the hypertrophy in both fat depots examined (Figure 2A–C).

2.4. Impact of Pistachio Consumption on Adipose and Hepatic Tissue Inflammation

The presence of Crown Like Structures CLS as an index of macrophage infiltration was evaluated and quantified in VAT and SAT. As shown in Figure 3, more crown-like structures were detected in HFD mice, as compared to the lean animals. Interestingly, in HFD-P mice, the CLS density was significantly lower in comparison to the HFD adipose tissues (Figure 3A,B). Furthermore, RT-PCR analysis revealed significantly higher levels of IL-1β, TNF-α, F4-80, and CCL2 mRNA in HFD mouse VAT and SAT than in the lean mice. However, pistachio-diet reduced the increase of the pro-inflammatory cytokines and the macrophage infiltration markers in both adipose tissue depots (Figure 3C).

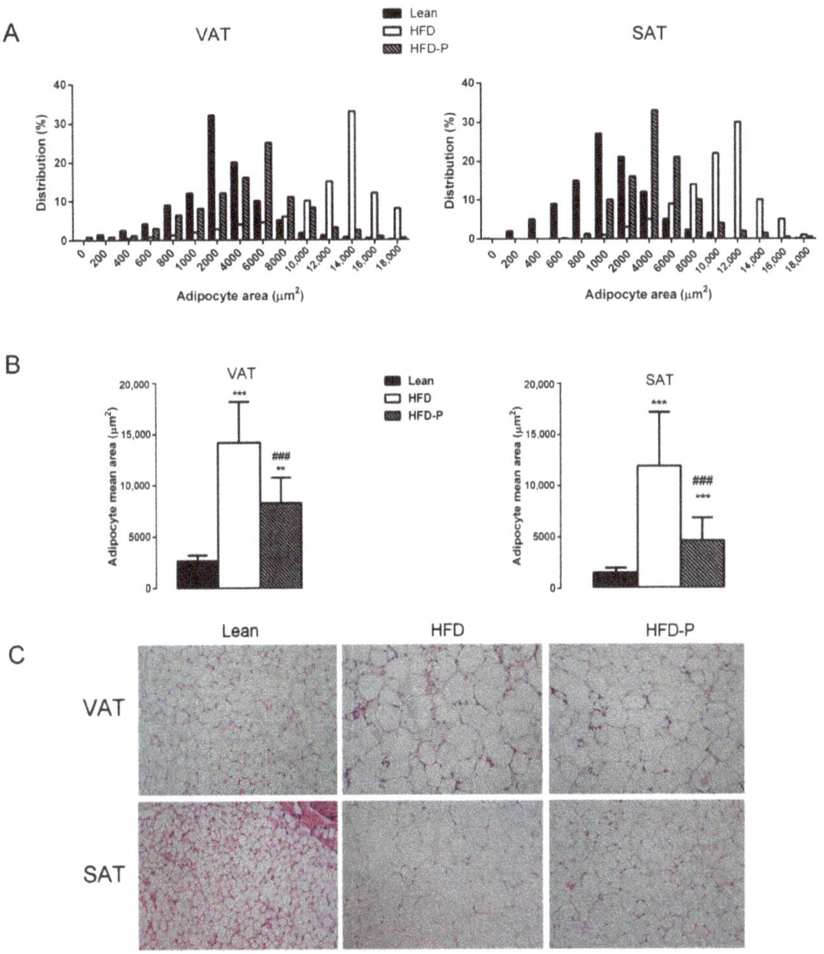

Figure 2. Effects of pistachio consumption on adipocyte morphology. (**A**) Adipocyte size distribution (%) and (**B**) adipocyte mean area (μm^2) of the epididymal visceral adipose tissues (VAT) and subcutaneous adipose tissue (SAT) in lean, HFD, and HFD-P mice. (**C**) Adipose tissue staining (H&E staining, magnification 10×) in the lean, HFD, and HFD-P mice. Data are expressed as mean ± SEM; (n = 8/group). Compared to the lean mice (** $p < 0.01$; *** $p < 0.001$); Compared to the HFD mice (### $p < 0.001$).

As previously reported [27], pistachio consumption counteracted the hepatic steatosis development consequent to HFD (Figure 4A). HFD mice showed higher infiltration of inflammatory cells in the liver compared to the STD animals. Nevertheless, infiltration was reduced in HFD-P livers in comparison to the HFD ones (Figure 4A,B). Moreover, pistachio intake significantly prevented the increase in hepatic mRNA levels of IL-1β and CCL2 observed in the HFD liver, as compared to the STD animals (Figure 4C).

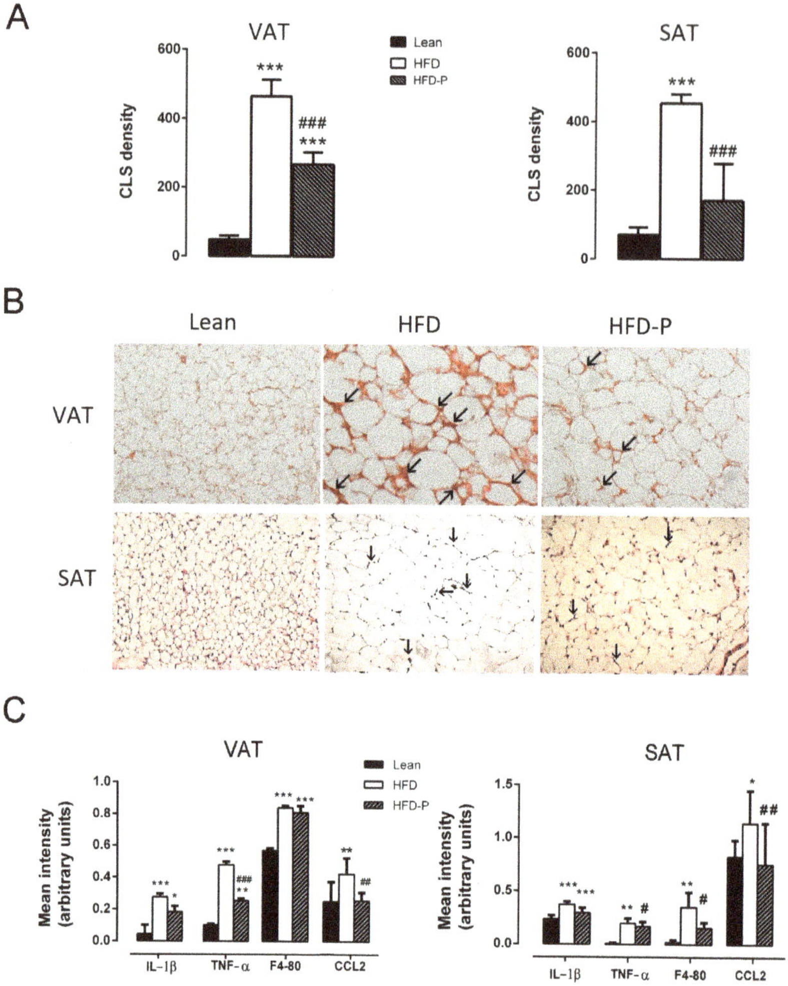

Figure 3. Effects of pistachio consumption on Crown Like Structures CLS density. (**A**) Representative results of the density of MAC-2 positive CLS stained in epididymal visceral adipose tissue (VAT) and subcutaneous adipose tissue (SAT) of the three groups of animals (CLS number/10.000 adipocytes). (**B**) VAT and SAT immunohistochemistry (IHC) analysis for MAC-2 positive macrophages forming CLS (arrows) in the lean, HFD, and HFD-P animals (magnification 10×). (**C**) Effect of Pistachio consumption on IL-1β, TNF-α, F4-80, and CCL2 mRNA expression in VAT and SAT of the lean, HFD, and HFD-P mice. Data are expressed as mean ± SEM; (n = 8/group). * $p < 0.05$ compared to the lean mice (* $p < 0.05$; ** $p < 0.01$; *** $p < 0.001$); # $p < 0.05$ compared to the HFD mice (# $p < 0.05$; ## $p < 0.01$; ### $p < 0.001$).

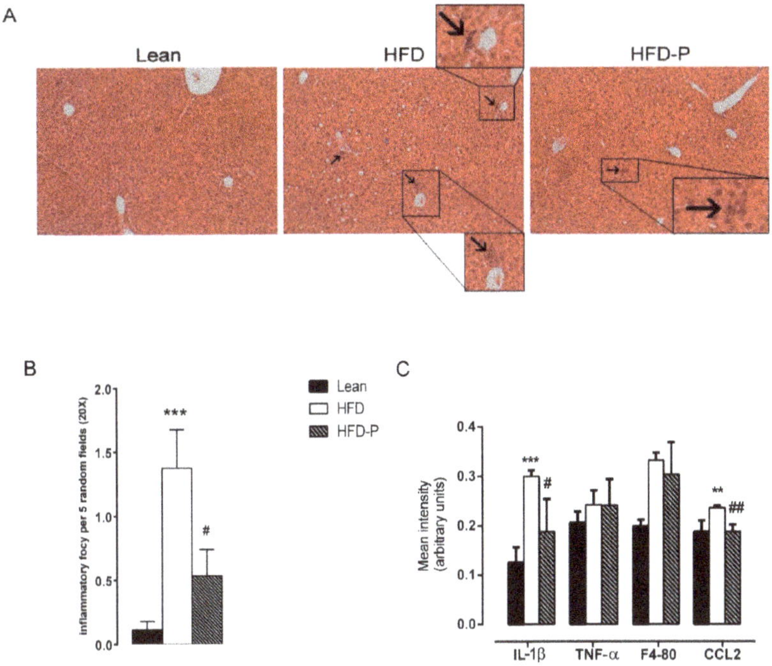

Figure 4. Effect of pistachio consumption on liver inflammation. (**A**) Liver histology of the lean, HFD, and HFD-P mice was examined by H&E staining. Arrows indicate the points of inflammatory foci (magnification 10×). (**B**) Quantification of inflammatory foci per 5 random fields under 20× magnification. (**C**) mRNA levels of IL-1β, TNF-α, F4-80, and CCL2 in the livers of the lean, HFD, and HFD-P mice (**B**). Data are represented by the means ± SEM. (n = 8/group). * $p < 0.05$ compared to the lean mice (** $p < 0.01$; *** $p < 0.001$); # $p < 0.05$ compared to the HFD mice (# $p < 0.05$; ## $p < 0.01$).

2.5. *Impact of Pistachio Consumption on Gut Microbial Community*

To examine the changes of the gut microbiota in response to the pistachio diet in obese HFD mice, we analyzed the microbial composition in the feces of mice fed STD, HFD, and HFD-P, through Next-Generation Sequencing (NGS) analysis. After 16 weeks of HFD feeding, a decrease in the phyla *Bacteroidetes* and an increase in the phyla *Firmicutes* and *Proteobacteria* relative to STD were observed both in the HFD group and in the HFD-P mice (Figure 5A). The ratio of *Firmicutes* to *Bacteroidetes* was significantly higher in the HFD group than in the lean mice group, consistent with the microbial changes of the two phyla in the mice with HFD-induced obesity. Although this value was also an index of dysbiosis in the HFD-P group, it was significantly improved by pistachio intake (Figure 5B). Interestingly, *Tenericutes* abundance of the HFD-P mice was significantly increased in comparison with the HFD control mice; on the contrary, the pistachio diet significantly reduced the *Proteobacteria* abundance (Figure 5A).

At the genus level, a pistachio diet significantly altered the abundances of 10 genera in a positive direction, as compared to the HFD animals. In particular, an abundance of *Parabacteroides*, *Dorea*, *Allobaculum*, *Turicibacter*, *Lactobacillus*, and *Anaeroplasma* genera was observed, while *Oscillospira*, *Desulfovibrio*, *Coprobacillus*, and *Bilophila* abundance was reduced in comparison to the HFD mice (Figure 6).

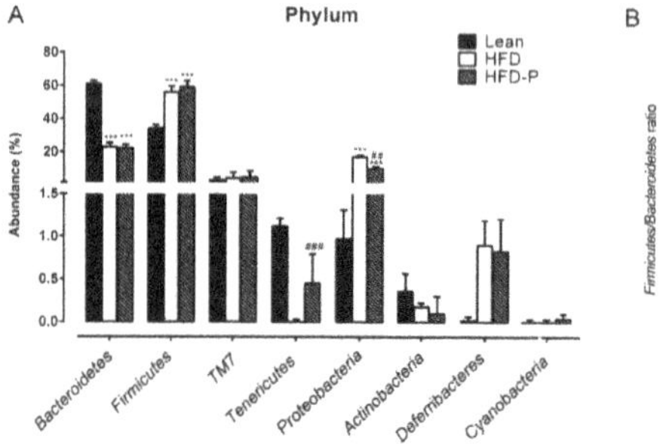

Figure 5. 16S rDNA sequencing of bacterial DNA in the feces of the lean, HFD, and HFD-P mice, in order to discriminate the intestinal microbial profile. (**A**) Graphic representation of the relative abundance (%) of the gut microbiota phyla composition of the three groups of animals. (**B**) Ratio of *Firmicutes* to *Bacteroidetes* in the lean, HFD, and HFD-P. Data are expressed as mean ± SEM; (n = 8/group). (*** $p < 0.001$); hash denotes significant difference when compared to the HFD group (# $p < 0.05$; ## $p < 0.01$; ### $p < 0.001$).

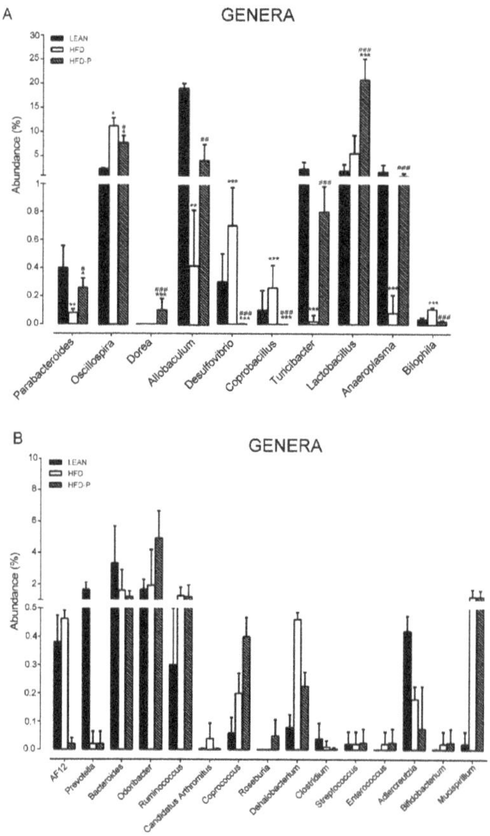

Figure 6. Genus level taxonomic distributions of the microbial communities in the feces of the lean, HFD, and HFD-P mice. (**A**) Genera abundance (%) was significantly modified by the pistachio intake. (**B**) Genera abundance (%) was not modified by the pistachio intake. Data are expressed as means ± SEM; (n = 8/group). (* $p < 0.05$; ** $p < 0.01$; *** $p < 0.001$); hash denotes significant difference compared to the HFD group (# $p < 0.05$; ## $p < 0.01$; ### $p < 0.001$).

2.6. Impact of Pistachio Consumptiom on the Intestinal Barrier

Barrier integrity in small intestine sections was evaluated in an Ussing chamber system through conductance measurements of mucosal preparations from all experimental groups. Conductance values in the duodenal sections from animals fed HFD were significantly higher than those from the STD group (about 60% increase). Notably, the conductance values from the HFD-P group were significantly lower than the HFD group and were very similar to the lean group (Figure 7).

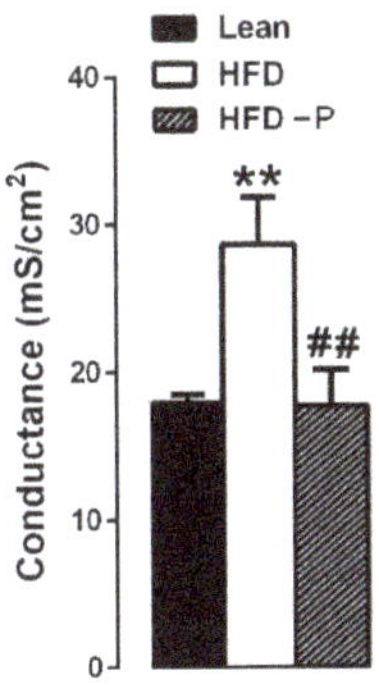

Figure 7. Effect of pistachio consumption on the conductance of isolated duodenal sections from the lean, HFD, and HFD-P mice through the Ussing chambers technique. Data are expressed as mean ± SEM; (n = 8/group). (** $p < 0.01$); hash denotes significant difference compared with the HFD (## $p < 0.01$).

3. Discussion

The present study provided evidence that regular pistachio intake in HFD-fed obese mice ameliorates systemic and metabolic tissue inflammation, positively modulates the gut microbial composition, and increases the intestinal barrier function.

Previous in vitro and in vivo studies have examined the antioxidant, anti-inflammatory, and anti-apoptotic potential of pistachio [21,34–38]. In particular, the pistachio properties were tested on carrageenan or LPS-induced acute inflammatory response [39,40], inflammatory bowel disease, colitis [24,41–43], cancer [44–46], and allergic inflammation in the asthmatic model [47]. To our knowledge, the present study was the first to explore the anti-inflammatory effects of pistachios in mice with HFD-induced obesity.

Obesity is characterized by a chronic low-degree inflammation. In fact, excessive calorie intake increases fat accumulation and the lipotoxicity activates the production of cytokines and the cells involved in innate immunity. This production promotes a chronic, low-grade inflammatory status, induces recruitment, and activation of mature immune cells and other cells, such as macrophages and adipocytes, respectively, which modify the tissue and reinforce the inflammatory response [12,48].

We previously reported that a pistachio-based diet exerts beneficial effects in HFD obese mice. In fact, it reduces the dyslipidemia and hepatic steatosis, and is able to prevent and improve visceral fat mass accumulation in HFD mice through a redistribution towards the subcutaneous fat depot, which is indicative of a healthier profile [27]. The present work not only confirms that the pistachio diet modifies fat depots, as suggested by the morphological analysis of visceral and subcutaneous adipose tissue, but also reduces the obesity-linked inflammatory status.

First, we highlighted that a pistachio diet significantly prevents the increase of pro-inflammatory cytokines, TNF-α, and IL-1β induced by HFD in the systemic circulation. Furthermore, we provided evidence that visceral and subcutaneous adipose tissue and liver inflammation induced by obesity were strongly prevented by pistachio intake. Various inflammatory mediators are involved in adipose tissue and liver inflammation. In the adipose tissue, a paracrine loop linking fatty acids, TNF-α and CCL2 establishes a vicious cycle between adipocytes and macrophages that aggravates inflammation [46]. In the liver, the increased influx of fatty acids induces lipotoxic injury and activation of inflammatory response. Accordingly, an abundant expression of pro-inflammatory cytokines IL-1β and TNF-α is often associated with Non-Alcoholic Fatty Liver Disease (NAFLD) [49].

We found that HFD-P mice exhibit lower levels of TNF-α, F4-80, and CCL2 as well as minor macrophage infiltration, detected as CLS density in adipose tissues, in comparison to HFD animals. Additionally, in the liver, we found a reduction of IL-1β and CCL2 mRNA levels and a decreased number of inflammatory foci, in comparison to the HFD mice. These changes would favor an anti-inflammatory microenvironment that is able to counteract the biochemical dysfunctions occurring in adipose tissue or in the liver of HFD mice.

Obesity and metabolic disorders are complex processes that also involve crosstalk between the gut microbiota and host metabolism [50]. The gut microbiota might induce inflammation in visceral adipose tissue via the LPS and TLR4 signaling pathways with an increased macrophage infiltration and release of a variety of pro-inflammatory mediators, which in turn recruit additional macrophages to further propagate the chronic inflammatory status [51,52]. Therefore, in attempt to elucidate an eventual contribution of the gut microbiota to the beneficial pistachio effects, we investigated the profiling changes of the gut microbiota composition in mice, performing 16S rDNA sequencing through NGS analyses. Indeed, pistachio consumption can modify human gut microbiota composition by increasing the number of potentially beneficial butyrate-producing bacteria [53]. Our results demonstrated that the microbial communities were influenced by different type of diets. Analysis at the phylum level indicated that the fecal microbiota was dominated by seven major phyla: *Firmicutes, Bacteroidetes, Proteobacteria, TM7, Deferribacteres, Actinobacteria,* and *Tenericutes*. We observed a dramatic reduction in *Bacteroidetes* abundance and a marked increase in *Firmicutes*, in the HFD group, in accordance with the increased *Firmicutes* to *Bacteroidetes* ratio identified in obese humans and mice [54,55]. However, although the pistachio diet failed to maintain the *Firmicutes/Bacteroidetes* proportion observed in STD mice, the ratio value in HFD-P was significantly lower than the HFD group, suggesting a pistachio protective effect against dysbiosis.

Interestingly, compared to the HFD control mice, we found a significant increase in the *Tenericutes* abundance and a significant decrease in *Proteobacteria* abundance in the HFD-P mice. Bacteria from the *Tenericutes* phylum have been found to be positively associated with the modulation of the immune system induced by high-polyphenol content food such as cocoa [56], and lower counts of these bacteria were found in the intestinal inflammation induced by dextran sodium sulfate [57]. Therefore, a more relative abundance of *Tenericutes* induced by the pistachio diet could provide some beneficial effects in the intestinal integrity. In addition, several reports have endorsed an abundance of *Proteobacteria* in the gut microbiota as a potential marker for obesity-related metabolic disorders in both humans and rodents [6,58]. Therefore, the lower level of *Proteobacteria* in the HFD-P-fed mice than the HFD mice could be indicative of less severe health conditions.

At the genus level, *Lactobacillus* was significantly increased in the HFD-P, in comparison to the other groups. The relative abundance of *Lactobacillus* caused by pistachio intake can be interpreted as a positive effect because *Lactobacillus* is a well-known probiotic that has been associated with reduced colitis in several models of inflammatory bowel diseases [59], and has been shown to have protective effect in the intestinal barrier function and steatosis [60–62].

Interestingly, pistachio intake was found to improve the abundance of other genera that are usually associated with a positive impact on host health, such as *Parabacteroides, Dorea, Allobaculum, Turicibacter,* and *Anaeroplasma*. *Parabacteroides* is a genus predominantly found in the gut of healthy individuals, which is negatively correlated with body weight gain, liver steatosis, and epididymal fat accumulation [63]. The *Allobaculum* genus has been associated with a better mucus layer in the colon [64], suggesting that its decrease reflects the alteration of the mucus layer in HFD. Thus, a pistachio diet might prevent this alteration. Moreover, *Allobaculum* and *Dorea* are among the major producers of butyrate, an important fuel for epithelial colonocytes that have been shown to help maintain normal differentiation. Thus, an increase in the amount of butyrate generated in the gut might be an indication of improved health. Accordingly, butyrate-producing probiotics reduce NAFLD progression in rats [65] and attenuate HFD-induced steatohepatitis in mice, by improving intestinal permeability [66–69].

The decreased *Turicibacter* abundance in HFD mice, which was prevented by pistachio intake, fits well with previous data showing a depletion of *Turicibacter* in animal models of inflammatory bowel disease and confirms the hypothesis that *Turicibacter* is an anti-inflammatory taxon [70–72]. Recent data report that *Anaeroplasma* abundance is significantly decreased in obese rats, while the increased abundance is related to a reduction in fat accumulation and expression of inflammatory factors in the liver [73].

Another interesting effect of pistachio intake on gut microbiota concerns the decrease of genera associated with inflammation, such as *Desulfovibrio*, *Coprobacillus*, *Oscillosphira*, and *Bilophila*. *Desulfovibrio* is a genus responsible for 60% of the total hydrogen sulfide (H_2S) production in the colon. H_2S inhibits the mitochondrial respiration of colonic epithelial cells [74], reducing the diffusion of oxygen, and then subtracting energy that is useful to the beta-oxidation of butyrate [75]. Thus, it is likely that the reduction of H_2S-producing bacteria by pistachio enhances the output of short chain fatty acids (SCFAs), such as butyrate, improving intestinal health and inflammation [76]. *Coprobacillus* has been reported to be negatively correlated with most of the features of obesity in obese rats [77,78]. An abundance of *Oscillospira* has been associated with systemic inflammation and altered intestinal permeability [79,80] and diets rich in polyphenols improve HFD-induced liver steatosis by reducing *Oscillospira* abundance [81,82]. *Bilophila* abundance seems to be related with colon inflammation [82]. A recent work reports that the treatment with phenolic compounds alleviate obesity-related inflammation in HFD-mice, by inhibiting the expansion of the *Bilophila* bacteria genus [76].

The changes in microbiota composition might be due to the different components of the pistachios, such as fatty acids, flavonoids, or fiber. Pistachios might exhibit prebiotic effects by enriching potentially beneficial microbes, such as lactic acid bacteria.

Therefore, taken together, these results suggest that the gut microbial alterations observed in HFD-P mice might be associated with pistachio metabolic and anti-inflammatory benefits.

It is interesting to note that an increased intestinal conductance was observed in the small intestine of HFD mice, in comparison with lean or HFD-P mice, suggesting that HFD induces a decrease in the intestinal epithelial integrity and an increased ability of ions and small molecules to permeate through the paracellular pathway. According to our data, several studies report an increased gut permeability in the HFD mice [83,84]. The intestinal conductance value in the HFD-P group was similar to the lean group, suggesting that the pistachio diet is able to prevent an increase in permeability and, thus, showing the protective action of the pistachio diet on the intestinal barrier functions. Cani and collaborators [85] provided evidence that the development of metabolic endotoxemia and the linked metabolic disorders induced by high-fat feeding are associated with an increased intestinal permeability. Therefore, it is likely that the modulation of gut bacteria associated with increased intestinal barrier functions are involved in the anti-inflammatory effects of pistachio diet.

4. Materials and Methods

4.1. Animals and Diets

The procedures were performed in accordance with the conventional guidelines for animal experimentation (Italian D.L. No. 26/2014 and subsequent variations) and the recommendations of the European Economic Community (2010/63/UE). The experimental protocols were approved by the animal welfare committee of the Istituto Zooprofilattico Sperimentale della Sicilia "A. Mirri" (Palermo, Italy) and authorized by the Ministry of Health (Rome, Italy; Authorization Number 349/2016-PR date of approval: 1 April 2016)).

Four-week-old male C57BL/6J (B6) mice, purchased from Harlan Laboratories (San Pietro al Natisone Udine, Italy) were housed in a room with controlled temperature and dark–light cycles, with free access to water and food. After acclimatization (1 week), the animals were weighed and divided into three groups. (1) Lean group—control animals were fed the standard diet (STD; 4RF25 Mucedola, Milan, Italy) for 16 weeks; (2) High-fat diet (HFD) group—obese animals fed HFD (PF4215, Mucedola, Milan, Italy) for 16 weeks. (3) HFD-P group—obese animals fed HFD supplemented with pistachio for 16 weeks. HFD-P was custom designed and prepared by Mucedola S.r.l (PF4215/C; R&S 34/16). It was obtained by substituting 20% of the caloric intake from HFD with pistachio (180 g/kg of HFD). The HFD and HFD-P were stored in vacuum containers at 4 °C. The energy densities of the diets are shown in Table 2.

Table 2. Composition and energy densities of the STD, HFD, and HFD-P groups.

Ingredient (g/kg)	STD	HFD	HFD-P
Total Energy, Kcal/g	3.5	6	6
Protein, %	20	20	20
Carbohydrate, %	70	20	20
Fat, %	10	60	60

STD—standard diet. HFD—high fat diet. HFD-P—HFD supplemented with pistachio.

Pistachio nuts belong to *Pistacia vera* L. species and were purchased by Pistachio Valle del Platani Association and Pistacchio di Raffadali (Agrigento-AG, Sicily, Italy). As previously described [31], during the 16 weeks of the experiment, changes in body weight and food-intake were measured weekly and results from the different groups of animals were compared. At the end, the animals were sacrificed by cervical dislocation; the blood was collected immediately by intracardiac puncture, and the plasma was isolated by centrifugation at 3000 rpm at 4 °C for 15 min and stored at −80 °C, until analysis. The liver, adipose tissue, and small intestine were rapidly removed; a part of each tissue was fixed in 4% neutral formalin solution for histological analysis and another part was stored at −80 °C for biomolecular analysis. Five-centimeter segments of the small intestine were taken for the Ussing chamber assays.

4.2. Plasma Biomarker Analysis

IL-1β and TNF-α were quantified by a commercial ELISA Kit (Cloud-Clone Corp, Wuhan, Hubei), based on the manufacturer's instructions. The levels of triglyceride and total cholesterol in the serum were evaluated by using the automatic biochemical analyzer (ILab 600, Instrumentation Laboratory, Milano, Italy).

4.3. Liver and Adipose Tissues Histology and Immunohistochemistry

Hepatic, visceral (epididymal), and subcutaneous white adipose tissues (WAT) were fixed with 4% formaldehyde solution for 24 h and embedded in paraffin. Then, 5 μm sections were prepared and stained with hematoxylin and eosin (H&E) for morphological examination. The number of liver inflammatory foci was calculated by counting the inflammatory cell aggregates in the hepatic lobules, per 5 random fields at a magnification of 20×. Hepatic inflammatory foci were defined as aggregates of inflammatory cells that accumulate in the liver during chronic inflammation [86,87]. The number of adipocytes per microscopical field (density) was determined at a magnification of 20×. The mean surface area of the adipocytes (μm^2) was calculated using the image analyzer software (Visilog 6, Courtaboeuf, France). Each adipocyte was manually delineated and 700–1000 adipocytes per condition were assessed.

Images of the H&E liver and WAT sections were captured using an optical microscope (Leica DMLB, Meyer instruments, Houston, Texas) equipped with a DS-Fi1 camera (Nikon, Florence, Italy), and were analyzed at 10× and 20× magnification.

For the immunohistochemistry, deparaffinized sections were treated with 3% hydrogen peroxide to inactivate the endogenous peroxidase followed by a rinse in PBS for 5 min. Subsequently, the sections were incubated with the primary antibody Mac-2 at 4 °C overnight (1:2800, Cedarlane, Ontario, Canada CL8942AP). After PBS washing, the sections were incubated with the biotinylated secondary antibody (Anti-Mouse IgG/Rabbit IgG) (1:400, Vector Laboratories, BA-4001) for 30 min. Histochemical reactions were performed using the Vector's Vectastain ABC Kit (Vector Laboratories, Burlingame, CA, USA) and diaminobenzidine as a substrate (Sigma, Milano, Italia). Crown-like structures (CLS) were counted as a measure of adipose tissue inflammation and were expressed as number of CLS/10,000 adipocytes.

4.4. *Reverse Transcription Polymerase Chain Reaction (RT-PCR)*

RNA was extracted from liver, epididymal, and subcutaneous adipose tissue, using the RNeasy plus Mini Kit (Qiagen, Valencia, CA, USA), according to the manufacturer's protocol. The extraction from adipose tissues was performed after a preliminary step of lysis using Triazol. Two nanograms of the total RNA were used for cDNA synthesis with High Capacity cDNA Reverse Transcription (Applied Biosystems, Waltham, MA, USA). The target cDNA was amplified using genetic-specific primers, as listed in Table 3. The amplification cycles included denaturation at 95 °C for 45 s, annealing at 52 °C for 45 s, and elongation at 72 °C for 45 s. After 40 cycles, the PCR products were separated by electrophoresis on a 1.8% agarose gel for 45 min at 85 V. The gels were stained with 1 mg/mL ethidium bromide and visualized with ultraviolet (UV) light, using E-Gel GelCapture (Thermo Fisher Scientific, Monza, Italy). The expression levels of the gene targets, normalized to the endogenous reference (β-actin), were analyzed using the E-Gel GelQuant Express Analysis Software (Thermo Fisher Scientific, Monza, Italy).

Table 3. Oligonucleotide sequence of primers for RT-PCR.

Gene	Forward Primer	Reverse Primer	Size (bp)
IL-1β	5′-CAGGATGAGGACATGAGCACC-3′	5′-CTCTGCAGACTCAAACTCCAC-3′	450
TNF-α	5′-AGCCCACGTCGTAGCAAACCA-3′	5′-GCAGGGGCTCTTGACGGCAG-3′	260
F4-80	5′-GCCACGGGGCTATGGGATGC-3′	5′-TCCCGTACCTGACGGTTGAGCA-3′	360
CCL2	5′-TCTGTGCTGACCCCAAGAAGG-3′	5′-TGGTTGTGGAAAAGGTAGTGGAT-3′	183
β-actin	5′-GGATCCCCGCCCTAGGCACCAGGGT-3′	5′-GGAATTCGGCTGGGGTGTTGAAGGTCTCAAA-3′	289

4.5. *Gut Microbiota Composition*

Six hours before the sacrifice, the mice were kept individually in a clean cage without food and the stool samples were collected from each mouse for gut microbiota analysis, using an autoclaved tube. Bacteria DNA was extracted from stool samples (200 mg per mouse) using the QIAamp DNA Stool Handbook kit (QIAGEN, Milan-Italy), following the manufacturer's protocol. The extracted DNA was used for the metagenomic study carried out by the BMR Genomics company s.r.l. (Padova, Italy).

For the NGS sequencing, the V3–V5 regions of the 16S rRNA gene were amplified. After confirming that all V3–V5 amplicons had good levels of concentration, purity, and integrity, a massive sequencing was carried out utilizing the Illumina MiSeq platform (San Diego, CA, USA). Reference-based UCLUST algorithm (Qiime1.9.1) was used to pick the OTUs at 97% of similarity against Greengenes v13.8 database. OTUs were collected in the biom file and filtered at 0.005% abundance to eliminate spurious OTUs that were present at a low frequency.

4.6. *Ussing Chamber Measurements*

Intestinal barrier integrity was evaluated in an Ussing chamber system. A segment of small intestine was excised from freshly sacrificed mice and transferred to an ice-cold oxygenated Krebs solution containing (mM) NaCl 119, KCl 4.5, MgSO4 2.5, NaHCO3 25, KH2PO4 1.2, CaCl2 2.5, and glucose 11.1. The segment was cut longitudinally along the mesenteric border and mounted in an Ussing chamber. The Ussing chambers contained a hydrated mixture of 5% CO^2/95% O^2 (v/v). The Ussing chamber system were filled with 10 mL Krebs solution, maintained at 37 °C, and continuously bubbled with the short-current (Isc), i.e., the current generated by the ionic transport through the epithelium. The transepithelial potential difference was continuously monitored under open circuit conditions, using a DVC 1000 amplifier (DVC 1000, World Precision Instruments, Sarasota, FL, USA) and was recorded through filled agar electrodes. The conductance was calculated according to Ohm's law, using the potential difference and current (Isc) values. Tissues whose conductance increased during the course of the experiment (calculated every 15 min) were considered damaged and were excluded from the data analysis.

4.7. Statistical Analyses

Results are shown as means ± the standard error of the mean (SEM). The letter 'n' indicates the number of animals. Statistical analyses were performed using the Prism Version 6.0 Software (Graph Pad Software, Inc., San Diego, CA, USA). The comparison between the groups was performed by Analysis of Variance (ANOVA), followed by Bonferroni's post-test. A p-value ≤ 0.05 was considered statistically significant.

5. Conclusions

The present study demonstrated that chronic intake of pistachio exerts beneficial effects in obese mice by alleviating inflammation in adipose tissues and liver, and impacting the gut microbiome composition. In particular, it enhances the abundance of beneficial bacteria genera, such as *Lactobacillus*, *Dorea*, *Allobaculum*, and inhibited the growth of bacteria associated with obesity-related comorbidities and inflammation, such as *Desulfovibrio* and *Bilophila*.

Author Contributions: Conceptualization, A.A.; methodology, A.A. and S.T.; formal analysis, S.T. and A.A.; investigation, S.T., G.F.C., R.P., S.B., and M.V.; resources, V.F.; data curation, A.A. and F.M.; writing—original draft preparation, S.T.; writing—review and editing, A.A. and F.M.; supervision, A.A. and F.M.; project administration, G.C.; funding acquisition, V.F. All authors have read and agreed to the published version of the manuscript.

Funding: This research was funded by Istituto Zooprofilattico Sperimentale della Sicilia "A. Mirri", Via Gino Marinuzzi 3, 90129 Palermo (Italy).

Acknowledgments: The authors would like to thank the associations "Pistachio Valle del Platani" and "Pistacchio di Raffadali" for supplying pistachios and the Istituto Zooprofilattico Sperimentale della Sicilia "A. Mirri" for providing administrative support.

Conflicts of Interest: The authors declare no conflict of interest.

Abbreviations

HFD	High-fat diet
STD	Standard diet
HFD-P	HFD supplemented with pistachios
TNF-α	Tumor Necrosis Factor alfa
IL-1β	Interleukin 1 beta
CCL-2	Chemokine (C-C motif) Ligand 2
LPS	Lipopolysaccharides
VAT	Visceral Adipose Tissues
SAT	Subcutaneous Adipose Tissues
CLS	Crown Like Structures
H&E	Hematoxylin and Eosin
SEM	Standard Error of the Mean
TLR4	Toll-Like Receptor 4
NAFLD	Non-Alcoholic Fatty Liver Disease
H_2S	Hydrogen Sulfide

References

1. Hruby, A.; Hu, F.B. The epidemiology of obesity: A big picture. *Pharmacoeconomics* **2015**, *33*, 673–689. [CrossRef] [PubMed]
2. Eckel, R.H.; Grundy, S.M.; Zimmet, P.Z. The metabolic syndrome. *Lancet* **2005**, *365*, 1415–1428. [CrossRef]
3. Trayhurn, P.; Wood, I.S. Adipokines: Inflammation and the pleiotropic role of white adipose tissue. *Br. J. Nutr.* **2004**, *92*, 347–355. [CrossRef] [PubMed]
4. Lackey, D.E.; Olefsky, J.M. Regulation of metabolism by the innate immune system. *Nat. Rev. Endocrinol.* **2015**, *12*, 15–28. [CrossRef]
5. Cani, P.D.; Osto, M.; Geurts, L.; Everard, A. Involvement of gut microbiota in the development of low-grade inflammation and type 2 diabetes associated with obesity. *Gut Microbes* **2012**, *3*, 279–288. [CrossRef]
6. Ridaura, V.K.; Faith, J.J.; Rey, F.E.; Cheng, J.; Duncan, A.E.; Kau, A.L.; Griffin, N.W.; Lombard, V.; Henrissat, B.; Bain, J.R.; et al. Gut microbiota from twins discordant for obesity modulate metabolism in mice. *Science* **2013**, *341*. [CrossRef]
7. Álvarez-Mercado, A.I.; Navarro-Oliveros, M.; Robles-Sánchez, C.; Plaza-Díaz, J.; Sáez-Lara, M.J.; Muñoz-Quezada, S.; Fontana, L.; Abadía-Molina, F. Microbial population changes and their relationship with human health and disease. *Microorganisms* **2019**, *7*, 68. [CrossRef]
8. Gentile, C.L.; Weir, T.L. The gut microbiota at the intersection of diet and human health. *Science* **2018**, *362*, 776–780. [CrossRef]
9. Myles, I.A. Fast food fever: Reviewing the impacts of the Western diet on immunity. *Nutr. J.* **2014**, *13*, 61. [CrossRef]
10. Moreira, A.P.B.; Texeira, T.F.S.; Ferreira, A.B.; do Carmo Gouveia Peluzio, M.; de Cássia Gonçalves Alfenas, R. Influence of a high-fat diet on gut microbiota, intestinal permeability and metabolic endotoxaemia. *Br. J. Nutr.* **2012**, *108*, 801–809. [CrossRef]
11. Cani, P.D.; Amar, J.; Iglesias, M.A.; Poggi, M.; Knauf, C.; Bastelica, D.; Neyrinck, A.M.; Fava, F.; Tuohy, K.M.; Chabo, C.; et al. Metabolic endotoxemia initiates obesity and insulin resistance. *Diabetes* **2007**, *56*, 1761–1772. [CrossRef] [PubMed]
12. Clemente-Postigo, M.; Oliva-Olivera, W.; Coin-Aragüez, L.; Ramos-Molina, B.; Giraldez-Perez, R.M.; Lhamyani, S.; Alcaide-Torres, J.; Perez-Martinez, P.; El Bekay, R.; Cardona, F.; et al. Metabolic endotoxemia promotes adipose dysfunction and inflammation in human obesity. *Am. J. Physiol. Metab.* **2019**, *316*, E319–E332. [CrossRef] [PubMed]
13. Tian, Y.; Su, L.; Wang, J.; Duan, X.; Jiang, X. Fruit and vegetable consumption and risk of the metabolic syndrome: A meta-analysis. *Public Health Nutr.* **2018**, *21*, 756–765. [CrossRef] [PubMed]
14. Rizzo, N.S.; Sabaté, J.; Jaceldo-Siegl, K.; Fraser, G.E. Vegetarian dietary patterns are associated with a lower risk of metabolic syndrome: The adventist health study 2. *Diabetes Care* **2011**, *34*, 1225–1227. [CrossRef] [PubMed]
15. Roopchand, D.E.; Carmody, R.N.; Kuhn, P.; Moskal, K.; Rojas-Silva, P.; Turnbaugh, P.J.; Raskin, I. Dietary polyphenols promote growth of the gut bacterium akkermansia muciniphila and attenuate high-fat diet induced metabolic syndrome. *Diabetes* **2015**, *64*, 2847–2858. [CrossRef] [PubMed]
16. Anhê, F.F.; Roy, D.; Pilon, G.; Dudonn'e, S.; Matamoros, S.; Varin, T.V.; Garofalo, C.; Moine, Q.; Desjardins, Y.; Levy, E.; et al. A polyphenol-rich cranberry extract protects from diet-induced obesity, insulin resistance and intestinal inflammation in association with increased Akkermansia spp. Population in the gut microbiota of mice. *Gut* **2015**, *64*, 872–883. [CrossRef] [PubMed]
17. De Souza, R.G.M.; Schincaglia, R.M.; Pimentel, G.D.; Mota, J.F. Nuts and human health outcomes: A systematic review. *Nutrients* **2017**, *9*, 1311. [CrossRef]
18. Dreher, M.L. Pistachio nuts: Composition and potential health benefits. *Nutr. Rev.* **2012**, *70*, 234–240. [CrossRef]
19. Terzo, S.; Baldassano, S.; Caldara, G.F.; Ferrantelli, V.; Lo Dico, G.; Mulè, F.; Amato, A. Health benefits of pistachios consumption. *Nat. Prod. Res.* **2019**, *33*, 715–726. [CrossRef]
20. Paterniti, I.; Impellizzeri, D.; Cordaro, M.; Siracusa, R.; Bisignano, C.; Gugliandolo, E.; Carughi, A.; Esposito, E.; Mandalari, G.; Cuzzocrea, S. The anti-inflammatory and antioxidant potential of pistachios (*Pistacia vera* L.) in vitro and in vivo. *Nutrients* **2017**, *9*, 915. [CrossRef]

21. Gentile, C.; Tesoriere, L.; Butera, D.; Fazzari, M.; Monastero, M.; Allegra, M.; Livrea, M.A. Antioxidant activity of Sicilian pistachio (*Pistacia vera* L. var. Bronte) nut extract and its bioactive components. *J. Agric. Food Chem.* **2007**, *55*, 643–648. [CrossRef] [PubMed]
22. Ahmad, N.S.; Waheed, A.; Farman, M.; Qayyum, A. Analgesic and anti-inflammatory effects of pistacia integerrima extracts in mice. *J. Ethnopharmacol.* **2010**, *129*, 250–253. [CrossRef] [PubMed]
23. Esmat, A.; Al-Abbasi, F.A.; Algandaby, M.M.; Moussa, A.Y.; Labib, R.M.; Ayoub, N.A.; Abdel-Naim, A.B. Anti-inflammatory activity of pistacia khinjuk in different experimental models: Isolation and characterization of its flavonoids and galloylated sugars. *J. Med. Food* **2012**, *15*, 278–287. [CrossRef] [PubMed]
24. Naouar, M.S.; Mekki, L.Z.; Charfi, L.; Boubaker, J.; Filali, A. Preventive and curative effect of pistacia lentiscus oil in experimental colitis. *Biomed. Pharmacother.* **2016**, *83*, 577–583. [CrossRef]
25. Bisignano, C.; Filocamo, A.; Faulks, R.M.; Mandalari, G. In vitro antimicrobial activity of pistachio (*Pistacia vera* L.) polyphenols. *FEMS Microbiol. Lett.* **2013**, *341*, 62–67. [CrossRef]
26. Smeriglio, A.; Denaro, M.; Barreca, D.; Calderaro, A.; Bisignano, C.; Ginestra, G.; Bellocco, E.; Trombetta, D. In vitro evaluation of the antioxidant, cytoprotective, and antimicrobial properties of essential oil from *Pistacia vera* L. Variety bronte hull. *Int. J. Mol. Sci.* **2017**, *18*, 1212. [CrossRef]
27. Terzo, S.; Caldara, G.F.; Ferrantelli, V.; Puleio, R.; Cassata, G.; Mulè, F.; Amato, A. Pistachio consumption prevents and improves lipid dysmetabolism by reducing the lipid metabolizing gene expression in diet-induced obese mice. *Nutrients* **2018**, *10*, 1857. [CrossRef]
28. Baldassano, S.; Amato, A.; Cappello, F.; Rappa, F.; Mulè, F. Glucagon-like peptide-2 and mouse intestinal adaptation to a high-fat diet. *J. Endocrinol.* **2013**, *217*, 11–20. [CrossRef]
29. Baldassano, S.; Rappa, F.; Amato, A.; Cappello, F.; Mulè, F. GLP-2 as Beneficial factor in the glucose homeostasis in mice fed a high fat diet. *J. Cell. Physiol.* **2015**, *230*, 3029–3036. [CrossRef]
30. Rossmeis, M.; Rim, J.S.; Koza, R.A.; Kozak, L.P. Variation in type 2 diabetes–related traits in mouse strains susceptible to diet-induced obesity. *Diabetes* **2003**, *52*, 1958–1966. [CrossRef]
31. Amato, A.; Caldara, G.F.; Nuzzo, D.; Baldassano, S.; Picone, P.; Rizzo, M.; Mulè, F.; Di Carlo, M. NAFLD and atherosclerosis are prevented by a natural dietary supplement containing curcumin, silymarin, guggul, chlorogenic acid and inulin in mice fed a high-fat diet. *Nutrients* **2017**, *9*, 492. [CrossRef] [PubMed]
32. Oppi, S.; Lüscher, T.F.; Stein, S. Mouse models for atherosclerosis research—Which is my line? *Front. Cardiovasc. Med.* **2019**, *6*. [CrossRef] [PubMed]
33. Nuzzo, D.; Picone, P.; Baldassano, S.; Caruana, L.; Messina, E.; Marino Gammazza, A.; Cappello, F.; Mulè, F.; Di Carlo, M. Insulin resistance as common molecular denominator linking obesity to Alzheimer's disease. *Curr. Alzheimer Res.* **2015**, *12*, 723–735. [CrossRef] [PubMed]
34. Yayeh, T.; Hong, M.; Jia, Q.; Lee, Y.C.; Kim, H.J.; Hyun, E.; Kim, T.W.; Rhee, M.H. Pistacia chinensis inhibits no production and upregulates ho-1 induction via pi-3k/akt pathway in lps stimulated macrophage cells. *Am. J. Chin. Med.* **2012**, *40*, 1085–1097. [CrossRef]
35. Mehla, K.; Balwani, S.; Kulshreshtha, A.; Nandi, D.; Jaisankar, P.; Ghosh, B. Ethyl gallate isolated from pistacia integerrima linn. Inhibits cell adhesion molecules by blocking ap-1 transcription factor. *J. Ethnopharmacol.* **2011**, *137*, 1345–1352. [CrossRef]
36. Zhang, J.; Kris-Etherton, P.M.; Thompson, J.T.; Vanden Heuvel, J.P. Effect of pistachio oil on gene expression of ifn-induced protein with tetratricopeptide repeats 2: A biomarker of inflammatory response. *Mol. Nutr. Food Res.* **2010**, *54*, S83–S92. [CrossRef]
37. Gentile, C.; Perrone, A.; Attanzio, A.; Tesoriere, L.; Livrea, M.A. Sicilian pistachio (*Pistacia vera* L.) nut inhibits expression and release of inflammatory mediators and reverts the increase of paracellular permeability in IL-1β-exposed human intestinal epithelial cells. *Eur. J. Nutr.* **2015**, *54*, 811–821. [CrossRef]
38. Hong, M.Y.; Groven, S.; Marx, A.; Rasmussen, C.; Beidler, J. Anti-inflammatory, antioxidant, and hypolipidemic effects of mixed nuts in atherogenic diet-fed rats. *Molecules* **2018**, *23*, 3126. [CrossRef]
39. Ben Khedir, S.; Mzid, M.; Bardaa, S.; Moalla, D.; Sahnoun, Z.; Rebai, T. In vivo evaluation of the anti-inflammatory effect of pistacia lentiscus fruit oil and its effects on oxidative stress. *Evid.-Based Complement. Altern. Med.* **2016**, *2016*. [CrossRef]
40. Gentile, C.; Allegra, M.; Angileri, F.; Pintaudi, A.M.; Livrea, M.A.; Tesoriere, L. Polymeric proanthocyanidins from sicilian pistachio (*Pistacia vera* L.) nut extract inhibit lipopolysaccharide-induced inflammatory response in raw 264.7 cells. *Eur. J. Nutr.* **2012**, *51*, 353–363. [CrossRef]

41. Kim, H.J.; Neophytou, C. Natural anti-inflammatory compounds for the management and adjuvant therapy of inflammatory bowel disease and its drug delivery system. *Arch. Pharm. Res.* **2009**, *32*, 997–1004. [CrossRef] [PubMed]
42. Papalois, A.; Gioxari, A.; Kaliora, A.C.; Lymperopoulou, A.; Agrogiannis, G.; Papada, E.; Andrikopoulos, N.K. Chios mastic fractions in experimental colitis: Implication of the nuclear factor kappab pathway in cultured ht29 cells. *J. Med. Food* **2012**, *15*, 974–983. [CrossRef] [PubMed]
43. Tafti, L.D.; Shariatpanahi, S.M.; Damghani, M.M.; Javadi, B. Traditional persian topical medications for gastrointestinal diseases. *Iran. J. Basic Med. Sci.* **2017**, *20*, 222–241. [PubMed]
44. Catalani, S.; Palma, F.; Battistelli, S.; Nuvoli, B.; Galati, R.; Benedetti, S. Reduced cell viability and apoptosis induction in human thyroid carcinoma and mesothelioma cells exposed to cidofovir. *Toxicol. In Vitro* **2017**, *41*, 49–55. [CrossRef]
45. Spyridopoulou, K.; Tiptiri-Kourpeti, A.; Lampri, E.; Fitsiou, E.; Vasileiadis, S.; Vamvakias, M.; Bardouki, H.; Goussia, A.; Malamou-Mitsi, V.; Panayiotidis, M.I.; et al. Dietary mastic oil extracted from pistacia lentiscus var. Chia suppresses tumor growth in experimental colon cancer models. *Sci. Rep.* **2017**, *7*, 3782. [CrossRef]
46. Balan, K.V.; Prince, J.; Han, Z.; Dimas, K.; Cladaras, M.; Wyche, J.H.; Sitaras, N.M.; Pantazis, P. Antiproliferative activity and induction of apoptosis in human colon cancer cells treated in vitro with constituents of a product derived from pistacia lentiscus l. Var. Chia. *Phytomedicine* **2007**, *14*, 263–272. [CrossRef]
47. Qiao, J.; Li, A.; Jin, X.; Wang, J. Mastic alleviates allergic inflammation in asthmatic model mice by inhibiting recruitment of eosinophils. *Am. J. Respir. Cell Mol. Biol.* **2011**, *45*, 95–100. [CrossRef]
48. Lumeng, C.N.; Saltiel, A.R. Inflammatory links between obesity and metabolic disease. *J. Clin. Investig.* **2011**, *121*, 2111–2117. [CrossRef]
49. Haynes, P.; Liangpunsakul, S.; Chalasani, N. Nonalcoholic fatty liver disease in individuals with severe obesity. *Clin. Liver Dis.* **2004**, *8*, 535–547. [CrossRef]
50. Marchesi, J.R.; Adams, D.H.; Fava, F.; Hermes, G.D.; Hirschfield, G.M.; Hold, G.; Quraishi, M.N.; Kinross, J.; Smidt, H.; Tuohy, K.M.; et al. The gut microbiota and host health: A new clinical frontier. *Gut* **2016**, *65*, 330–339. [CrossRef]
51. Olefsky, J.M.; Glass, C.K. Macrophages, inflammation, and insulin resistance. *Annu. Rev. Physiol.* **2010**, *72*, 219–246. [CrossRef] [PubMed]
52. Caesar, R.; Tremaroli, V.; Kovatcheva-Datchary, P.; Cani, P.D.; Bäckhed, F. Crosstalk between gut microbiota and dietary lipids aggravates WAT inflammation through TLR Signaling. *Cell Metab.* **2015**, *22*, 658–668. [CrossRef] [PubMed]
53. Ukhanova, M.; Wang, X.; Baer, D.J.; Novotny, J.A.; Fredborg, M.; Mai, V. Effects of almond and pistachio consumption on gut microbiota composition in a randomised cross-over human feeding study. *Br. J. Nutr.* **2014**, *111*, 2146–2152. [CrossRef] [PubMed]
54. Ley, R.E.; Bäckhed, F.; Turnbaugh, P.; Lozupone, C.A.; Knight, R.D.; Gordon, J.I. Obesity alters gut microbial ecology. *Proc. Natl. Acad. Sci. USA* **2005**, *102*, 11070–11075. [CrossRef] [PubMed]
55. Delzenne, N.M.; Neyrinck, A.M.; Cani, P.D. Modulation of the gut microbiota by nutrients with prebiotic properties: Consequences for host health in the context of obesity and metabolic syndrome. *Microb. Cell Factories* **2010**, *10*. [CrossRef]
56. Camps-Bossacoma, M.; Massot-Cladera, M.; Abril-Gil, M.; Franch, A.; Pérez-Cano, F.J.; Castell, M. Cocoa diet and antibody immune response in preclinical studies. *Front. Nutr.* **2017**, *4*, 28. [CrossRef]
57. Nagalingam, N.A.; Kao, J.Y.; Young, V.B. Microbial ecology of the murine gut associated with the development of dextran sodium sulfate-induced colitis. *Inflamm. Bowel Dis.* **2011**, *17*, 917–926. [CrossRef]
58. De Filippo, C.; Cavalieri, D.; Di Paola, M.; Ramazzotti, M.; Poullet, J.B.; Massart, S.; Collini, S.; Pieraccini, G.; Lionetti, P. Impact of diet in shaping gut microbiota revealed by a comparative study in children from Europe and rural Africa. *Proc. Natl. Acad. Sci. USA* **2010**, *107*, 14691–14696. [CrossRef]
59. Zhu, Q.; Gao, R.; Wu, W.; Qin, H. The role of gut microbiota in the pathogenesis of colorectal cancer. *Tumour Biol.* **2013**, *34*, 1285–1300. [CrossRef]
60. Zhang, B.; Lu, X.L.; Song, Y.H.; Shi, H.T.; Li, J.; Geng, Y. Changes in the intestinal microenvironment during development of alcoholic fatty liver disease and related effects of probiotic therapy. *Zhonghua Gan Zang Bing Za Zhi* **2012**, *20*, 848–852.
61. Guarner, F.; Perdigon, G.; Corthier, G.; Salminen, S.; Koletzko, B.; Morelli, L. Should yoghurt cultures be considered probiotic? *Br. J. Nutr.* **2005**, *93*, 783–786. [CrossRef] [PubMed]

62. Ritze, Y.; Bárdos, G.; Claus, A.; Ehrmann, V.; Bergheim, I.; Schwiertz, A.; Bischoff, S.C. Lactobacillus rhamnosus GG protects against non-alcoholic fatty liver disease in mice. *PLoS ONE* **2014**, *9*. [CrossRef]
63. Carbajo-Pescador, S.; Porras, D.; García-Mediavilla, M.V.; Martínez-Flórez, S.; Juarez-Fernández, M.; Cuevas, M.J.; Mauriz, J.L.; González-Gallego, J.; Nistal, E.; Sánchez-Campos, S. Beneficial effects of exercise on gut microbiota functionality and barrier integrity, and gut-liver crosstalk in an in vivo model of early obesity and non-alcoholic fatty liver disease. *Dis. Models Mech.* **2019**, *12*. [CrossRef] [PubMed]
64. Jakobsson, H.E.; Rodríguez-Piñeiro, A.M.; Schütte, A.; Ermund, A.; Boysen, P.; Bemark, M.; Sommer, F.; Bäckhed, F.; Hansson, G.C.; Johansson, M.E. The composition of the gut microbiota shapes the colon mucus barrier. *EMBO Rep.* **2015**, *16*, 164–177. [CrossRef] [PubMed]
65. Endo, H.; Niioka, M.; Kobayashi, N.; Tanaka, M.; Watanabe, T. Butyrate-producing probiotics reduce nonalcoholic fatty liver disease progression in rats: New insight into the probiotics for the gut-liver axis. *PLoS ONE* **2013**, *8*. [CrossRef] [PubMed]
66. Xu, J.; Mahowald, M.A.; Ley, R.E.; Lozupone, C.A.; Hamady, M.; Martens, E.C.; Henrissat, B.; Coutinho, P.M.; Minx, P.; Latreille, P.; et al. Evolution of symbiotic bacteria in the distal human intestine. *PLoS Biol.* **2007**, *5*. [CrossRef] [PubMed]
67. Hamer, H.M.; Jonkers, D.; Venema, K.; Vanhoutvin, S.; Troost, F.J.; Brummer, R.J. Review article: The role of butyrate on colonic function. *Aliment. Pharmacol. Ther.* **2008**, *27*, 104–119. [CrossRef] [PubMed]
68. Yang, C.; Deng, Q.; Xu, J.; Wang, X.; Hu, C.; Tang, H.; Huang, F. Sinapic acid and resveratrol alleviate oxidative stress with modulation of gut microbiota in high-fat diet-fed rats. *Food Res. Int.* **2019**, *116*, 1202–1211. [CrossRef]
69. Zhou, D.; Chen, Y.W.; Zhao, Z.H.; Yang, R.X.; Xin, F.Z.; Liu, X.L.; Pan, Q.; Zhou, H.; Fan, J.G. Sodium butyrate reduces high-fat diet-induced non-alcoholic steatohepatitis through upregulation of hepatic GLP-1R expression. *Exp. Mol. Med.* **2018**, *50*. [CrossRef]
70. Liu, W.; Crott, J.W.; Lyu, L.; Pfalzer, A.C.; Li, J.; Choi, S.W.; Yang, Y.; Mason, J.B.; Liu, Z. Diet- and Genetically-induced Obesity Produces Alterations in the Microbiome, Inflammation and Wnt Pathway in the Intestine of Apc+/1638N Mice: Comparisons and Contrasts. *J. Cancer* **2016**, *7*, 1780–1790. [CrossRef]
71. Johnson, A.M.; Costanzo, A.; Gareau, M.G.; Armando, A.M.; Quehenberger, O.; Jameson, J.M.; Olefsky, J.M. High fat diet causes depletion of intestinal eosinophils associated with intestinal permeability. *PLoS ONE* **2015**, *10*. [CrossRef] [PubMed]
72. Jiao, N.; Baker, S.S.; Nugent, C.A.; Tsompana, M.; Cai, L.; Wang, Y.; Buck, M.J.; Genco, R.J.; Baker, R.D.; Zhu, R.; et al. Gut microbiome may contribute to insulin resistance and systemic inflammation in obese rodents: A meta-analysis. *Physiol. Genom.* **2018**, *50*, 244–254. [CrossRef] [PubMed]
73. Liang, Y.; Zhang, Y.; Deng, Y.; Liang, S.; He, Y.; Chen, Y.; Liu, C.; Lin, C.; Han, L.; Tu, G.; et al. Chaihu-Shugan-San. Decoction modulates intestinal microbe dysbiosis and alleviates chronic metabolic inflammation in NAFLD rats via the NLRP3 inflammasome pathway. *Evid.-Based Complement. Alternat. Med.* **2018**, *2018*. [CrossRef] [PubMed]
74. Beaumont, M.; Andriamihaja, M.; Lan, A.; Khodorova, N.; Audebert, M.; Blouin, J.M.; Grauso, M.; Lancha, L.; Benetti, P.H.; Benamouzig, R.; et al. Detrimental effects for colonocytes of an increased exposure to luminal hydrogen sulfide: The adaptive response. *Free Radic. Biol. Med.* **2016**, *93*, 155–164. [CrossRef] [PubMed]
75. Byndloss, M.X.; Olsan, E.E.; Rivera-Chávez, F.; Tiffany, C.R.; Cevallos, S.A.; Lokken, K.L.; Torres, T.P.; Byndloss, A.J.; Faber, F.; Gao, Y.; et al. Microbiota-activated PPAR-γ signaling inhibits dysbiotic Enterobacteriaceae expansion. *Science* **2017**, *357*, 570–575. [CrossRef]
76. Guo, J.; Han, X.; Zhan, J.; You, Y.; Huang, W. Vanillin alleviates high fat diet-induced obesity and improves the gut microbiota composition. *Front. Microbiol.* **2018**, *9*. [CrossRef]
77. Li, X.; Wang, H.; Wang, T.; Zheng, F.; Wang, H.; Wang, C. Dietary wood pulp-derived sterols modulation of cholesterol metabolism and gut microbiota in high-fat-diet-fed hamsters. *Food Funct.* **2019**, *10*, 775–785. [CrossRef]
78. Wang, Y.; Fei, Y.; Liu, L.; Xiao, Y.; Pang, Y.; Kang, J.; Wang, Z. Polygonatum odoratum polysaccharides modulate gut microbiota and mitigate experimentally induced obesity in rats. *Int. J. Mol. Sci.* **2018**, *19*, 3587. [CrossRef]
79. Thevaranjan, N.; Puchta, A.; Schulz, C.; Naidoo, A.; Szamosi, J.C.; Verschoor, C.P.; Loukov, D.; Schenck, L.P.; Jury, J.; Foley, K.P.; et al. Age-associated microbial dysbiosis promotes intestinal permeability, systemic inflammation, and macrophage dysfunction. *Cell Host Microbe* **2017**, *21*, 455–466. [CrossRef]

80. Santisteban, M.M.; Qi, Y.; Zubcevic, J.; Kim, S.; Yang, T.; Shenoy, V.; Cole-Jeffrey, C.T.; Lobaton, G.O.; Stewart, D.C.; Rubiano, A.; et al. Hypertension-linked pathophysiological alterations in the gut. *Circ. Res.* **2017**, *120*, 312–323. [CrossRef]
81. Wu, S.; Hu, R.; Nakano, H.; Chen, K.; Liu, M.; He, X.; Zhang, H.; He, J.; Hou, D.X. Modulation of gut microbiota by lonicera caerulea L. berry polyphenols in a mouse model of fatty liver induced by high fat diet. *Molecules* **2018**, *23*, 3213. [CrossRef] [PubMed]
82. David, L.A.; Maurice, C.F.; Carmody, R.N.; Gootenberg, D.B.; Button, J.E.; Wolfe, B.E.; Ling, A.V.; Devlin, A.S.; Varma, Y.; Fischbach, M.A.; et al. Diet rapidly and reproducibly alters the human gut microbiome. *Nature* **2014**, *505*, 559–563. [CrossRef] [PubMed]
83. Devkota, S.; Wang, Y.; Musch, M.W.; Leone, V.; Fehlner-Peach, H.; Nadimpalli, A.; Antonopoulos, D.A.; Jabri, B.; Chang, E.B. Dietary-fat-induced taurocholic acid promotes pathobiont expansion and colitis in Il10-/-mice. *Nature* **2012**, *487*, 104–108. [CrossRef] [PubMed]
84. Cani, P.D.; Lecourt, E.; Dewulf, E.M.; Sohet, F.M.; Pachikian, B.D.; Naslain, D.; De Backer, F.; Neyrinck, A.M.; Delzenne, N.M. Gut microbiota fermentation of prebiotics increases satietogenic and incretin gut peptide production with consequences for appetite sensation and glucose response after a meal. *Am. J. Clin. Nutr.* **2009**, *90*, 1236–1243. [CrossRef] [PubMed]
85. Cani, P.D.; Bibiloni, R.; Knauf, C.; Waget, A.; Neyrinck, A.M.; Delzenne, N.M.; Burcelin, R. Changes in gut microbiota control metabolic endotoxemia-induced inflammation in high-fat diet-induced obesity and diabetes in mice. *Diabetes* **2008**, *57*, 1470–1481. [CrossRef]
86. Kleiner, D.E.; Brunt, E.M. Nonalcoholic fatty liver disease: Pathologic patterns and biopsy evaluation in clinical research. *Semin. Liver Dis.* **2012**, *32*, 3–13. [CrossRef] [PubMed]
87. Wu, N.; Siow, Y.L.; Karmin, O. Induction of hepatic cyclooxygenase-2 by hyperhomocysteinemia via nuclear factor-kappaB activation. *Am. J. Physiol. Regul. Integr. Comp. Physiol.* **2009**, *297*, 1086–1094. [CrossRef]

International Journal of
Molecular Sciences

MDPI

Article

Chitosan Ameliorates DSS-Induced Ulcerative Colitis Mice by Enhancing Intestinal Barrier Function and Improving Microflora

Jia Wang, Cuili Zhang, Chunmei Guo and Xinli Li *

Department of Biotechnology, College of Basic Medical Sciences, Dalian Medical University, Dalian 116044, China; wangjia@dmu.edu.cn (J.W.); biot_zhangcl@sina.com (C.Z.); 18340841956@139.com (C.G.)
* Correspondence: lixinlibio@hotmail.com; Tel.: +86-0411-8611-0296

Received: 25 October 2019; Accepted: 12 November 2019; Published: 15 November 2019

check for updates

Abstract: Ulcerative colitis (UC) has been identified as one of the inflammatory diseases. Intestinal mucosal barrier function and microflora play major roles in UC. Modified-chitosan products have been consumed as effective and safe drugs to treat UC. The present work aimed to investigate the effect of chitosan (CS) on intestinal microflora and intestinal barrier function in dextran sulfate sodium (DSS)-induced UC mice and to explore the underlying mechanisms. KM (Kunming) mice received water/CS (250, 150 mg/kg) for 5 days, and then received 3% DSS for 5 days to induce UC. Subsequently, CS (250, 150 mg/kg) was administered daily for 5 days. Clinical signs, body weight, colon length, and histological changes were recorded. Alterations of intestinal microflora were analyzed by PCR-DGGE, expressions of TNF-α and tight junction proteins were detected by Western blotting. CS showed a significant effect against UC by the increased body weight and colon length, decreased DAI (disease activity index) and histological injury scores, and alleviated histopathological changes. CS reduced the expression of TNF-α, promoted the expressions of tight junction proteins such as claudin-1, occludin, and ZO-1 to maintain the intestinal mucosal barrier function for attenuating UC in mice. Furthermore, *Parabacteroides*, *Blautia*, *Lactobacillus*, and *Prevotella* were dominant organisms in the intestinal tract. *Blautia* and *Lactobacillus* decreased with DSS treatment, but increased obviously with CS treatment. This is the first time that the effect of original CS against UC in mice has been reported and it is through promoting dominant intestinal microflora such as *Blautia*, mitigating intestinal microflora dysbiosis, and regulating the expressions of TNF-α, claudin-1, occludin, and ZO-1. CS can be developed as an effective food and health care product for the prevention and treatment of UC.

Keywords: chitosan; ulcerative colitis; tight junction protein; intestinal microflora

1. Introduction

Ulcerative colitis (UC) has been identified as one of the modern inflammatory diseases [1]. It is a chronic and nonspecific inflammatory bowel disease (IBD), which presents with abdominal pain, diarrhea and bloody mucopurulent stool [2], and associated with a high risk of colon cancer if not treated in a timely manner.

The pathogenesis of UC is not clearly understood, and it is generally considered to be caused by multiple factors [3]. Among these, intestinal mucosal barrier function and microflora play major roles in which UC occurs and develops [4]. The intestinal mucosal barrier is the first barrier against a hostile environment, mainly formed by the tight junctions (TJs) of epithelial cells. TJs consist of transmembrane proteins (occludins and claudins) and accessory proteins (zonula occludens) for preventing the spread of pathogens and harmful antigens across the epithelium. ZOs (zonula occludens), occludin, and claudins are thought to be important integral membrane proteins which participate in TJ structural integrity by binding to a actin-cytoskeleton [5]. In addition, there is a strong relationship between intestinal microflora and intestinal barrier function. Intestinal microflora dysbiosis decreases the intestinal mucosal barrier function and increases bacterial translocation, and intestinal pathogenic bacteria damage structural barriers by changing intestinal TJ proteins [6]. Thus, the research works focusing on effective methods to regulate intestinal mucosal barrier function and balance intestinal microflora for treatment of UC are important.

Most drugs for UC treatment interfere with metabolism and immune responses, often causing some serious adverse reactions. Alternative treatments, including probiotics and nutritional supplements, have been given more attention due to fewer side effects [7]. A range of dietary oligosaccharides, such as lactosucrose, galactooligosaccharides, fructooligosaccharides, and isomaltooligosaccharides, have nutritional supplement properties [8,9]. Chitosan (CS) is a polymer obtained by the deacetylation of chitin extracted from the exoskeletons of crustaceans [10], which possesses diverse biological and pharmacological effects, including antitumor, anti-inflammatory, antioxidant, anticoagulant, wound healing, antimicrobial, anti-obesity, and antidiabetic activities [11–13]. As a main marine natural compound, CS is a much sought after bioresource due to its therapeutic value, stability, biodegradability, biocompatibility and low toxicity. In recent years, modified-chitosan products have been consumed as effective and safe drugs to treat UC [14,15]. To avoid unintended absorption of the drug or its degradation products in the gastrointestinal tract, rectal administration was used. However, intestinal microflora has an important role in chronic human diseases, it should not be ignored. Our previous work [16] also indicated that CS was a potential food supplement for protecting intestinal microflora. *Lactobacillus* was promoted with a CS treatment in an antibiotics-induced intestinal dysbiosis mice model. In the present study, the curative effect and mechanism of original CS were evaluated in a DSS-induced UC mice model by the alterations of intestinal microflora, and the expressions of TNF-α and tight junction proteins. We attempted to explore the association between intestinal microflora and UC, and to provide a novel insight into the mechanisms of CS.

2. Results

2.1. Effects of CS on Body Weight and DAI in DSS-Induced UC Mice

On the first 5 days of the experiment, the mice in each group showed a steady increase in body weight. Subsequently, the mice were treated with 3% DSS for 5 days to induce UC. The control mice still showed a steady increase in body weight. But the DSS-alone treatment group had a significantly decreased body weight and increased DAI score ($p < 0.01$) compared with control mice. Both CSH and CSL treatment groups reduced the body weight loss, and attenuated the increased DAI score (Figure 1A,B). These results indicated that CS effectively relieved DSS-induced UC symptoms.

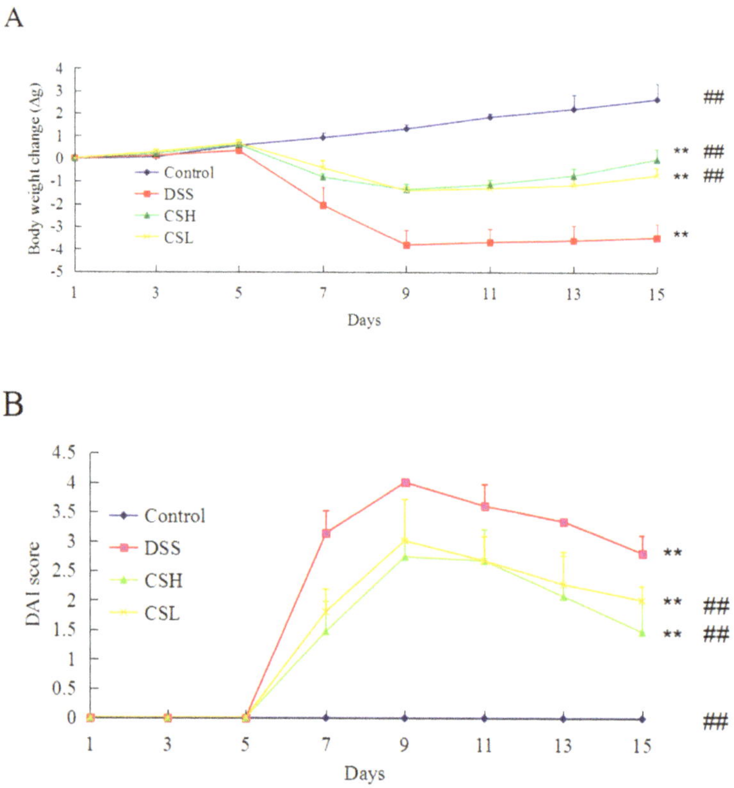

Figure 1. Effects of CS on body weight (**A**) and the disease activity index (DAI) (**B**) in DSS-induced UC mice. Control: normal mice; DSS: mice treated with 3% DSS alone; CSH: mice treated with DSS plus chitosan (250 mg/kg); CSL: mice treated with DSS plus chitosan (150 mg/kg). Values are expressed as mean ± SD ($n = 10$). ** $p < 0.01$ versus control; ## $p < 0.01$ versus DSS-alone.

2.2. Effects of CS on Colon Length and Histopathology in DSS-Induced UC Mice

Colon length was shortened in all DSS-treated mice. The colon length of the DSS-only treatment group showed a significant reduction compared with the control group ($p < 0.01$). Both CSH and CSL alleviated the effects of DSS on colon length shortening (Figure 2A,B).

As shown in Figure 2C, intact colonic epithelial cells and crypt structure, and complete goblet cells were observed in the control group. Severe lesions were present in all DSS-treated groups, with loss of colonic epithelial cells, distortion of crypt structure, and massive inflammatory cell infiltration. However, compared with the DSS-only treatment group, the colons of CSH-treated mice showed ameliorated structural damage, exhibited less inflammatory cell infiltration and only mild evidence of crypt distortion. In addition, both CSH and CSL treatment groups resulted in a significant reduction of the histological injury scores caused by DSS (Figure 2D; $p < 0.01$). Taken together, our results suggested that CS significantly protected colon tissue and attenuated DSS-induced tissue morphological changes.

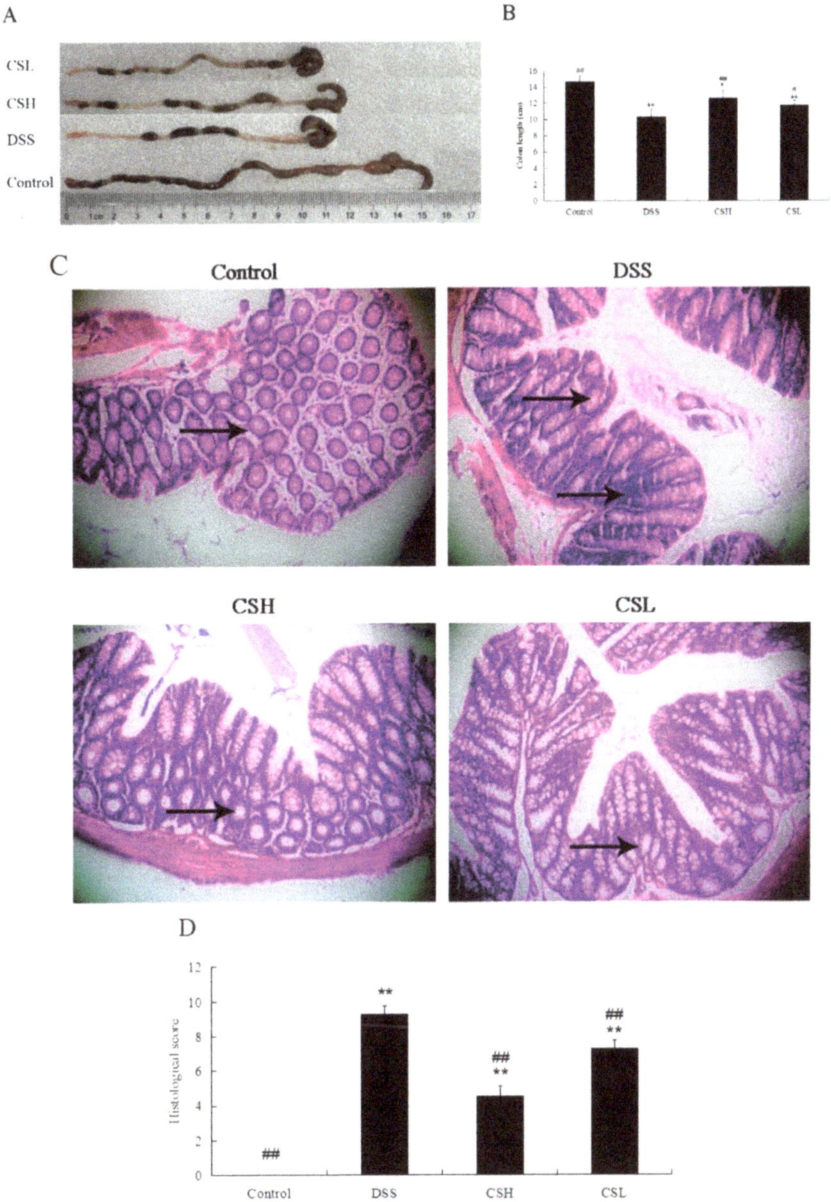

Figure 2. Effects of CS on colon length and histopathology in DSS-induced UC mice. Representative colons (**A**). Colon length (**B**). Histopathology (magnification 200×) (**C**). Histopathological scores (**D**). Arrows indicated the inflammatory infiltration, mucosal erosion, and damage of crypts. Control: normal mice; DSS: mice treated with 3% DSS alone; CSH: mice treated with DSS plus chitosan (250 mg/kg); CSL: mice treated with DSS plus chitosan (150 mg/kg). Values are expressed as mean ± SD ($n = 10$). * $p < 0.05$ and ** $p < 0.01$ versus control; # $p < 0.05$ and ## $p < 0.01$ versus DSS-alone.

2.3. Effects of CS on Expressions of TNF-α, Claudin-1, Occludin, and ZO-1

Western blotting analysis (Figure 3) showed that the expression levels of claudin-1, occludin, and ZO-1 were significantly decreased in the DSS-alone treatment group. Meanwhile, the expression of TNF-α increased significantly in this group. Compared with the DSS-alone treatment group, the expressions of claudin-1, occludin, and ZO-1 increased, but expression of TNF-α decreased in both CSH and CSL treatment groups significantly. Furthermore, as shown in Figure 2D, CS at the dose of 250 mg/kg (CSH) significantly increased the expressions of claudin-1 and occludin, which were even better than the effects produced by CS at the dose of 150 mg/kg (CSL). We proposed that the effects of CS on DSS-induced UC mice were related to the regulation of the colonic mucosal barrier function, where the expressions of ZO-1, occludin, and claudin-1 play important roles in maintaining the intestinal mucosal barrier function.

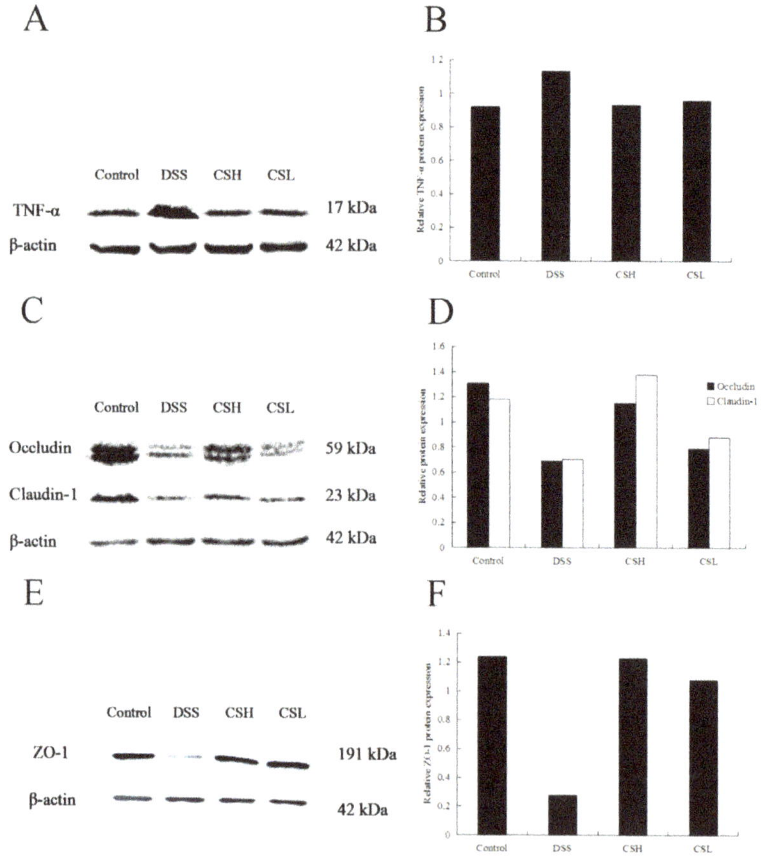

Figure 3. Effects of CS on protein expression of TNF-α, claudin-1, occludin, and ZO-1. Control: normal mice; DSS: mice treated with 3% DSS alone; CSH: mice treated with DSS plus chitosan (250 mg/kg); CSL: mice treated with DSS plus chitosan (150 mg/kg). Values are expressed as mean ± SD (n = 10).

2.4. PCR-DGGE Analysis

The dominant intestinal microflora of experimental groups at different time intervals was examined by PCR-DGGE analysis (Figure 4A). The gene sequencing results were showed in Table 1. Mice pretreated with water/CS for 5 days, and then received 3% DSS for 5 days to induce UC. On the

11th day of the experiment, *Lactobacillus johnsonii* (band d) was found in all groups (D11-Group 1–4). But the intensities of *Blautia* sp. (band c) and *Lactobacillus ruminis* (band e) weakened in the DSS-alone treatment group (D11-Group 2). Especially, *Blautia* sp. (band c) almost disappeared in the DSS-alone treatment group, but this band existed in both CSH and CSL pretreated groups. The intensity of *Blautia* sp. (band c) in the CSH group (D11-Group 3) was stronger than that in CSL group (D11-Group 4). In addition, *Parabacteroides distasonis* (band a) and *Prevotella intermedia* (band f) existed in all DSS-treated groups, but did not exist in the control group. On the 15th day of the experiment, the mice had stopped taking DSS and continued to receive water/CS for 5 days. *Blautia* sp. (band c) still decreased in the DSS-alone treatment group, almost disappearing. *Parabacteroides distasonis* (band a) did not exist in all groups. Other bands had no significant change at different time intervals (D11 and D15). The bacterium from the genera of *Parabacteroides, Blautia, Lactobacillus*, and *Prevotella* were dominant organisms in the intestinal tract of mice. We further proposed that the alterations of dominant intestinal microflora play causal roles in UC. Specifically, *Blautia* and *Lactobacillus* decreased in the DSS-alone treatment group, which showed that the intestinal microflora balance was disturbed by DSS. CS treatment mitigated intestinal microflora dysbiosis, *Blautia* and *Lactobacillus* increased with CS treatment, and the effect produced by CSH was better than the actions produced by CSL.

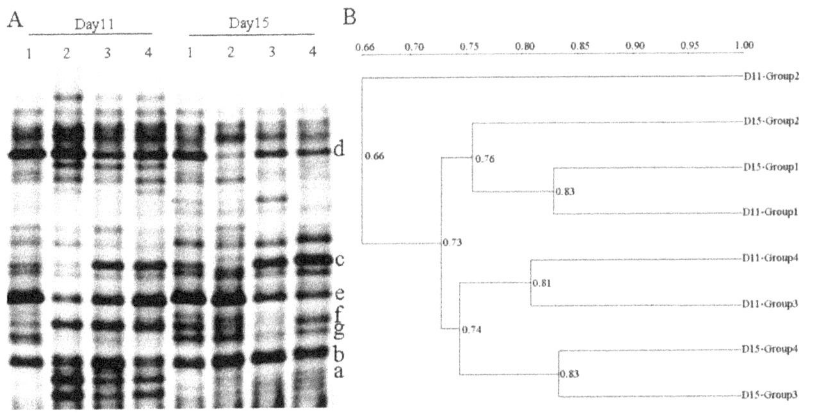

Figure 4. Representative denaturing gradient gel electrophoresis (DGGE) profiles (**A**) and unweighted pair group method using arithmetic average (UPGMA) dendrograms (**B**) of intestinal microflora at different time intervals. D11: the 11th day of the experiment. D11-Group 1: mice treated with normal water; D11-Group 2: mice treated with 3% DSS alone; D11-Group 3: mice treated with DSS plus chitosan (250 mg/kg); D11-Group 4: mice treated with DSS plus chitosan (150 mg/kg). D15: the 15th day of the experiment. D15-Group 1: mice treated with normal water; D15-Group 2: mice treated with normal water; D15-Group 3: mice treated with chitosan (250 mg/kg); D15-Group 4: mice treated with chitosan (150 mg/kg). Bands marked with letters a–g were excised and proceeded for sequencing.

Figure 4B displayed that different groups formed the statistically significant UPGMA clustering profiles. There were two main clusters in the dendrogram, the first was D11-Group 2 (related the DSS-alone treatment group (D11)), the second was the remaining seven groups. The minimum bacterial similarity index between cluster one and cluster two was 0.66, which suggested that the intestinal microflora community of DSS-induced UC mice was seriously damaged. In cluster two, there were also two secondary clusters. The maximum bacterial similarity index between D11-Group 1 and D15-Group 1 was 0.83, and D15-Group 2 possessed high similarity (0.76) to the above two groups, which suggested that the intestinal microflora community of normal mice was little changed at D11 and D15. Moreover, the similarity between the CS-treated groups at D11 and D15 was 0.74, which suggested that the treatment cycle of CS had little effect on the intestinal microflora community.

Table 1. Sequences of bands a–g based on the BLAST database.

Selected Band	Most Similar Sequence Relative (GenBank Accession Number)	Bacteria Genus	Identity (%)
a	*Parabacteroides distasonis* (NZ 009615.1)	*Parabacteroides*	83
b	*Parabacteroides* sp. (NC CP015402.2)		91
c	*Blautia* sp. (NZ CP015405.2)	*Blautia*	97
d	*Lactobacillus johnsonii* (NZ 022909.1)	*Lactobacillus*	92
e	*Lactobacillus ruminis* (NZ 015975.1)		91
f	*Prevotella intermedia* (NZ CP019301.1)	*Prevotella*	87
g	*Prevotella dentalis* (NZ 019968.1)		89

The richness (S), diversity index (H') and evenness score (E) decreased in all DSS-treated groups (Table 2). Compared to the control group, the diversity index was lower in the DSS-only treatment group ($p < 0.01$), but increased in CS-treated groups with no significant difference. It appeared that intestinal microflora community was changed by DSS with S, H', and E decreasing. However, the intestinal microflora community of the mice treated with CS was considerably ameliorated, indicating that CS showed a significant effect on the intestinal microflora in mice.

Table 2. Microflora diversity indexes analysis.

Group	S	H'	E
D11-Group1	20.50 ± 1.29	3.0679 ± 0.1258	1.0165 ± 0.0439
D11-Group2	18.50 ± 0.58 *	2.7446 ± 0.0469 **	0.9410 ± 0.0259 *
D11-Group3	20.25 ± 0.96	2.9627 ± 0.0473	0.9852 ± 0.0111
D11-Group4	18.75 ± 0.50 *	2.9097 ± 0.0270	0.9929 ± 0.0218
D15-Group1	19.75 ± 0.96	2.9585 ± 0.0480	0.9922 ± 0.0226
D15-Group2	20.25 ± 0.50	2.8867 ± 0.0244 *	0.9597 ± 0.0145 *
D15-Group3	19.25 ± 0.96	2.9241 ± 0.0270	0.9889 ± 0.0267
D15-Group4	20.00 ± 0.82	2.9269 ± 0.0409	0.9775 ± 0.0279

Results are expressed as mean ± SD ($n = 10$). * $p < 0.05$ and ** $p < 0.01$ versus control (D11-Group 1). $H' = -\sum (p_i)(\ln p_i)$, p_i was the proportion of the bands in the track, $p_i = n_i/\sum n_i$, n_i was the average density of peak i in the densitometric curve. E = H'/ln S, S was the number of bands. D11: the 11th day of the experiment. D11-Group 1: mice treated with normal water; D11-Group 2: mice treated with 3% DSS alone; D11-Group 3: mice treated with DSS plus chitosan (250 mg/kg); D11-Group 4: mice treated with DSS plus chitosan (150 mg/kg). D15: the 15th day of the experiment. D15-Group 1: mice treated with normal water; D15-Group 2: mice treated with normal water; D15-Group 3: mice treated with chitosan (250 mg/kg); D15-Group 4: mice treated with chitosan (150 mg/kg).

3. Discussions

Increasing evidence [17] has demonstrated that intestinal mucosal barrier dysfunction is critical in UC development. As a key component of the intestinal mucosal barrier, TJ proteins seal the gaps between adjacent intestinal epithelial cells and keeps substances such as antigens and microbes contained in the lumen. They play important roles in the maintenance of intestinal permeability, tissue differentiation, and homeostasis. Previous reports have demonstrated that dietary threonine maintained intestinal barrier function by modulating intestinal TJ proteins synthesis [18]. Fermented *Pueraria lobata* extract ameliorated DSS-induced inflammation in the colon, and recovered the disrupted intestinal barrier through retrieving the expression and architecture of TJ proteins [19]. Expressions of ZO-1, claudin-3, and occludin decreased in enterotoxigenic *Escherichia coli* K88-infected intestinal mucosa damaged pig [20]. Thus, regulation of TJs to keep epithelial barrier integrity in UC is vitally important. Excessive tumor necrosis factor-alpha (TNF-α) expands a local or systemic inflammation, which triggers a disturbance of both TJ proteins and intestinal mucous barrier functions [21], and is associated with UC severity. TNF-α has long been recognized as the key inflammatory mediator in colon inflammation [22]. Blockade of TNF-α activity has proven to be an effective way of inhibiting inflammation. BaweiXileisan (a traditional Chinese compound medicine) inhibited TNF-α expression and improved the mucosa barrier function in DSS-induced UC mice [23]. Hydroxynaphthoquinone mixtures exerted their anti-inflammatory actions through inhibiting TNF-α, down-regulating nuclear factor-κB (NF-κB) signaling. In clinics, TNF-α blockers such as infliximab, adalimumab, and certolizumab pegol have been successfully used for the treatment of IBD patients [24]. In the present study, CS showed significant effects against UC in mice by the increased body weight and colon length, decreased DAI scores and histological injury scores, and alleviated histopathological changes. CS was effective in reducing the expression of TNF-α, and promoting the expressions of TJ proteins such as claudin-1, occludin, and ZO-1 to maintain intestinal mucosal barrier function for attenuating UC in mice, and to this effect, a high dose of CS was better than a low dose.

Probiotics/prebiotics have been suggested as a useful integrative treatment of inflammatory bowel disease, for their ability to alter the expression of epithelial tight junctions. For example, *Lactobacillus rhamnosus* and *Lactobacillus plantarum* strengthened intestinal barrier function, promoted TJ integrity, and protected against experimental necrotizing enterocolitis [25]. A probiotic mixture of *Bacillus subtilis* and *Enterococcus faecium* improved gut microbiota, ameliorated permeability of the intestinal epithelial cell barrier, and enhanced intestinal integrity through up-regulating the expressions of occludin, ZO-1, and JAM-A in heat stress-induced laying hens [26]. Likewise, prebiotic supplements lead to changes in the intestinal microflora, and further improves patients' well-being and health. Alpinetin decreases intestinal inflammation, and regulates the expression of tight junctions in UC mice. Purple potato extract could be used as a supportive dietary therapeutic strategy for improving gut epithelial health through improving gut epithelial differentiation and barrier function [27]. Accordingly, the effects of probiotic/prebiotics treatment in DSS-induced UC are more important. We were interested in examining whether CS has an impact on the intestinal microflora in UC mice. In the present work, both *Blautia* and *Lactobacillus* decreased with DSS administration for 5 days, and *Blautia* almost disappeared after DSS withdrawal, which showed that the intestinal microflora balance may seriously be disturbed by DSS. However, the CSH group showed higher counts for *Blautia*, and higher diversity index, which demonstrated that a high dose of CS better mitigated intestinal microflora dysbiosis and had beneficial effects on *Blautia*. Additionally, *Blautia* is one of the major intestinal microbes often found in human fecal samples [28]. There is a strong relationship between decreased levels of genus *Blautia* and diseases. Increasing the ratio of *Blautia* in the intestine might be beneficial for health [29]. An increase of *Blautia coccoides* through the intake of Japanese koji might be one mechanism explaining Japanese longevity. So the effect of supplements on health, which increased the ratio of *Blautia* in the intestinal microflora, is of significant concern. Our findings showed that CSH improved *Blautia* and *Lactobacillus*. *Blautia* and *Lactobacillus* can protect the intestine by producing antibacterial substances such as lactic acid and short-chain fatty acids, competing for the nutrients and intestinal adhesion sites to inhibit pathogenic bacteria, and preventing cell apoptosis so as to enhance intestinal barrier function [30]. Moreover, *Blautia coccoides* decreases NF-κB activity in Caco-2 cells [31]. Most anti-inflammatory drugs activate the NF-κB signaling pathway, which in turn promotes and controls the expression of TNF-α and other cytokines. As expected, the expression of TNF-α was reduced with CS treatment. Thus, *Blautia* may act as a key regulator in the pathogenesis of UC.

DSS causes a change in the intestinal microflora composition and induces intestinal barrier dysfunction in mice. Intestinal microflora dysbiosis is further associated with UC and a reversal occurred by CS treatment. We hypothesize that the expressions of TNF-α and TJ proteins may be affected by dominant intestinal microflora such as *Blautia*. CS has prebiotic-like effects which can induce microbial competition and reduce the populations of non-beneficial intestinal microflora. Further studies are required to decipher the role of dominant intestinal microflora in the development of UC.

4. Materials and Methods

4.1. Material and Reagents

Chitosan (M_V = 21.70 × 10^4 Da, DD ≥ 95%) was purchased from Jinan Haidebei Marine Biological Engineering Co., Ltd. (Jinan, China). A voucher specimen (No. CS 201701) was deposited in the Department of Biotechnology, Dalian Medical University, China. DSS (MW 36–50 kD) was purchased from MP Biomedicals (Santa Ana, CA, USA). Stool DNA extract kit was purchased from ForeGenen (Chengdu, China). Polymerase Chain Reaction primers GC-357f (CGCCCGGGGCGCGCCCCGGGCGGGGCGGGGGACGGGGGGCCTACGGGAGGCAGCAG), 518r (ATTACCGCGGCTGCTGG) and 357f (CCTACGGGAGGCAGCAG) were synthesized by TaKaRa Biotechnology Co., Ltd. (Dalian, China). A PCR Mix kit was purchased from Beijing TransGen Biotech Co., Ltd. (Beijing, China). Antibodies against TNF-α, ZO-1, claudin-1, occludin, β-actin, and HRP-conjugated affinipure goat anti-rabbit IgG (H+L) were obtained from Proteintech Group Inc. (Chicago, IL, USA). The enhanced chemiluminescence (ECL) kit was from Amersham Life Science, Inc. (Arlington Heights, IL, USA). All other chemical reagents used were of analytical grade.

4.2. Animals and Experimental Design

Male KM mice weighing 20 ± 2 g were provided by the Experimental Animal Center of Dalian Medical University, Dalian, China (Quality certificate number: SCXK (Liao) 2013–0003; 20 May 2013). All experimental procedures were approved by the Animal Care and Use Committee of Dalian Medical University and performed in strict accordance with the People's Republic of China Legislation Regarding the Use and Care of Laboratory Animals (Approval number: SYXK (Liao) 2013–0006; 18 November 2013). The mice were kept under standardized conditions at a temperature of 22–24 °C, and 20% humidity with a 12 h light/dark cycle, and they had free access to standard diet and water ad libitum. After acclimatization for one week, 40 mice were randomly divided into four groups (n = 10). Two groups (Group 1 and Group 2) received normal drinking water only, and the other two groups received CS at doses of 250 (high dose of CS, CSH) or 150 (low dose of CS, CSL) mg/kg by oral gavage for 5 days. On the 6th day, all animals except Group 1 received 3% DSS dissolved in drinking water for 5 days to induce UC [19]. From Day 11 to 15, Group 1 and Group 2 received normal drinking water, and served as control and DSS groups, respectively. The other two groups received CS at doses of 250 or 150 mg/kg by oral gavage. All mice were sacrificed 12 h later after the last administration. The fecal samples were collected on the 11th and 15th day, respectively, and stored at −80 °C for intestinal microflora analysis. The length of the colons was measured and then washed instantly using ice-cold physiological saline. One part of colon tissue was rapidly divided and fixed in 10% formalin for pathological examination, and the remaining parts were stored at −80 °C for Western blotting assay.

4.3. Evaluation of Disease Activity Index

The mice were checked daily for UC based on body weight, gross rectal bleeding, and stool consistency. A disease activity index (DAI) score was calculated according to a described method [32] to assess the disease severity.

4.4. Colon Histopathology

The length of colon was measured. Then a 0.5 cm colon segment was fixed in 10% formalin for 24 h, paraffin embedded, sliced into 5 μm sections, and stained with hematoxylin-eosin (H & E) for histopathological examination. Each sample was observed at 200× magnification. Histological scores were given on a scale as described previously [33].

4.5. Western Blotting Assay

Total protein was extracted from the colon samples using a RIPA lysis buffer with protease inhibitors in a proportion of 1:100. The BCA assay kit was used to quantitate protein. Equal amounts of protein (50 μg) were separated by 10% SDS-PAGE gel using 100 V for 2 h and then transferred to a nitrocellulose membrane by semi-dry apparatus for 40 min for β-actin, 25 min for TNF-α, 30 min for claudin-1, 60 min for occludin, and 180 min for ZO-1, respectively. The membranes were blocked with 5% non-fat milk for 2 h at room temperature and then incubated with primary antibodies against TNF-α, claudin-1, occludin, ZO-1, and β-actin, respectively, at a 1:500 dilution overnight at 4 °C. The next day, the membranes were incubated with secondary antibody at a 1:5000 dilution for 2 h at room temperature after washed with TBST for three times. Then, the protein bands were visualized using an ECL kit by Bio-rad ChemiDoc XRS plus an image analyzer (Bio-Rad, Hercules, CA, USA) after TBST washing, as previously described. β-actin was used as internal reference.

4.6. Polymerse Chain Rection (PCR)-Denaturing Gradient Gel Electrophoresis (DGGE) Analysis

Total bacterial DNA was extracted from fecal samples with a Stool DNA kit. For 16S rRNA gene analysis, primers GC-357f and 518r were used to amplify the V3 region. PCR-DGGE analysis was performed with the methods reported in our previous study [34]. Some separated and strong bands were cut out and eluted in 20 μL sterile water at 4 °C overnight. The eluted DNA was reamplified using 357f and 518r primers with the same PCR program. Idiographic sequences were attained by TaKaRa Biotechnology (Dalian, China) Co., Ltd. The results were compared directly with those in GeneBank by Blast search (http://blast.ncbi.nlm.nih.gov/Blast.cgi).

4.7. Statistical Analysis

SPSS version 17.0 was used for analysis. p values were determined using Student's t-test, p-value < 0.01 was considered significant. DGGE gels and Western blotting gels were analyzed by using Quantity One 4.6.2 gel analysis software (Bio-Rad, Hercules, CA, USA). The Shannon–Wiener index of diversity (H') was used to determine the diversity of the bacterial community. The evenness (E) which reflected uniformity of bacterial species distribution was also computed.

5. Conclusions

CS can effectively reduce symptoms in a mouse model of DSS-induced UC and improve intestinal mucosal barrier function and affect intestinal microflora. A high dose of CS better regulates the expressions of TNF-α and TJ proteins such as claudin-1, occludin, and ZO-1. Moreover, the intestinal microflora composition of UC was distinct from controls, and CS treatment can mitigate intestinal microflora dysbiosis. These findings provide novel insights into the mechanisms of CS as a potential agent to ameliorate the severity of UC. Thus, CS can be developed as an effective food and health care product for the prevention of UC and restoration of intestinal microflora balance.

Author Contributions: J.W. analyzed the data and drafted the manuscript. C.L.Z. and C.M.G. participated in the animal experiments and performed the PCR-DGGE analysis. X.L.L. performed the Western blotting analysis, drew the figures and modified the manuscript. All authors have read and approved the final manuscript.

Funding: This work was supported by Foundation of Liaoning Educational Committee of China (LZ2019059), National Natural Science Foundation of China (31900517), and College of Basic Medical Sciences of Dalian Medical University.

Conflicts of Interest: The authors declare no conflict of interest.

References

1. Adams, S.M.; Bornemann, P.H. Ulcerative colitis. *Am. Fam. Physician* **2013**, *87*, 699–705. [PubMed]
2. Zhang, Z.L.; Fan, H.Y.; Yang, M.Y.; Zhang, Z.K.; Liu, K. Therapeutic effect of a hydroxynaphthoquinone fraction on dextran sulfate sodium-induced ulcerative colit. *World J. Gastroenterol.* **2014**, *20*, 15310. [CrossRef] [PubMed]
3. Actis, G.C.; Pellicano, R.; Rosina, F. Inflammatory bowel disease: Traditional knowledge holds the seeds for the future. *World J. Gastrointest. Pharmacol. Ther.* **2015**, *6*, 10–16. [CrossRef] [PubMed]
4. Li, C.; Cai, Y.Y.; Yan, Z.X. Brain-derived neurotrophic factor preserves intestinal mucosal barrier function and alters gut microbiota in mice. *Kaohsiung J. Med. Sci.* **2018**, *34*, 134–141. [CrossRef]
5. Yang, F.J.; Wang, A.N.; Zeng, X.F.; Hou, C.L.; Liu, H.; Qiao, S.Y. *Lactobacillus reuteri* I5007 modulates tight junction protein expression in IPEC-J2 cells with LPS stimulation and in newborn piglets under normal conditions. *BMC Microbiol.* **2015**, *15*, 32–42. [CrossRef]
6. Ren, X.; Zhu, Y.; Gamallat, Y.; Ma, S.; Chiwala, G.; Meyiah, A.; Xin, Y. *E. coli* O124 K72 alters the intestinal barrier and the tight junctions proteins of guinea pig intestine. *Biomed. Pharmacother.* **2017**, *94*, 468–473. [CrossRef]
7. Vargas Robles, H.; Castro Ochoa, K.F.; Nava, P.; Silva Olivares, A.; Shibayama, M.; Schnoor, M. Analyzing beneficial effects of nutritional supplements on intestinal epithelial barrier functions during experimental colitis. *J. Vis. Exp.* **2017**, *2017*, e55095. [CrossRef]
8. Lamsal, B.P. Production, health aspects and potential food uses of dairy prebiotic galactooligosaccharides. *J. Sci. Food Agric.* **2012**, *92*, 2020–2028. [CrossRef]
9. Peshev, D.; Ende, W.V. Fructans: Prebiotics and immunomodulators. *J. Funct. Foods* **2014**, *8*, 348–357. [CrossRef]
10. Diaz, Y.M.; Laverde, G.V.; Gamba, L.R.; Wandurraga, H.M.; Ferrom, C.A. Biofilm inhibition activity of compounds isolated from two *Eunicea* species collected at the Caribbean Sea. *Rev. Bras. Farmacogn.* **2015**, *25*, 605–611. [CrossRef]
11. Rubini, D.; Farisa Banu, S.; Veda Hari, B.N.; Ramya Devi, D.; Gowrishankar, S.; Karutha Pandian, S.; Nithyanand, P. Chitosan extracted from marine biowaste mitigates staphyloxanthin production and biofilms of Methicillin-resistant *Staphylococcus aureus*. *Food Chem. Toxicol.* **2018**, *118*, 733–744. [CrossRef] [PubMed]
12. Xiao, D.; Wang, Y.; Liu, G.; He, J.; Qiu, W.; Hu, X.; Feng, Z.; Ran, M.; Nyachoti, C.M.; Kim, S.W.; et al. Effects of chitosan on intestinal inflammation in weaned pigs challenged by enterotoxigenic *Escherichia coli*. *PLoS ONE* **2014**, *9*, e104192. [CrossRef] [PubMed]
13. Kim, J.G.; Jo, S.H.; Ha, K.S.; Kim, S.C.; Kim, Y.C.; Apostolidis, E.; Kwon, Y.I. Effect of long-term supplementation of low molecular weight chitosan oligosaccharide (GO2KA1) on fasting blood glucose and HbA1c in db/db mice model and elucidation of mechanism of action. *BMC Complement Altern. Medi.* **2014**, *14*, 272–278. [CrossRef] [PubMed]
14. Huang, Y.; Guo, J.; Gui, S. Orally targeted galactosylated chitosan poly(lactic-co-glycolic acid) nanoparticles loaded with TNF-α siRNA provide a novel strategy for the experimental treatment of ulcerative colitis. *Eur. J. Pharm. Sci.* **2018**, *125*, 232–243. [CrossRef] [PubMed]
15. Xu, J.; Tam, M.; Samaei, S.; Lerouge, S.; Barralet, J.; Stevenson, M.M.; Cerruti, M. Mucoadhesive chitosan hydrogels as rectal drug delivery vessels to treat ulcerative colitis. *Acta Biomaterialia.* **2017**, *48*, 247–257. [CrossRef] [PubMed]
16. Liu, L.K.; Wang, Y.; Kong, M.; Li, X.L. Prebiotic-like effects of water soluble chitosan on the intestinal microflora in mice. *Int. J. Food Eng.* **2018**, *14*, 1–8. [CrossRef]
17. Tan, Y.; Zheng, C. Effects of alpinetin on intestinal barrier function, inflammation and oxidative stress in dextran sulfate sodium-induced ulcerative colitis mice. *Am. J. Med. Sci.* **2018**, *55*, 377–386. [CrossRef]
18. Azzam, M.M.; Zou, X.T.; Dong, X.Y.; Xie, P. Effect of supplemental L-threonine on mucin 2 gene expression and intestine mucosal immune and digestive enzymes activities of laying hens in environments with high temperature and humidity. *Poult. Sci.* **2011**, *90*, 2251–2256. [CrossRef]
19. Choi, S.; Woo, J.K.; Jang, Y.S.; Kang, J.H.; Jang, J.E.; Yi, T.H.; Park, S.Y.; Kim, S.Y.; Yoon, Y.S.; Oh, S.H. Fermented *Pueraria lobata* extract ameliorates dextran sulfate sodium-induced colitis by reducing pro-inflammatory cytokines and recovering intestinal barrier function. *Lab. Anim. Res.* **2016**, *32*, 151–159. [CrossRef]

20. Li, H.H.; Li, Y.P.; Zhu, Q.; Qiao, J.Y.; Wang, W.J. Dietary supplementation with *Clostridium butyricum* helps to improve the intestinal barrier function of weaned piglets challenged with enterotoxigenic *Escherichia coli* K88. *J. Appl. Microbiol.* **2018**, *125*, 964–975. [CrossRef]
21. He, L.X.; Wang, J.B.; Sun, B.; Zhao, J.; Li, L.; Xu, T.; Li, H.; Sun, J.Q.; Ren, J.; Liu, R.; et al. Suppression of TNF-α and free radicals reduces systematic inflammatory and metabolic disorders: Radioprotective effects of ginseng oligopeptides on intestinal barrier function and antioxidant defense. *J. Nutr. Biochem.* **2017**, *40*, 53–61. [CrossRef] [PubMed]
22. Rosenstiel, P.; Fantini, M.; Bräutigam, K.; Kühbacher, T.; Waetzig, G.H.; Seegert, D.; Schreiber, S. TNF-α and IFN-γ regulate the expression of the NOD2 (CARD15) gene in human intestinal epithelial cells. *Gastroenterology* **2003**, *124*, 1001–1009. [CrossRef] [PubMed]
23. Pan, Y.; Ouyang, Q. Effects of Bawei XileiSan on mice with oxazolone-induced colitis and the mechanisms. *J. Chin. Integr. Med.* **2010**, *8*, 568–574. [CrossRef] [PubMed]
24. Löwenberg, M.; D'Haens, G. Novel targets for inflammatory bowel disease therapeutics. *Curr. Gastroenterol. Rep.* **2013**, *15*, 311–316. [CrossRef]
25. Blackwood, B.P.; Yuan, C.Y.; Wood, D.R.; Nicolas, J.D.; Grothaus, J.S.; Hunter, C.J. Probiotic *Lactobacillus* species strengthen intestinal barrier function and tight junction integrity in experimental necrotizing enterocolitis. *J. Probiotics Health.* **2017**, *5*, 59–78. [CrossRef]
26. Zhang, P.; Yan, T.; Wang, X.; Kuang, S.; Xiao, Y.; Lu, W.W.; Bi, D.R. Probiotic mixture ameliorates heat stress of laying hens by enhancing intestinal barrier function and improving gut microbiota. *Ital. J. Anim. Sci.* **2017**, *16*, 292–300. [CrossRef]
27. Sun, X.; Du, M.; Navarre, D.A.; Zhu, M.J. Purple potato extract promotes intestinal epithelial differentiation and barrier function by activating AMP-activated protein kinase. *Mol. Nutr. Food Res.* **2018**, *62*, 1700536. [CrossRef]
28. Park, S.K.; Kim, M.S.; Bae, J.W. *Blautia faecis* sp. nov., isolated from human faeces. *Int. J. Syst. Evol. Microbiol.* **2013**, *63*, 599–603. [CrossRef]
29. Hamajima, H.; Matsunaga, H.; Fujikawa, A.; Sato, T.; Mitsutake, S.; Yanagita, T.; Nagao, K.; Nakayama, J.; Kitagaki, H. Japanese traditional dietary fungus koji *Aspergillus oryzae* functions as a prebiotic for *Blautia coccoides* through glycosylceramide: Japanese dietary fungus koji is a new prebiotic. *Springerplus* **2016**, *5*, 1321–1330. [CrossRef]
30. Jiang, D.; Kang, A.; Yao, W.; Lou, J.; Zhang, Q.; Bao, B.; Cao, Y.; Yu, S.; Guo, S.; Zhang, Y.; et al. *Euphorbia kansui* fry-baked with vinegar modulates gut microbiota and reduces intestinal toxicity in rats. *J. Ethnopharmacol.* **2018**, *226*, 26–35. [CrossRef]
31. Lakhdari, O.; Tap, J.; Béguet-Crespel, F.; Le Roux, K.; de Wouters, T.; Cultrone, A.; Nepelska, M.; Lefèvre, F.; Doré, J.; Blottière, H.M. Identification of NF-κB modulation capabilities within human intestinal commensal bacteria. *J. Biomed. Biotechnol.* **2011**, *2011*, 282356. [CrossRef]
32. Chaudhary, G.; Mahajan, U.B.; Goyal, S.N.; Ojha, S.; Patil, C.R.; Subramanya, S.B. Protective effect of *Lagerstroemia speciosa* against dextran sulfate sodium induced ulcerative colitis in C57BL/6 mice. *Am. J. Transl. Res.* **2017**, *9*, 1792–1800.
33. Yan, H.; Wang, H.; Zhang, X.; Li, X.; Yu, J. Ascorbic acid ameliorates oxidative stress and inflammation in dextran sulfate sodium-induced ulcerative colitis in mice. *Int. J. Clin. Exp. Med.* **2015**, *8*, 20245–20253.
34. Li, X.L.; Zhang, C.L.; Li, W.L.; Wu, D.C.; Liu, J.J.; Tang, L.; Xin, Y. In vivo effects on the intestinal microflora of *Physalis alkekengi* var. *francheti* extracts. *Fitoterapia* **2013**, *87*, 43–48. [CrossRef]

International Journal of
Molecular Sciences

MDPI

Article

The Constipation-Relieving Property of D-Tagatose by Modulating the Composition of Gut Microbiota

Yu-Xuan Liang [1], Peng Wen [1], Yu Wang [2], Dong-Mei OuYang [1], Da Wang [1], Yu-Zhong Chen [2], Ya Song [1], Jie Deng [1], Yuan-Ming Sun [1,*] and Hong Wang [1,*]

[1] Guangdong Provincial Key Laboratory of Food Quality and Safety/College of Food Science, South China Agricultural University, Guangzhou 510642, China; 1402972616@163.com (Y.-X.L.); wenpeng100@163.com (P.W.); ouydm2019@163.com (D.-M.O.); wangda134@foxmail.com (D.W.); songya_1990@163.com (Y.S.); scaudengjie@163.com (J.D.)

[2] Guangzhou Institute for Food Inspection, Guangzhou 511400, China; xxwangyu@163.com (Y.W.); Kittychenyuzhong@gmail.com (Y.-Z.C.)

* Correspondence: ymsun@scau.edu.cn (Y.-M.S.); gzwhongd@163.com (H.W.)

Received: 30 September 2019; Accepted: 12 November 2019; Published: 14 November 2019

check for updates

Abstract: D-tagatose, a monosaccharide as well as a dietary supplement, has been reported as having a wide range of applicability in the food industry, however, the prebiotic activity, anticonstipation effects, and related mechanisms are still unclear. In this study, using the loperamide-induced constipation Kunming mice as the animal model, the effects of D-tagatose for the prevention of constipation were evaluated by gastrointestinal transit experiment and defecation experiment. Furthermore, the underlying mechanism was clarified by evaluating the change of the biochemical indicators and analyzing 16S rRNA amplicon of gut microbiota among the different mice groups. The results showed that the gastrointestinal transit rate, fecal number, and weight in six hours were significantly enhanced after the administration of D-tagatose. In addition, D-tagatose significantly increased the serum levels of acetylcholine (Ach) and substance P (SP), whereas the serum levels of nitric oxide (NO) were significantly decreased. Moreover, the 16S rRNA sequencing analysis revealed that the changes in the gut microbiota caused by constipation were restored by D-tagatose treatment. In conclusion, this study indicated that the administration of D-tagatose as a dietary supplement can effectively prevent and relieve constipation in Kunming mice, and it is a promising prebiotic candidate with constipation-relieving properties.

Keywords: D-tagatose; prebiotic; constipation; gut microbiota; 16S rRNA; neurotransmitter

1. Introduction

Constipation is one of the most common health issues worldwide [1], and it is characterized by several symptoms such as infrequent bowel movements, small, dry and hard feces, and prolonged gastrointestinal transit time [2]. The cause of constipation is considered multifactorial [3]. Previous studies have shown that constipation is associated with imbalances in the composition of the gut microbiota [4,5]. For example, some common probiotics (*Lactobacilli* and *Bifidobacteria*) were generally reduced in constipation, while pathogenic bacteria (*Methanogenic archaea* and *Clostridia*) were increased in constipation [1,6,7]. Additionally, the enteric nervous system (ENS) plays a key role in the regulation of gastrointestinal motility, and the excitability of which is regulated by the levels of neurotransmitters [8]. Currently, the common way to relieve constipation is by using osmotic, secretagogues, and prokinetic drugs [9–11], however, the administration of these drugs is often accompanied by some side effects, such as irritancy and dependence [12]. Hence, an effective and easy-to-implement strategy to relieve constipation is desirable.

Recently, dietary supplements of natural products are regarded as a promising alternative to relieve constipation because of their safety and effectiveness [5]. For instance, some flavonoids are able to change the structure of epithelial cells and smooth muscle cells, or effectively regulate the quantitative alternations of the interstitial cells of Cajal and aquaporin 3 to relieve constipation in mice [13,14]. In addition, different doses of fructo-oligosaccharide, galacto-oligosaccharide, and isomalto-oligosaccharide have been proven to have the potential to regulate intestinal microbiota to relieve constipation [15]. Nevertheless, there are few studies on the ease of constipation by monosaccharides.

Tagatose, a kind of monosaccharide with a molecular weight of 180.16 Da, is recognized as a safe food ingredient by the China Food and Drug Administration (CFDA) and USA Food and Drug Administration (FDA) [16,17]. It has been reported that D-tagatose possesses low small intestinal digestibility (15~20%) [18,19], however, it can be utilized by the intestinal microflora to increase the concentration of short-chain fatty acids (SCFAs), lower the intestinal pH, and make the intestinal environment more harmonious [15,19,20]. A previous study found that the abundance of *Bacteroides*, *Lactobacillus*, and *Akkermansia* was significantly associated with the tagatose treatment in the colitis murine model [21]. In view of the aforementioned cause of constipation, we hypothesized that these characteristics of D-tagatose may confer beneficial effects on relieving constipation.

The aim of this study were as follows: (1) to evaluate the preventive effects of D-tagatose on the gastrointestinal transit and defecation status in mice and (2) to clarify D-tagatose underlying mechanisms of anticonstipation by analyzing the composition of gut microbiota and the indicators of serum, including acetylcholine (Ach), nitric oxide (NO), substance P (SP), and vasoactive intestinal peptide (VIP), in slow transit constipation mice.

2. Results

2.1. Effects of D-Tagatose on the Gastrointestinal Transit Rate

The effect of D-tagatose with different doses on gastrointestinal transit rates of mice is shown in Figure 1. Constipation symptoms were effectively induced by loperamide, indicating the constipation model was established successfully. The lowest gastrointestinal transit rate was in the model group, and the highest was in the high doses D-tagatose group. Although there was no significant difference between the low dose D-tagatose group and the model group, the medium and high dose D-tagatose groups showed significant acceleration on the gastrointestinal transit ($p < 0.01$). In particular, there was no significant difference in the gastrointestinal transit rate between the high-dose tagatose group (83.73%) and the blank group (83.42%), indicating that high-dose tagatose could restore gastrointestinal peristalsis to normal levels. These results indicate that the administration of D-tagatose could improve the intestinal peristalsis of constipated mice in a dose-dependent manner.

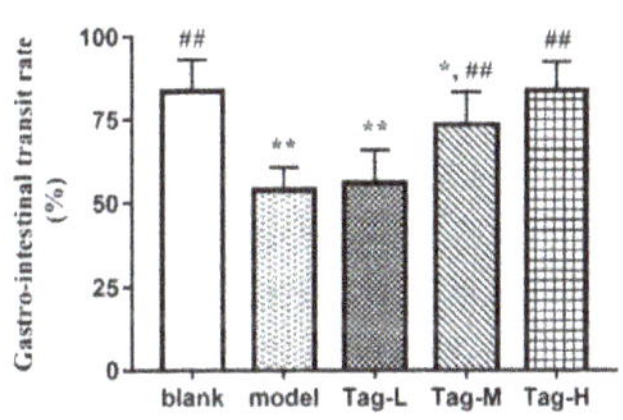

Figure 1. Effects of D-tagatose on the gastrointestinal transit rate of mice. Blank group (days 1–7, distilled water administration period and day 8, distilled water administration but no induction of constipation); model group (days 1–7, distilled water administration period and day 8, distilled water administration and induction of constipation); tagatose group (days 1–7, tagatose administration period and day 8, tagatose administration and induction of constipation, Tag-L mice treated with 0.43 g/kg body weight (BW) tagatose, Tag-M mice treated with 0.85 g/kg BW tagatose, and Tag-H mice treated with 1.70 g/kg BW tagatose). *, compared with the blank group, $p < 0.05$; **, compared with the blank group, $p < 0.01$; and ##, compared with the model group, $p < 0.01$.

2.2. *Effects of* D*-Tagatose on the Defecation Status*

The effect of different doses of D-tagatose on the defecation status of mice is shown in Figure 2. In comparison to the model group, low and medium doses of D-tagatose treatment led to a significant increase in the fecal number and weight in six hours ($p < 0.05$), while in the high doses of D-tagatose group, the significant difference was observed only in the fecal weight in six hours. Regardless of the time under the action of low-, medium- and high-tagatose groups did not reach the level of the blank group, however, a decreased shorten time of the first black stool defecation (shortening rate to 28.53%, 31.09%, and 26.41%, respectively) than that of the model group was achieved. Overall, the administration of D-tagatose could shorten the average time of defecation; nevertheless, there were no significant differences between all D-tagatose groups and model group in the time to the first blank stool defecation.

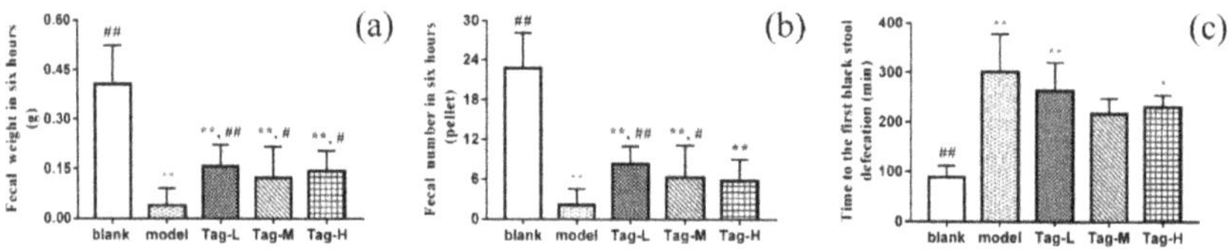

Figure 2. Effects of D-tagatose on the defecation status of mice. (**a**) Fecal weight in six hours, (**b**) fecal number in six hours, and (**c**) time to the first black stool defecation. Blank group (days 1–7, distilled water administration period and day 8, distilled water administration but no induction of constipation); model group (days 1–7, distilled water administration period and day 8, distilled water administration and induction of constipation); and tagatose group (days 1–7, tagatose administration period and day 8, tagatose administration and induction of constipation, Tag-L mice treated with 0.43 g/kg BW tagatose, Tag-M mice treated with 0.85 g/kg BW tagatose, and Tag-H mice treated with 1.70 g/kg BW tagatose). *, compared with the blank group, $p < 0.05$; **, compared with the blank group, $p < 0.01$; #, compared with the model group, $p < 0.05$; ##, compared with the model group, $p < 0.01$.

2.3. *Effects of* D*-Tagatose on Serum Neurotransmitter*

The effects of D-tagatose on relieving constipation were further evaluated by measuring serum parameters in mice, including excitability neurotransmitters (Ach and SP), and the inhibitory neurotransmitters (NO and VIP). As shown in Table 1, the levels of Ach and SP were significantly increased in the D-tagatose treatment group ($p < 0.05$), while the levels of NO exhibited a significant decrease ($p < 0.05$). No statistical differences were observed in the levels of VIP between the D-tagatose group and model group. These results show that D-tagatose treatment effectively increased the levels of the excitability neurotransmitters (Ach and SP) and decreased the levels of the inhibitory neurotransmitters (NO) to speed up the motility of the gastrointestinal tract.

Table 1. Effects of D-tagatose on serum parameters in mice.

Group	Ach (pg/L)	NO (μmol/L)	VIP (pg/L)	SP (pg/L)
blank	38.17 ± 10.17 $^{\#}$	30.71 ± 3.81	18.19 ± 3.75 $^{\#\#}$	31.97 ± 12.81
model	27.82 ± 7.28 *	26.99 ± 4.77	26.19 ± 4.53 **	24.76 ± 6.23
Tag-H	37.98 ± 11.58 $^{\#}$	15.60 ± 2.76 $^{*,\#\#}$	24.07 ± 2.50	49.11 ± 11.25 $^{\#\#}$

Blank group (days 1–14, distilled water administration period and days 15–28, distilled water administration period); model group (days 1–14, induction of constipation period and days 15–28, distilled water administration period); and tagatose group (days 1–14, induction of constipation period and days 15–28, tagatose administration period). Tag-H, mice treated with 1.70 g/kg BW tagatose). *, compared with the blank group, $p < 0.05$; **, compared with the blank group, $p < 0.01$; #, compared with the model group, $p < 0.05$; and ##, compared with the model group, $p < 0.01$. Acetylcholine (Ach); nitric oxide (NO); substance P (SP); vasoactive intestinal peptide (VIP).

2.4. Effect of D-Tagatose on the Composition of Rectum Microbiota

To assess the influence of D-tagatose in the gut microbiome, we collected rectal content of three groups of mice, and for each group of 10 mice took six samples (a total of 18 rectum samples) and a community structure analysis was performed. All of the sequences were clustered with representative sequences, and a 97% sequence identity was obtained. Based on alpha and beta diversity measurement, overall association tests were then conducted.

From Figure 3, regarding the indexes of alpha diversity (the species richness, Chao-1 and Shannon, and the observed species), significant differences were observed between the blank and model group ($p < 0.05$). After the administration of D-tagatose, the above three indexes exhibited a significant increase in comparison with the model group ($p < 0.05$), while no significant differences were observed between the blank and D-tagatose group indicating the general diversity and richness of gut microbiota in constipated mice restored to normal. In addition, PCoA based on unweighted uniFrac matrixes revealed the beta diversity of the intestinal flora in mice. As shown in Figure 3, we can see that the data points shifted from the lower of the score plot to the upper in all of the D-tagatose treatment in comparison with the model group. Combined with the results of alpha diversity measurement, we can conclude that the administration of D-tagatose could restore the microbial richness and diversity in constipated mice.

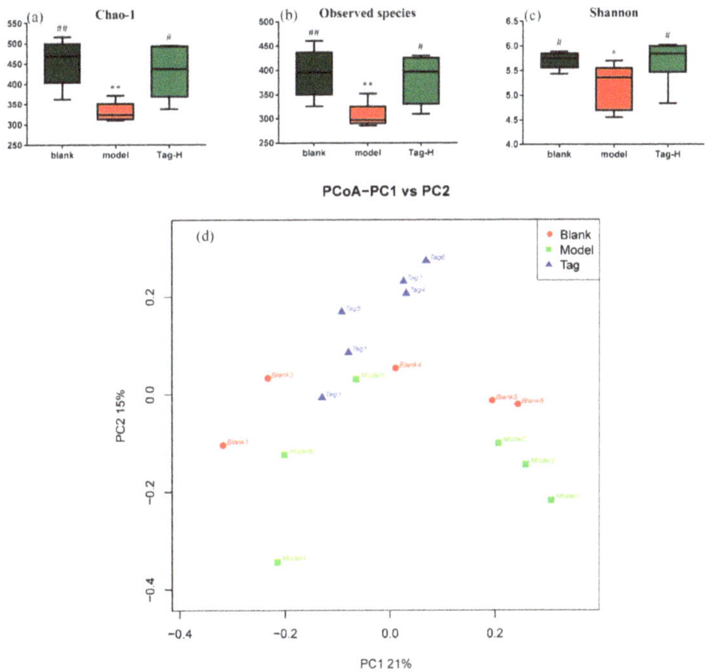

Figure 3. Alpha diversity indexes and beta diversity indexes in mice. (**a**) Chao-1 (alpha diversity index), (**b**) observed species (alpha diversity index), (**c**) Shannon (alpha diversity index), and (**d**) PCoA plots based on unweighted uniFrac metrics (beta diversity index). Blank group (days 1–14, distilled water administration period and days 15–28, distilled water administration period); model group (days 1–14, induction of constipation period and days 15–28, distilled water administration period); tagatose group (days 1–14, induction of constipation period and days 15–28, tagatose administration period, Tag-H mice treated with 1.70 g/kg BW tagatose). *, compared with the blank group, $p < 0.05$; **, compared with the blank group, $p < 0.01$; $^{\#}$, compared with the model group, $p < 0.05$; and $^{\#\#}$, compared with the model group, $p < 0.01$.

2.5. The Relative Abundance of Rectum Microbiota

To further investigate the effect of tagatose treatment on the difference in the composition of the intestinal flora, the levels of phylum, family, and genus were further analyzed. All of the effective reads were clustered into 11 phyla, 30 families, and 35 genera using the Ribosomal Database Project (RDP) classifier. At the phylum level (Figure 4), the major dominant phyla in the rectum contents of mice were Bacteroidetes and Firmicutes, followed by Proteobacteria, Verrucomicrobia, Actinobacteria, Deferribacteres, Tenericutes, Saccharibacteria, Cyanobacteria and an unclassified group denoted as "other". The core bacterium was regarded as the bacterium with the highest average relative abundance. At the phylum level, the core bacterium was Bacteroidetes, which were 77.97%, 82.90%, and 69.31% in the blank, model, and tagatose groups, respectively. In order to characterize the effect of tagatose on the difference of phylum level, the specific phylum levels with their relative abundance higher than 0.1% were selected and analyzed. As shown in Figure 4b, there were significant differences between Bacteroidetes and Proteobacteria. After treatment with loperamide, the mean abundance of Bacteroidetes in the model group was increased by 2.37% as compared with the blank group. However, under the action of tagatose, Bacteroidetes were then decreased by 14.01% as compared with the model group ($p < 0.05$), and there was no significant difference as compared with the blank group. By contrast, an increase in the level of Proteobacteria after the D-tagatose treatment was also observed as compared with the model group. These results indicated that D-tagatose could improve the change of gut microbiota in the levels of Bacteroidetes and Proteobacteria caused by loperamide.

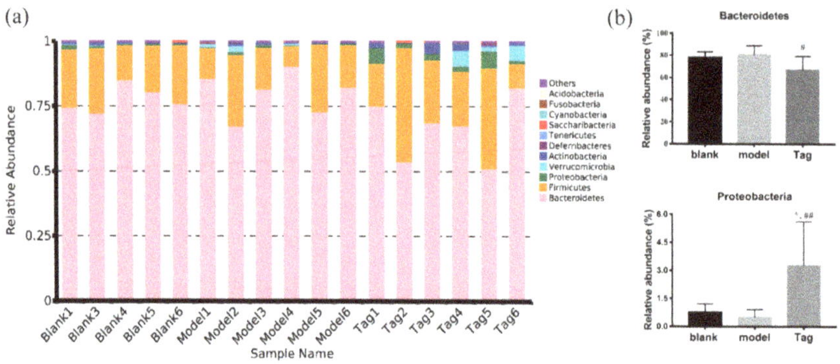

Figure 4. Community abundance on phyla level. (**a**) Changes in the level of phyla in different groups and (**b**) changes of the abundance of selected phyla in different groups (relative abundance more than >0.1%). Blank group (days 1–14, distilled water administration period and days 15–28, distilled water administration period); model group (days 1–14, induction of constipation period and days 15–28, distilled water administration period); and tagatose group (days 1–14, induction of constipation period and days 15–28, tagatose administration period, Tag-H, mice treated with 1.7 g/kg BW tagatose). *, compared with the blank group, $p < 0.05$; #, compared with the model group, $p < 0.05$; and ##, compared with the model group, $p < 0.01$.

At the family level, the major dominant families in the rectum contents of mice were BacteroidalesS24-7group (phylum Bacteroidetes), Prevotellaceae (phylum Bacteroidetes), Bacteroidaceae (phylum Bacteroidetes), and Lachnospiraceae (phylum Firmicutes), followed by twenty-six families, which are shown in Figure 5. It can be seen that the core bacterium was BacteroidalesS24-7group, and its abundance in the blank, model, and tagatose group were 45.81%, 49.68%, and 33.55%, respectively. Similarly, the specific family levels with their relative abundance higher than 0.1% were selected and analyzed, and results showed that six families' levels revealed significant differences (Figure 6). In comparison to the blank group, the structure of intestinal flora was changed in the model group. For example, the mean abundance of BacteroidalesS24-7group and Lactobacillaceae increased in the model group, while the mean abundance of Lachnospiraceae, Porphyromonadaceae, Corynebacteriaceae, and Erysipelotrichaceae reduced in the model group. However, after administration of D-tagatose, the trends of six families had been changed. The relative abundance of Erysipelotrichaceae, Lachnospiraceae, Porphyromonadaceae, and Corynebacteriaceae was significantly increased as compared with the model group ($p < 0.05$), while the relative abundance of BacteroidalesS24-7group significantly reduced ($p < 0.01$). These results reveal that the administration of D-tagatose could effectively relieve the change of gut microbiota caused by loperamide in the above mentioned six families.

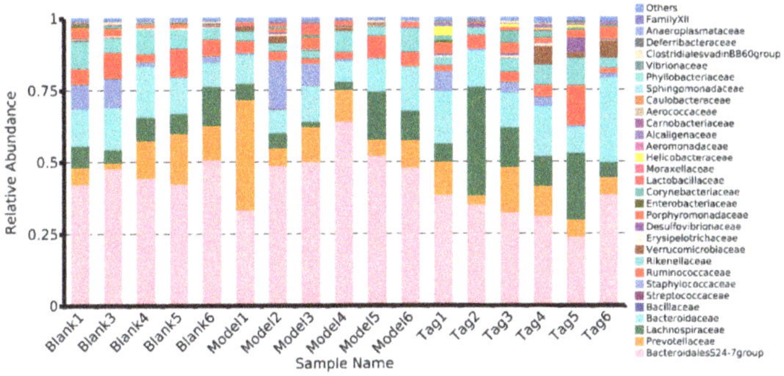

Figure 5. Changes in the level of families in different groups. Blank group (days 1–14, distilled water administration period and days 15–28, distilled water administration period); model group (days 1–14, induction of constipation period and days 15–28, distilled water administration period); and tagatose group (days 1–14, induction of constipation period and days 15–28, tagatose administration period, Tag-H, mice treated with 1.70 g/kg BW tagatose).

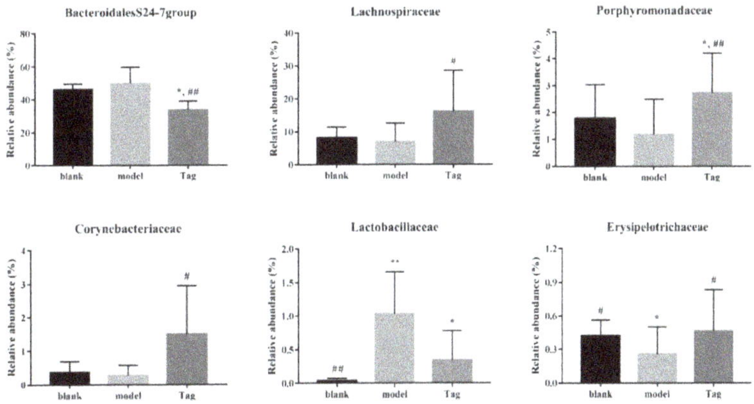

Figure 6. Changes of the abundance of selected families in different groups (relative abundance more than >0.1%). Blank group (days 1–14, distilled water administration period and days 15–28, distilled water administration period); model group (days 1–14, induction of constipation period and days 15–28, distilled water administration period); and tagatose group (days 1–14, induction of constipation period and days 15–28, tagatose administration period, Tag-H mice treated with 1.70 g/kg BW tagatose). *, compared with the blank group, $p < 0.05$; **, compared with the blank group, $p < 0.01$; $^{\#}$, compared with the model group, $p < 0.05$; and $^{\#\#}$, compared with the model group, $p < 0.01$.

At the genus level (Figure 7), the major dominant genera in the rectum contents of mice were *Bacteroides* (phylum Bacteroidetes, family Bacteroidaceae) and *Alloprevotella* (phylum Bacteroidetes, family Prevotellaceae), followed by thirty-three genera. As shown in Figure 7, the core bacterium of the genus level was *Bacteroides* with the levels of 13.42%, 10.86%, and 18.12% in the blank, model, and tagatose groups, respectively. Similar to the analysis of phylum and family levels, we found that twelve genus levels exhibited significant differences (Figure 8). Among them, after treatment with loperamide, the average abundance of *Bacteroides, Alistipes, Parabacteroides, Eubacterium xylanophilum group, Paenalcaligenes, Facklamia, Atopostipes, Clostridium innocuum group,* and *RuminococcaceaeNK4A-214group* were reduced, while the average abundance of *Lactobacillus* was increased. Interestingly, the trends of ten genera had been reversed with the administration of D-tagatose. As presented in Figure 8, the relative abundance of *Bacteroides, Alistipes, Desulfovibrio, RuminococcaceaeTCG-004,* and *RuminococcaceaeNK4A-214group* was significantly increased in the D-tagatose group ($p < 0.05$), in particular, the relative abundance of *Parabacteroides, Paenalcaligenes, Facklamia, Atopostipes,* and *Clostridium innocuum group* highly significantly increased in the D-tagatose group ($p < 0.01$). These results indicate the beneficial effect of D-tagatose on the levels of ten genera, which were influenced by loperamide.

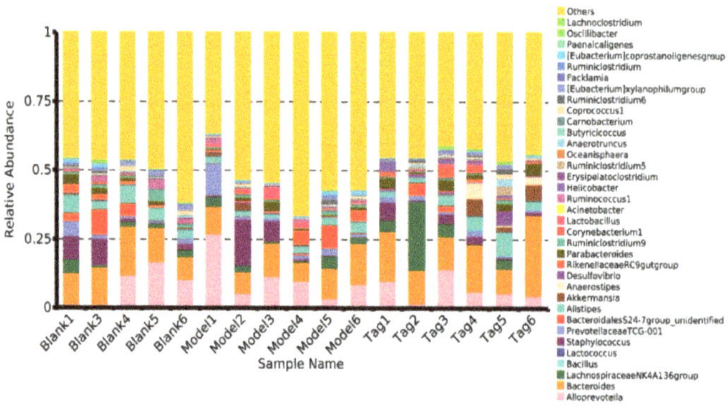

Figure 7. Changes in the level of genera in different groups. Blank group (days 1–14, distilled water administration period and days 15–28, distilled water administration period); model group (days 1–14, induction of constipation period and days 15–28, distilled water administration period); and tagatose group (days 1–14, induction of constipation period and days 15–28, tagatose administration period, Tag-H mice treated with 1.70 g/kg BW tagatose).

2.6. Correlation between Intestinal Microflora and Biological Indexes

The correlation of the top 40 genera and the biological indexes were performed by Spearman correlation analysis (Figure 9). Among them, 13 genera significantly exhibited a correlation with the biological indexes of constipation. As the indicators of constipation, for instance, the gastrointestinal transit rate was positively correlated with *Bacteroides* and *Enterobacter* ($p < 0.05$). Conversely, it was negatively correlated with *FamilyXIII AD3011 group* and *Eubacterium coprostanoligenes group* ($p < 0.05$). The fecal number in six hours and fecal weight in six hours were negatively correlated with *RuminococcaceaeTCG−004, Lactobacillus,* and *Eubacterium coprostanoligenes group* ($p < 0.05$), while they were positively related to *Enterobacter* ($p < 0.01$). In addition, the time to the first black stool defecation was negatively correlated with *Enterobacter* ($p < 0.05$). For neurotransmitters, there was a significant negative correlation between *Enterobacter* and VIP ($p < 0.05$), and *Helicobacter, Corynebacterium1, Desulfovibrio* were also negatively with the NO ($p < 0.05$). *Akkermansia* and *Eubacterium coprostanoligenes group* were negatively correlated with the SP ($p < 0.05$), while *Helicobacter* was positively correlated with the SP ($p < 0.05$). These results indicate that biological indicators of constipation, such as intestinal movement, defecation, and level of the related neurotransmitter, are significantly correlated with specific bacteria. Thus, changes in the abundance of specific bacteria may be beneficial to alleviating or aggravating constipation.

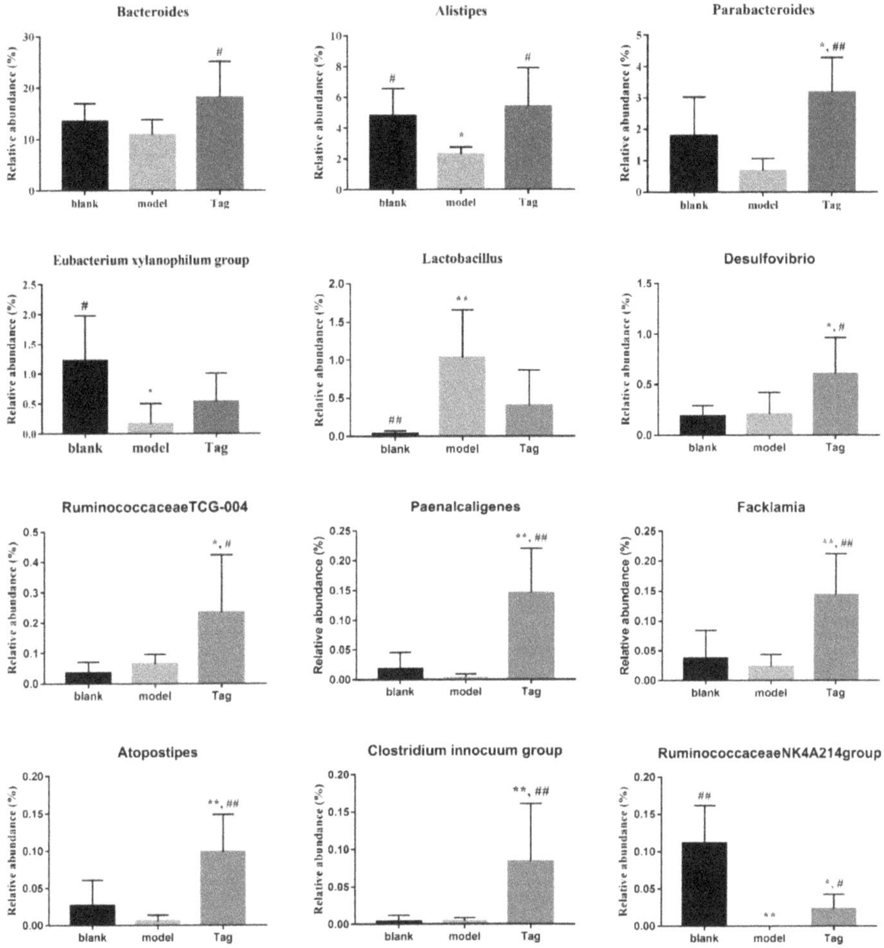

Figure 8. Changes of the abundance of selected genera in different groups (relative abundance more than > 0.1%). Blank group (days 1–14, distilled water administration period and days 15–28, distilled water administration period); model group (days 1–14, induction of constipation period and days 15–28, distilled water administration period); and tagatose group (days 1–14, induction of constipation period and days 15–28, tagatose administration period, Tag-H mice treated with 1.70 g/kg BW tagatose). *, compared with the blank group, $p < 0.05$; **, compared with the blank group, $p < 0.01$; #, compared with the model group, $p < 0.05$; ##, compared with the model group, $p < 0.01$.

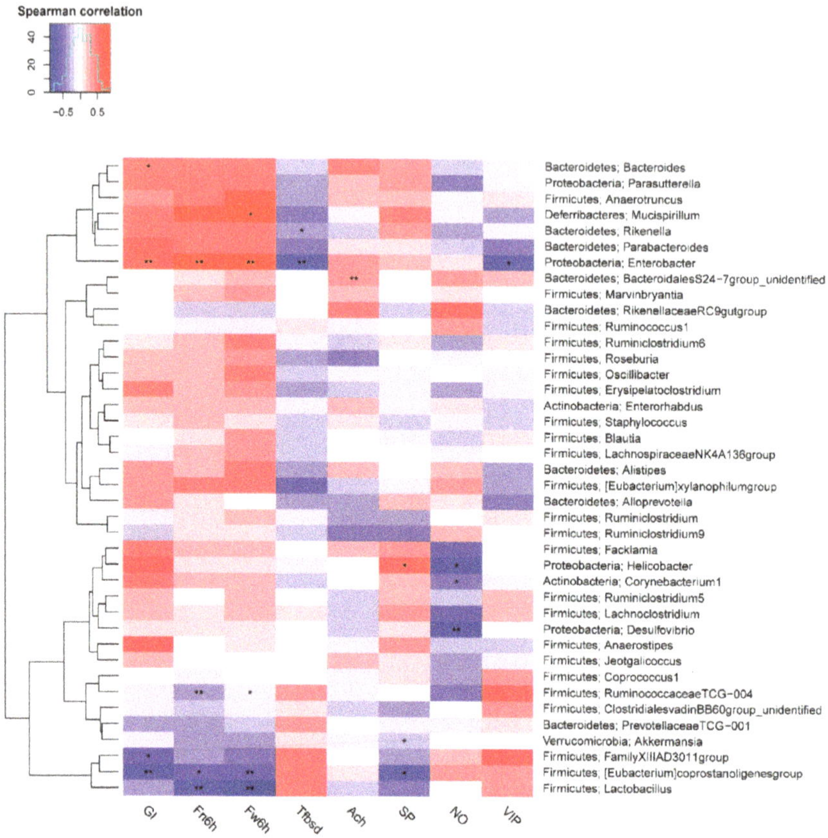

Figure 9. Spearman correlation analysis of top 40 genera and biological indicators. *, $p < 0.05$, **, $p < 0.01$. GI, gastrointestinal transit rate; Fn6h, fecal number in six hours; Fw6h, fecal weight in six hours; Tfbsd, time to the first black stool defecation; Ach, acetylcholine; SP, substance P; NO, nitric oxide; and VIP, vasoactive intestinal peptide.

3. Discussion

Many studies have shown that the indigestible carbohydrates of a supplement possess beneficial effects on gut health [22–24]. These carbohydrate supplements are mainly polysaccharides and oligosaccharides, while there are few studies about monosaccharides. D-tagatose is a monosaccharide with 92% sweetness of sucrose but only 38% of the calorie, the primary features of which include a lower absorption rate in the small intestine but utilization by the intestinal flora [18,19,25]. Importantly, D-tagatose has not been studied in relieving constipation. Hence, based on the indigestible properties of tagatose and the activity of other similar functional carbohydrates in the management of relieving constipation, the effect of D-tagatose on the constipation was investigated using a constipation animal model.

To establish the model, loperamide was used to induce constipation, the primary mechanism of which can be attributed to the inhibition of the release of Ach and prostaglandins, thus, leading to the reduced frequency of bowel motility and prolonged bowel movement [1]. The gastrointestinal transit, time to the first black stool defecation, as well as fecal weight and number in unit time are key indices with which to evaluate the function of the gastrointestinal transit [1,26]. The movement of gastrointestinal contents is mainly promoted by peristalsis, which can be detected by measuring the gastrointestinal transit rate and time to the first black stool defecation [3]. In addition, the fecal weight and number in six hours are also considered as direct indicators for reflecting the status of constipation. This study revealed that D-tagatose administration effectively accelerated gastrointestinal peristalsis to prevent constipation in a dose-dependent manner (Figure 1). This may be related to the lower intestinal absorption of tagatose, allowing it to pass through the small intestine faster, as well as, in the defecation experiment, the administration of tagatose could enhance the fecal number and weight in six hours ($p < 0.05$). In addition, it can be seen that both the GI transit rate and the time to the first stool defecation were improved after D-tagatose treatment. For the GI transit rate, it was shown to be restored upon tagatose treatment. However, for the time to first stool, although a higher level still existed as compared with the blank group, there was a reduction of 28.53%, 31.09%, and 26.41% by the treatment of the low-, medium- and high-tagatose, respectively as compared with the model group (Figure 2c). Different measurement methods and doses of loperamide may be responsible for the different results of the gastrointestinal transit and the time to the first black stool defecation. In a word, these results may be explained by the restoring of gut diversity, and thus the improvement of gut microbiota to promote the intestinal peristalsis. Combined with the Spearman correlation analysis (Figure 9), the parameters of relieving constipation were significantly correlated with the specific bacterium (such as *Bacteroides*). This was consistent with the previous reports that the indexes of constipation were significantly associated with the level of the specific bacterium [3,27].

Neurotransmitter also plays an important role in gastrointestinal hormones [28]. A couple of neurotransmitters have been identified to be associated with the motor activity of the gastrointestinal tract, such as Ach, NO, SP, and VIP, which play important roles in the regulation of gastrointestinal motility [14,24,29]. Among them, Ach and SP are excitatory neurotransmitters that promote bowel motility and speed defecation, while NO and VIP are inhibitory neurotransmitters that relax the gastrointestinal tract and inhibit bowel movement [3,30]. Disruption of the balance between these excitatory and inhibitory neurotransmitters may lead to constipation. Our study revealed that the administration of D-tagatose effectively regulated the excitability of the ENS, as the levels of excitatory neurotransmitters, Ach and SP, were increased, while the level of inhibitory neurotransmitter NO was decreased (Table 1). These results are in line with the previous report, suggesting that the regulation in neurotransmitters plays a crucial role in intestinal peristalsis [3].

The 16S rRNA amplicon of gut microbiota was applied to investigate the underlying mechanism among the different mice groups. First, the overall association tests by the alpha and beta diversity measurements were conducted to provide a holistic view of the microbiota structure [26]. From Figure 2, we can see that the administration of D-tagatose changed alpha diversity based on species richness (Chao-1, observed species, and Shannon), and a significant change in the structure of the rectum microbiota was observed in a PCoA plot of beta diversity (Figure 2). Moreover, there were significant differences among the blank, model, and D-tagatose groups, suggesting that the diversity of general intestinal flora induced by loperamide-constipation was restored after D-tagatose treatment.

Bacteroidetes were the dominant phylum in our study. In this study, as compared with the model group, a significant decrease in the abundance of Bacteroidetes was exerted in the D-tagatose group (Figure 3, $p < 0.01$). Since a previous study reported that a lower abundance of Bacteroidetes was observed in healthy than constipated patients [31], and the intake of dietary fiber was beneficial to the decrease of Bacteroidetes [32], in this study, the results indicated that D-tagatose has a beneficial function as a dietary fiber to improve gut microbiota.

At the family level (Figure 4), this study revealed that the relative abundance of Lachnospiraceae, Porphyromonadaceae, and Erysipelotrichaceae were significantly increased after administration of D-tagatose ($p < 0.05$). It was reported that the increases of Lachnospiraceae, Porphyromonadaceae, and Erysipelotrichaceae were beneficial for the production of SCFAs, which play an important role in protecting the host from infection and maintaining the normal physiological function of the gut [33,34]. Furthermore, studies have shown that SCFAs can also decrease intestinal pH and promote colon motility [35,36]. For instance, Lachnospiraceae is a family of clostridia [37,38], which is associated with the production of butyric acid [39,40]. Erysipelotrichaceae is another important butyrate-producing member [41], and Porphyromonadaceae can alleviate colonic inflammation [42,43]. Therefore, we hypothesized that D-tagatose treatment could promote the proliferation of SCFAs-producing bacteria, thus accelerating the intestinal peristalsis. In addition, the Lactobacillaceae and Bifidobacteriaceaec contain some beneficial common genera of bacteria, such as *Lactobacillus* and *Bifidobacterium*; in this study, higher abundance of Lactobacillaceae was found in constipated groups than other groups, while Bifidobacteriaceae was not detected. Ren et al. reported that the polysaccharides of *Enteromorpha* could relieve loperamide-induced constipation; similarly, the highest level of Lactobacillaceae was found in constipated groups and *Bifidobacterium* was not detected in the treatment group [44]. The characteristics of constipation-relieving properties may not necessarily be shown in the increased level of Lactobacillaceae and Bifidobacteriaceae, and thus further study needs to investigate the cause of such a phenomenon.

At the genus level, previous studies reported that *Bacteroides* is the major portion of the mammalian gut microbiota, which plays an important effect on the processing of complex molecules to simpler ones [45–47]. As documented in the literature, the species of *Bacteroides* in the gut were devoted to the uptake and breakdown of indigestible carbohydrates [47–49]. As expected, *Bacteroides* was the dominant genus in the gut microbiota of mice, and its content was significantly increased after administration of D-tagatose (Figure 7, $p < 0.01$), suggesting that the higher levels of *Bacteroides* could be beneficial to produce more metabolite (SCFAs), as well as accelerate intestinal peristalsis [35,36]. In addition, the relative abundances of *Parabacteroide*, *RuminococcaceaeTCG-004*, and *RuminococcaceaeNK4A-214group* increased after administration of D-tagatose (Figure 7, $p < 0.05$), which were positively associated with the production of SCFAs [50–52]. In a word, it can be concluded that for the intestinal flora diversity was restored after tagatose administration. Regarding the composition, changes in core bacteria (reduction of Bacteroidetes and an increase in *Bacteroides*) were observed, which are conducive to forming a harmonious intestinal environment and promoting the relief of constipation.

4. Materials and Methods

4.1. Chemicals and Reagents

The D-tagatose was purchased from Jcantek Pharmaceuticals Ltd. (Wuxi, China), and the content of the D-tagatose was 98%. Loperamide hydrochloride (Xi'an Janssen Pharmaceutical Ltd., Xi'an, China) was purchased from Dashenlin drugstore. Enzyme-linked immunosorbent assay (ELISA) kits were used to measure the levels of Ach, SP and VIP were purchased from Jianglai industrial Ltd. (Shanghai, China). The kit for the determination of the level of NO was purchased from Nanjing Jiancheng Bioengineering Institute (Nanjing, China). The Microbial DNA was extracted from rectum contents samples using the E.Z.N.A.® Soil DNA Kit (Omega Bio-tek, Norcross, GA, USA). Amplicons were extracted and purified using the AxyPrep DNA Gel Extraction Kit (Axygen Biosciences, Union City, CA, USA). The methods were adopted to prepare loperamide hydrochloride and activated carbon meal solution [1].

4.2. Animals and Experimental Design

Specific Pathogen Free (SPF) male Kunming mice (initial weight 18~22 g) were purchased from Guangdong Medical Laboratory Animal Centre (Guangzhou, China); the mice were fed under standard conditions at a room temperature of 25 ± 2 °C and humidity of 50% ± 5% with a 12 h light-dark cycle. They were fed with standard commercial mouse food (containing 64% carbohydrate, 19% protein, and 17% fat), and water was provided ad libitum. All of the protocols were approved by the Ethics Committee of South China Agricultural University (SYXK2014-0136), and the approval number was 2017009.

4.2.1. The Gastrointestinal Transit Experiment

The methods were conducted by the Technical Standards for Testing and Assessment of Health Food formulated by the Chinese Ministry of Health. After 7 days of adaptive feeding, the 50 mice were randomly divided into five groups, each group consisting of ten animals. All animals were administered orally by gavage. A dose of 5 g/day is the recommended dietary allowance of D-tagatose for a human being weighing 60 kg, which is equivalent to 0.083 g/kg per day. The dose groups set for the mice were 5, 10, and 20 times the equivalent recommended dietary allowance of prebiotics for humans. The blank group and model group were given distilled water via gavage once per day for 7 consecutive days; the low, medium, and high dose D-tagatose groups received 0.6 mL of 0.43 g/kg BW, 0.85 g/kg BW, or 1.70 g/kg BW D-tagatose solutions, respectively, in the same manner as the blank and model groups for 7 days. Then, all groups fasted overnight for 16 h (water was not restricted) before measurement of the indicator. The model and D-tagatose groups were treated with loperamide (5 mg/kg BW, 0.6 mL) via gavage to induce constipation, and the blank group was treated with distilled water (0.6 mL). Thirty minutes later, all groups were treated with activated carbon meal solution containing the corresponding content. At the end of the experiment, the gastrointestinal transit rate (the rate of carbon powder propulsion through the small intestine) was measured by the methods in [53,54].

4.2.2. The Experiment of Defecation

The methods were conducted by the Technical Standards for Testing and Assessment of Health Food formulated by the Chinese Ministry of Health. The process for defecation experiment was the same with the gastrointestinal transit experiment, while the model and D-tagatose groups were treated with loperamide (10 mg/kg BW, 0.6 mL) according to the dose of Ren [44]. At the end of the experiment, all of the mice were used to examine defecation status including the time of first black stool defecation, fecal numbers in six hours, and fecal weight in six hours [55].

4.2.3. The Slow Transit Constipation Experiment

The experiment was conducted according to the method of Zhu with slight modifications [56]. After 7 days of adaptive feeding, the mice were randomly divided with ten mice into each of the three groups. All the animals were administered orally by gavage. Both the model and D-tagatose groups were treated with 10 mg/kg BW dose of loperamide for the first two weeks to induce the model of slow transit constipation mice, and the blank group was treated with distilled water. Then, for the following two weeks, the blank and model groups were treated with distilled water, and the D-tagatose group received 1.70 g/kg D-tagatose solutions, respectively. At the end of the experiment, the mice were killed with light ether anesthesia. Blood samples were collected and centrifuged at 3000 r/min at 4 °C for 15 min to obtain serum. The rectum contents of each mouse were collected and stored at −80 °C until analysis.

4.3. Determination of Ach, NO, SP, and VIP Levels in Serum

The levels of Ach, SP, and VIP in the serum were measured using ELISA kits, and the levels of NO in the serum were determined by commercially available diagnostic kits. All measurement steps were carried out according to the manufactuerers' instructions.

4.4. DNA Extraction, PCR, and 16S rDNA Sequencing

The microbial DNA was extracted from rectum contents samples using the E.Z.N.A.® Soil DNA Kit by the protocols. The V4-V5 region of the bacterial 16S ribosomal RNA gene was amplified by PCR using primers 515F 5′-GTGCCAGCMGCCGCGG-3′ and 907R 5′-CCGTCAATTCMTTTRAGTTT-3′, where the barcode is an eight-base sequence unique to each sample. The PCR reactions were performed in triplicate 20 μL mixtures containing 4 μL of 5× FastPfu Buffer, 2 μL of 2.5 mM dNTPs, 0.8 μL of each primer (5 μM), 0.4 μL of FastPfu Polymerase, and 10 ng of template DNA.

Amplicons were extracted from 2% agarose gels and purified using a AxyPrep DNA Gel Extraction Kit according to the instructions. Purified amplicons were pooled in equimolar, and sequenced using paired-end sequencing (2 × 250) on the HiSeq 2500 platform according to the standard protocols. The raw reads were deposited into the NCBI Sequence Read Archive (SRA) database (accession number: SRP041836).

4.5. Analysis of Sequences and Relative Abundance of Microbiota Community Members

The raw sequences were trimmed and filtered by using the QIIME software (Version1.9). The high-quality reads were merged to generate the 16S rDNA V4 fragment sequences using FLASH software. Chimera sequences were filtered out using the Gold database by UCHIME (Version 4.2.40). All quality-filtered sequences were then clustered into operational taxonomic units (OTUs) with a threshold of 97% sequence similarity, by utilizing UPARSE software (Version 7.0). The representative sequences for each OTU were taxonomically assigned to the Silva database (16S rDNA) and unite database (ITS) using the RDP classifier [57]. Then, OTUs were processed by removing chloroplast sequences, chondriosome sequences, and unclassified sequences. The normalized OTU abundance profile was generated by utilizing a standard sequence number corresponding to the sample with the least sequences.

On the basis of the normalized OTU abundance profile, the three alpha diversity indices (Chao1, observed species, and Shannon) were calculated to estimate the species diversity and richness of each sample [58]. Subsequently, the differences of samples in OTU-level were evaluated through the principal co-ordinates analysis (PCoA) based on Bray-Curtis by using R software [59].

4.6. Statistical Analysis

All statistics were analyzed by using GraphPad Prism 7. The data are presented as mean ± SD for each group. The differences between the mean values of the groups were analyzed by one-way analysis of variance using (least significant difference, LSD) multiple range tests. If the variance was not uniform, and the non-parametric test was used (Kruskal-Wallis). The analyses were performed using SPSS software (version 20.0; SPSS Inc., Chicago, IL, USA). A *p*-value of less than 0.05 was considered to indicate statistical significance.

5. Conclusions

In summary, the results showed that the key constipation indicators (the gastrointestinal transit rate, fecal number in six hours, and fecal weight in six hours) were significantly enhanced after the administration of D-tagatose, suggesting that different doses of D-tagatose can prevent constipation symptoms in mice. In addition, the administration of D-tagatose treatment was beneficial for restoring the diversity of gut microbiota, improving the composition of gut microbiota (mainly for phylum of Bacteroidetes), and influencing the ENS by regulating the balance between the inhibitory and excitatory neurotransmitters (Ach, SP, and NO). Therefore, this study indicates that the administration of D-tagatose as a dietary supplement could effectively relieve constipation in mice, and it is a promising prebiotic candidate for the food industry. Further studies are required to study the process change of defecation status and gut microbiota during the intervention and the relationship between the SCFAs and intestinal peristalsis.

Author Contributions: Y.-X.L., Y.-M.S., and H.W. conceived and designed the experiments; Y.-X.L., Y.W., D.-M.O., D.W., Y.-Z.C., Y.S., and J.D. performed the experiments; Y.-X.L. and H.W. analyzed the data; and Y.-X.L., P.W., and H.W. wrote the paper.

Funding: This research was funded by Key Realm R&D Program of GuangDong Province (No. 2019B020211002), National Natural Science Foundation of China (No. 31972157), Science and Technology Planning Project of Guangzhou City (201804020077), International Cooperation Program of SCAU (2019SCAUGH03) and Project Supported by Guangdong Province Universities and Colleges Pearl River Scholar Funded Scheme (2017).

Acknowledgments: We thank Huan Liu, Yong-Jian He, and Yong-Chun Chen, for directing the animal use and care at Laboratory Animal Center of South China Agricultural University.

Conflicts of Interest: The authors declare no conflict of interest.

Abbreviations

Ach	acetylcholine
SP	substance P
NO	nitric oxide
VIP	vasoactive intestinal peptide
ENS	enteric nervous system
FDA	USA Food and Drug Administration
CFDA	China Food and Drug Administration
SCFAs	short-chain fatty acids
SPF	specific pathogen free
ELISA	enzyme-linked immunosorbent assay
RDP	ribosomal database project
PCoA	principal co-ordinates analysis
SRA	sequence read archive
OTUs	operational taxonomic units

References

1. Wang, L.; Hu, L.; Xu, Q.; Yin, B.; Fang, D.; Wang, G.; Zhao, J.; Zhang, H.; Chen, W. Bifidobacterium adolescentis Exerts Strain-Specific Effects on Constipation Induced by Loperamide in BALB/c Mice. *Int. J. Mol. Sci.* **2017**, *18*, 318. [CrossRef] [PubMed]
2. Bharucha, A.E.; Pemberton, J.H.; Locke, G.R. American Gastroenterological Association technical review on constipation. *Gastroenterology* **2013**, *144*, 218–238. [CrossRef] [PubMed]
3. Hu, T.G.; Wen, P.; Fu, H.Z.; Lin, G.Y.; Liao, S.T.; Zou, Y.X. Protective effect of mulberry (*Morus atropurpurea*) fruit against diphenoxylate-induced constipation in mice through the modulation of gut microbiota. *Food Funct.* **2019**, *10*, 1513–1528. [CrossRef] [PubMed]
4. Zhao, Y.; Yu, Y.B. Intestinal microbiota and chronic constipation. *Springerplus* **2016**, *5*, 1130. [CrossRef]
5. Ohkusa, T.; Koido, S.; Nishikawa, Y.; Sato, N. Gut Microbiota and Chronic Constipation: A Review and Update. *Front. Med.* **2019**, *6*, 19. [CrossRef] [PubMed]
6. Zoppi, G.; Cinquetti, M.; Luciano, A.; Benini, A.; Muner, A.; Bertazzoni Minelli, E. The intestinal ecosystem in chronic functional constipation. *Acta Paediatr.* **1998**, *87*, 836–841. [CrossRef]
7. Kang, D.W.; DiBaise, J.K.; Ilhan, Z.E.; Crowell, M.D.; Rideout, J.R.; Caporaso, J.G.; Rittmann, B.E.; Krajmalnik-Brown, R. Gut microbial and short-chain fatty acid profiles in adults with chronic constipation before and after treatment with lubiprostone. *Anaerobe* **2015**, *33*, 33–41. [CrossRef]
8. Ge, X.; Zhao, W.; Ding, C.; Tian, H.; Xu, L.; Wang, H.; Ni, L.; Jiang, J.; Gong, J.; Zhu, W.; et al. Potential role of fecal microbiota from patients with slow transit constipation in the regulation of gastrointestinal motility. *Sci. Rep.* **2017**, *7*, 441. [CrossRef]
9. Nelson, A.D.; Camilleri, M.; Chirapongsathorn, S.; Vijayvargiya, P.; Valentin, N.; Shin, A.; Erwin, P.J.; Wang, Z.; Murad, M.H. Comparison of efficacy of pharmacological treatments for chronic idiopathic constipation: A systematic review and network meta-analysis. *Gut* **2017**, *66*, 1611–1622. [CrossRef]
10. Omer, A.; Quigley, E.M.M. An update on prucalopride in the treatment of chronic constipation. *Ther. Adv. Gastroenterol.* **2017**, *10*, 877–887. [CrossRef]
11. Chapman, R.W.; Stanghellini, V.; Geraint, M.; Halphen, M. Randomized clinical trial: Macrogol/PEG 3350 plus electrolytes for treatment of patients with constipation associated with irritable bowel syndrome. *Am. J. Gastroenterol.* **2013**, *108*, 1508–1515. [CrossRef] [PubMed]

12. Smith, B. Pathologic changes in the colon produced by anthraquinone purgatives. *Dis. Colon Rectum* **1973**, *16*, 455–458. [CrossRef] [PubMed]
13. Kim, J.E.; Lee, M.R.; Park, J.J.; Choi, J.Y.; Song, B.R.; Son, H.J.; Choi, Y.W.; Kim, K.M.; Hong, J.T.; Hwang, D.Y. Quercetin promotes gastrointestinal motility and mucin secretion in loperamide-induced constipation of SD rats through regulation of the mAChRs downstream signal. *Pharm. Biol* **2018**, *56*, 309–317. [CrossRef] [PubMed]
14. Yin, J.; Liang, Y.; Wang, D.; Yan, Z.; Yin, H.; Wu, D.; Su, Q. Naringenin induces laxative effects by upregulating the expression levels of c-Kit and SCF, as well as those of aquaporin 3 in mice with loperamide-induced constipation. *Int. J. Mol. Med.* **2018**, *41*, 649–658. [CrossRef]
15. Wang, L.; Hu, L.; Yan, S.; Jiang, T.; Fang, S.; Wang, G.; Zhao, J.; Zhang, H.; Chen, W. Effects of different oligosaccharides at various dosages on the composition of gut microbiota and short-chain fatty acids in mice with constipation. *Food Funct.* **2017**, *8*, 1966–1978. [CrossRef]
16. Roy, S.; Chikkerur, J.; Roy, S.C.; Dhali, A.; Kolte, A.P.; Sridhar, M.; Samanta, A.K. Tagatose as a Potential Nutraceutical: Production, Properties, Biological Roles, and Applications. *J. Food Sci.* **2018**, *83*, 2699–2709. [CrossRef]
17. Levin, G.V. Tagatose, the new GRAS sweetener and health product. *J. Med. Food* **2002**, *5*, 23–36. [CrossRef]
18. Laerke, H.N.; Jensen, B.B. D-tagatose has low small intestinal digestibility but high large intestinal fermentability in pigs. *J. Nutr.* **1999**, *129*, 1002–1009. [CrossRef]
19. Bertelsen, H.; Andersen, H.; Tvede, M. Fermentation of D-Tagatose by Human Intestinal Bacteria and Dairy Lactic Acid Bacteria. *Microb. Ecol. Health* **2001**, *13*, 87–95. [CrossRef]
20. Bertelsen, H. D-tagatose a novel low-calorie bulk sweetener with prebiotic properties. *World Rev. Nutr. Diet.* **1999**, *85*, 98–109.
21. Son, S.J.; Koh, J.H.; Park, M.R.; Ryu, S.; Lee, W.J.; Yun, B.; Lee, J.H.; Oh, S.; Kim, Y. Effect of the *Lactobacillus rhamnosus* strain GG and tagatose as a synbiotic combination in a dextran sulfate sodium-induced colitis murine model. *J. Dairy Sci.* **2019**, *102*, 2844–2853. [CrossRef] [PubMed]
22. Souza, D.D.S.; Tahan, S.; Weber, T.K.; Araujo-Filho, H.B.; de Morais, M.B. Randomized, Double-Blind, Placebo-Controlled Parallel Clinical Trial Assessing the Effect of Fructooligosaccharides in Infants with Constipation. *Nutrients* **2018**, *10*. [CrossRef] [PubMed]
23. Yu, T.; Zheng, Y.P.; Tan, J.C.; Xiong, W.J.; Wang, Y.; Lin, L. Effects of Prebiotics and Synbiotics on Functional Constipation. *Am. J. Med Sci.* **2017**, *353*, 282–292. [CrossRef] [PubMed]
24. Zhai, X.; Lin, D.; Zhao, Y.; Yang, X. Bacterial Cellulose Relieves Diphenoxylate-Induced Constipation in Rats. *J. Agric. Food Chem.* **2018**, *66*, 4106–4117. [CrossRef] [PubMed]
25. Hasibul, K.; Nakayama-Imaohji, H.; Hashimoto, M.; Yamasaki, H.; Ogawa, T.; Waki, J.; Tada, A.; Yoneda, S.; Tokuda, M.; Miyake, M.; et al. D-Tagatose inhibits the growth and biofilm formation of *Streptococcus mutans*. *Mol. Med. Rep.* **2018**, *17*, 843–851. [CrossRef] [PubMed]
26. Wang, L.; Hu, L.; Xu, Q.; Jiang, T.; Fang, S.; Wang, G.; Zhao, J.; Zhang, H.; Chen, W. Bifidobacteria exert species-specific effects on constipation in BALB/c mice. *Food Funct.* **2017**, *8*, 3587–3600. [CrossRef] [PubMed]
27. Fu, X.; Li, R.; Zhang, T.; Li, M.; Mou, H. Study on the ability of partially hydrolyzed guar gum to modulate the gut microbiota and relieve constipation. *J. Food Biochem.* **2019**, *43*. [CrossRef]
28. Schulz, S.; Röcken, C.; Mawrin, C.; Weise, W.; Höllt, V.; Schulz, S. Immunocytochemical identification of VPAC1, VPAC2, and PAC1 receptors in normal and neoplastic human tissues with subtype-specific antibodies. *Clin. Cancer Res.* **2004**, *10*, 8235–8242. [CrossRef]
29. Zhao, X.; Qian, Y.; Li, G.; Yi, R.; Park, K.Y.; Song, J.L. Lactobacillus plantarum YS2 (yak yogurt Lactobacillus) exhibited an activity to attenuate activated carbon-induced constipation in male Kunming mice. *J. Dairy Sci.* **2019**, *102*, 26–36. [CrossRef]
30. Baldassano, S.; Tesoriere, L.; Rotondo, A. Inhibition of the Mechanical Activity of Mouse Ileum by Cactus Pear (*Opuntia Ficus Indica*, L, Mill.) Fruit Extract and Its Pigment Indicaxanthin. *J. Agric. Food Chem.* **2010**, *58*, 7565–7571. [CrossRef]
31. Parthasarathy, G.; Chen, J.; Chen, X.; Chia, N.; O'Connor, H.M.; Wolf, P.G.; Gaskins, H.R.; Bharucha, A.E. Relationship Between Microbiota of the Colonic Mucosa vs Feces and Symptoms, Colonic Transit, and Methane Production in Female Patients With Chronic Constipation. *Gastroenterology* **2016**, *150*, 367–379. [CrossRef] [PubMed]

32. Wu, G.D.; Chen, J.; Hoffmann, C.; Bittinger, K.; Chen, Y.Y.; Keilbaugh, S.A.; Bewtra, M.; Knights, D.; Walters, W.A.; Knight, R.; et al. Linking long-term dietary patterns with gut microbial enterotypes. *Science* **2011**, *334*, 105–108. [CrossRef] [PubMed]
33. Roy, C.C.; Kien, C.L.; Bouthillier, L.; Levy, E. Short-chain fatty acids: Ready for prime time? *Nutr. Clin. Pract.* **2006**, *21*, 351–366. [CrossRef] [PubMed]
34. Ríos-Covián, D.; Ruas-Madiedo, P.; Margolles, A.; Gueimonde, M.; de Los Reyes-Gavilán, C.G.; Salazar, N. Intestinal Short Chain Fatty Acids and their Link with Diet and Human Health. *Front. Microbiol.* **2016**, *7*, 185. [CrossRef]
35. Holscher, H.D. Dietary fiber and prebiotics and the gastrointestinal microbiota. *Gut Microbes* **2017**, *8*, 172–184. [CrossRef]
36. Niwa, T.; Nakao, M.; Hoshi, S.; Yamada, K.; Inagaki, K.; Nishida, M.; Nabeshima, T. Effect of dietary fiber on morphine-induced constipation in rats. *Biosci. Biotechnol. Biochem.* **2002**, *66*, 1233–1240. [CrossRef]
37. Kittelmann, S.; Seedorf, H.; Walters, W.A.; Clemente, J.C.; Knight, R.; Gordon, J.I.; Janssen, P.H. Simultaneous amplicon sequencing to explore co-occurrence patterns of bacterial, archaeal and eukaryotic microorganisms in rumen microbial communities. *PLoS ONE* **2013**, *8*, e47879. [CrossRef]
38. Gosalbes, M.J.; Durbán, A.; Pignatelli, M.; Abellan, J.J.; Jiménez-Hernández, N.; Pérez-Cobas, A.E.; Latorre, A.; Moya, A. Metatranscriptomic approach to analyze the functional human gut microbiota. *PLoS ONE* **2011**, *6*, e17447. [CrossRef]
39. Meehan, C.J.; Beiko, R.G. A phylogenomic view of ecological specialization in the Lachnospiraceae, a family of digestive tract-associated bacteria. *Genome Biol. Evol.* **2014**, *6*, 703–713. [CrossRef]
40. Louis, P.; Young, P.; Holtrop, G.; Flint, H.J. Diversity of human colonic butyrate-producing bacteria revealed by analysis of the butyryl-CoA:acetate CoA-transferase gene. *Environ. Microbiol.* **2010**, *12*, 304–314. [CrossRef]
41. Estaki, M.; Pither, J.; Baumeister, P.; Little, J.P.; Gill, S.K.; Ghosh, S.; Ahmadi-Vand, Z.; Marsden, K.R.; Gibson, D.L. Cardiorespiratory fitness as a predictor of intestinal microbial diversity and distinct metagenomic functions. *Microbiome* **2016**, *4*, 42. [CrossRef] [PubMed]
42. Kverka, M.; Zakostelska, Z.; Klimesova, K.; Sokol, D.; Hudcovic, T.; Hrncir, T.; Rossmann, P.; Mrazek, J.; Kopecny, J.; Verdu, E.F.; et al. Oral administration of *Parabacteroides distasonis* antigens attenuates experimental murine colitis through modulation of immunity and microbiota composition. *Clin. Exp. Immunol.* **2011**, *163*, 250–259. [CrossRef] [PubMed]
43. Donaldson, G.P.; Lee, S.M.; Mazmanian, S.K. Gut biogeography of the bacterial microbiota. *Nat. Rev. Microbiol.* **2016**, *14*, 20–32. [CrossRef] [PubMed]
44. Ren, X.; Liu, L.; Gamallat, Y.; Zhang, B.; Xin, Y. Enteromorpha and polysaccharides from enteromorpha ameliorate loperamide-induced constipation in mice. *Biomed. Pharmacother.* **2017**, *96*, 1075–1081. [CrossRef]
45. Xu, J.; Gordon, J.I. Honor thy symbionts. *Proc. Natl. Acad. Sci. USA* **2003**, *100*, 10452–10459. [CrossRef]
46. Wexler, H.M. Bacteroides: The good, the bad, and the nitty-gritty. *Clin. Microbiol. Rev.* **2007**, *20*, 593–621. [CrossRef]
47. McNulty, N.P.; Wu, M.; Erickson, A.R.; Pan, C.; Erickson, B.K.; Martens, E.C.; Pudlo, N.A.; Muegge, B.D.; Henrissat, B.; Hettich, R.L.; et al. Effects of diet on resource utilization by a model human gut microbiota containing Bacteroides cellulosilyticus WH2, a symbiont with an extensive glycobiome. *PLoS Biol.* **2013**, *11*, e1001637. [CrossRef]
48. Martens, E.C.; Chiang, H.C.; Gordon, J.I. Mucosal glycan foraging enhances fitness and transmission of a saccharolytic human gut bacterial symbiont. *Cell Host Microbe* **2008**, *4*, 447–457. [CrossRef]
49. Tan, H.; Zhai, Q.; Chen, W. Investigations of *Bacteroides* spp. towards next-generation probiotics. *Food Res. Int.* **2019**, *116*, 637–644. [CrossRef]
50. Den Besten, G.; van Eunen, K.; Groen, A.K.; Venema, K.; Reijngoud, D.J.; Bakker, B.M. The role of short-chain fatty acids in the interplay between diet, gut microbiota, and host energy metabolism. *J. Lipid Res.* **2013**, *54*, 2325–2340. [CrossRef]
51. Barcenilla, A.; Pryde, S.E.; Martin, J.C.; Duncan, S.H.; Stewart, C.S.; Henderson, C.; Flint, H.J. Phylogenetic relationships of butyrate-producing bacteria from the human gut. *Appl. Environ. Microbiol.* **2000**, *66*, 1654–1661. [CrossRef] [PubMed]
52. Pryde, S.E.; Duncan, S.H.; Hold, G.L.; Stewart, C.S.; Flint, H.J. The microbiology of butyrate formation in the human colon. *FEMS Microbiol. Lett.* **2002**, *217*, 133–139. [CrossRef] [PubMed]

53. Verdu, E.F.; Bercik, P.; Collins, S.M. Effect of probiotics on gastrointestinal function: Evidence from animal models. *Ther. Adv. Gastroenterol.* **2009**, *2*, 31–35. [CrossRef] [PubMed]
54. Zhao, X.; Yi, R.; Qian, Y.; Park, K.Y. Lactobacillus plantarum YS-3 Prevents Activated Carbon-Induced Constipation in Mice. *J. Med. Food* **2018**, *21*, 575–584. [CrossRef]
55. Liu, C.-J.; Tang, X.-D.; Yu, J.; Zhang, H.-Y.; Li, X.-R. Gut microbiota alterations from different Lactobacillus probiotic-fermented yoghurt treatments in slow-transit constipation. *J. Funct. Foods* **2017**, *38*, 110–118. [CrossRef]
56. Zhu, F.; Xu, S.; Zhang, Y.; Chen, F.; Ji, J.; Xie, G. Total Glucosides of Paeony Promote Intestinal Motility in Slow Transit Constipation Rats through Amelioration of Interstitial Cells of Cajal. *PLoS ONE* **2016**, *11*, e0160398. [CrossRef]
57. Wang, Q.; Garrity, G.M.; Tiedje, J.M.; Cole, J.R. Naive Bayesian classifier for rapid assignment of rRNA sequences into the new bacterial taxonomy. *Appl. Environ. Microbiol.* **2007**, *73*, 5261–5267. [CrossRef]
58. Faith, D.P.; Baker, A.M. Phylogenetic diversity (PD) and biodiversity conservation: Some bioinformatics challenges. *Evol. Bioinform. Online* **2007**, *2*, 121–128. [CrossRef]
59. Somerfield, P.J. Identification of the Bray-Curtis similarity index: Comment on Yoshioka (2008). *Mar. Ecol. Prog. Ser.* **2008**, *372*, 303–306. [CrossRef]

MDPI
St. Alban-Anlage 66
4052 Basel
Switzerland
www.mdpi.com

International Journal of Molecular Sciences Editorial Office
E-mail: ijms@mdpi.com
www.mdpi.com/journal/ijms

www.ingramcontent.com/pod-product-compliance
Lightning Source LLC
Chambersburg PA
CBHW041949260926
51850CB00019B/2280
9783036589633